AF772374

Pillararenes

Monographs in Supramolecular Chemistry

Series Editors:
Professor Philip Gale, *University of Southampton, UK*
Professor Jonathan Steed, *Durham University, UK*

Titles in this Series:
1: Cyclophanes
2: Calixarenes
3: Crown Ethers and Cryptands
4: Container Molecules and Their Guests
5: Membranes and Molecular Assemblies: The Synkinetic Approach
6: Calixarenes Revisited
7: Self-assembly in Supramolecular Systems
8: Anion Receptor Chemistry
9: Boronic Acids in Saccharide Recognition
10: Calixarenes: An Introduction, 2nd Edition
11: Polymeric and Self Assembled Hydrogels: From Fundamental Understanding to Applications
12: Molecular Logic-based Computation
13: Supramolecular Systems in Biomedical Fields
14: Synthetic Receptors for Biomolecules: Design Principles and Applications
15: Polyrotaxane and Slide-Ring Materials
16: Boron: Sensing, Synthesis and Supramolecular Self-Assembly
17: Porous Polymers: Design, Synthesis and Applications
18: Pillararenes

How to obtain future titles on publication:
A standing order plan is available for this series. A standing order will bring delivery of each new volume immediately on publication.

For further information please contact:
Book Sales Department, Royal Society of Chemistry, Thomas Graham House, Science Park, Milton Road, Cambridge, CB4 0WF, UK
Telephone: +44 (0)1223 420066, Fax: +44 (0)1223 420247
Email: booksales@rsc.org
Visit our website at http://www.rsc.org/Shop/Books/

Pillararenes

Edited by

Tomoki Ogoshi
Kanazawa University, Kanazawa, Japan
Email: ogoshi@se.kanazawa-u.ac.jp

ROYAL SOCIETY OF CHEMISTRY

THE QUEEN'S AWARDS FOR ENTERPRISE: INTERNATIONAL TRADE 2013

Monographs in Supramolecular Chemistry No. 18

Print ISBN: 978-1-84973-970-2
PDF eISBN: 978-1-78262-232-1
ISSN: 1368-8642

A catalogue record for this book is available from the British Library

© The Royal Society of Chemistry 2016

All rights reserved

Apart from fair dealing for the purposes of research for non-commercial purposes or for private study, criticism or review, as permitted under the Copyright, Designs and Patents Act 1988 and the Copyright and Related Rights Regulations 2003, this publication may not be reproduced, stored or transmitted, in any form or by any means, without the prior permission in writing of The Royal Society of Chemistry or the copyright owner, or in the case of reproduction in accordance with the terms of licences issued by the Copyright Licensing Agency in the UK, or in accordance with the terms of the licences issued by the appropriate Reproduction Rights Organization outside the UK.Enquiries concerning reproduction outside the terms stated here should be sent to The Royal Society of Chemistry at the address printed on this page.

The RSC is not responsible for individual opinions expressed in this work.

The authors have sought to locate owners of all reproduced material not in their own possession and trust that no copyrights have been inadvertently infringed.

Published by The Royal Society of Chemistry,
Thomas Graham House, Science Park, Milton Road,
Cambridge CB4 0WF, UK

Registered Charity Number 207890

Visit our website at www.rsc.org/books

Printed in the United Kingdom by CPI Group (UK) Ltd, Croydon, CR0 4YY, UK

Preface

Macrocyclic compounds act as main players in the field of supramolecular chemistry because they form organized supramolecular assemblies by cavity-size dependent host–guest interactions. Therefore, development of the field of supramolecular chemistry has been pushed by the discovery of four key macrocycles, the cyclodextrins, crown ethers, calix[*n*]arenes and cucurbit[*n*]urils. In 2008, we introduced a new class of macrocycles: "Pillar[*n*]arenes". On account of their novel cross section of properties and great potential, pillar[*n*]arenes are becoming dominant in the field of macrocyclic chemistry and are opening up a new supramolecular chemistry, which cannot be achieved by the other four key players. The field called "pillar[*n*]arene chemistry" was started in 2008, and there was a small handful of players in the beginning. Fortunately, many chemists have joined the field of pillar[*n*]arene chemistry since 2012 and, now, pillar[*n*]arene chemistry attracts the attention of numerous researchers.

One important purpose of this book is to provide not only the fundamental properties of pillar[*n*]arenes but also their application for supramolecular materials. In Chapter 1, we first discuss the historical background of macrocyclic compounds, mainly the four key macrocyclic compounds. We discuss advantages of these macrocyclic compounds, and what are the desired properties of macrocyclic compounds to dominate the field of supramolecular chemistry. In Chapters 2–6, we discuss fundamental properties, characteristic features and advantages of pillar[*n*]arenes, such as their high yield synthesis and highly symmetrical structures (Chapter 2), versatile functionality (Chapter 3), conformation and planar chirality (Chapter 4), host–guest properties (Chapter 5) and assembled structures (Chapter 6). Based on these advantages and characteristic features, pillar[*n*]arenes have been applied in various fields of supramolecular materials, such as supramolecular polymers (Wang *et al.* in Chapter 7), interlocked molecules

Monographs in Supramolecular Chemistry No. 18
Pillararenes
Edited by Tomoki Ogoshi
© The Royal Society of Chemistry 2016
Published by the Royal Society of Chemistry, www.rsc.org

(Ogoshi *et al.* in Chapter 8), supramolecular assemblies (Zhao *et al.* in Chapter 9), hybrid materials (Yang *et al.* in Chapter 10), and biomedical applications (Hou *et al.* in Chapter 11). In the final Chapter 12, Stoddart *et al.* review pillar[*n*]arene-related macrocyclic compounds, including their new fantastic macrocycles asar[*n*]arenes. It is hoped that the book will contribute to the further development of pillararene chemistry and the quest to discover new and exciting macrocyclic compounds.

Finally, I would like to thank all the authors for their excellent contributions, and the Royal Society of Chemistry for their continuous support in publishing the book.

Tomoki Ogoshi
Kanazawa,
Japan

Contents

Monographs in Supramolecular Chemistry No. 18
Pillararenes
Edited by Tomoki Ogoshi
© The Royal Society of Chemistry 2016
Published by the Royal Society of Chemistry, www.rsc.org

CHAPTER 1

Historical Background of Macrocyclic Compounds

TOMOKI OGOSHI* AND TADA-AKI YAMAGISHI

Graduate School of Natural Science and Technology,
Kanazawa University, Kakuma-machi, Kanazawa 920-1192, Japan
*Email: ogoshi@se.kanazawa-u.ac.jp

1.1 Introduction

The concept of supramolecular chemistry was proposed by Lehn in the 1980s. He defined supramolecular chemistry as follows:[1]

> "'Chemistry beyond the molecule', bearing on the organized entities of higher complexity that result from the association of two or more chemical species held together by intermolecular forces. Its development requires the use of all resources of molecular chemistry combined with the designed manipulation of non-covalent interactions so as to form supramolecular entities, supramolecules possessing features as well defined as those of molecules themselves."

Host–guest chemistry using macrocyclic compounds as hosts absolutely adheres to Lehn's concept of supramolecular chemistry because these are organized entities constructed from two or more chemical species by designed intermolecular interactions. Thus, host–guest chemistry based on macrocyclic compounds was an important starting point of supramolecular chemistry. This fact was supported by the Nobel Prize for Chemistry in 1987

Monographs in Supramolecular Chemistry No. 18
Pillararenes
Edited by Tomoki Ogoshi
© The Royal Society of Chemistry 2016
Published by the Royal Society of Chemistry, www.rsc.org

being awarded jointly to Cram, Lehn and Pedersen "for their development and use of molecules with structure-specific interactions of high selectivity".

In addition to the ability of macrocyclic compounds to form host–guest complexes, macrocyclic compounds have attracted considerable interest because of their structural features as follows:

(1) Highly symmetrical structures: Unlike linear structures, macrocyclic structures have only repeating units and no terminal ends, therefore their structures have highly symmetrical architectures. Their beautiful cyclic structures have engaged scientists, and form the origins of their names in many cases.
(2) Oligomeric molecular weights with mono-dispersity: The molecular weights of macrocyclic compounds range from several hundreds into the thousands with mono-dispersity. Therefore, they can show interesting properties that cannot be observed in low molecular weight compounds and polymers with wide polydispersity.
(3) Building blocks with nano-size order: The size of macrocyclic compounds is approximately 1 nm; therefore, interesting supramolecular assemblies can be constructed using the nano-scale macrocyclic compounds like building blocks.
(4) Host–guest property: Owing to their cyclic structures, macrocyclic compounds possess a cavity in which they can accommodate guest molecules. Furthermore, in many cases, the complexation ability of macrocyclic compounds is higher than that of linear compounds with the same number of binding sites. Linear compounds need to alter their structure to form host–guest complexes during the complexation process, while cyclic compounds do not. Entropy loss during the complexation process using macrocyclic host compounds is therefore lower than that using linear host compounds.
(5) Functionality: When macrocyclic compounds have highly reactive groups, the introduction of functional substituents can be achieved and may contribute to changing various physical properties of the macrocyclic compounds and adding new functions.

In this chapter, we present the investigations of the major structural and physical properties of macrocyclic compounds in historical order. First, we compare the chemical and X-ray crystal structures and the physical properties of the four key classes of compounds, cyclodextrins, crown ethers, calix[*n*]arenes and cucurbit[*n*]urils, which have been dominant in the field of supramolecular chemistry. Other well-known and useful host molecules, and recently discovered compounds with great potential in the field of macrocyclic chemistry, are also described. Second, we discuss the desired properties of macrocyclic compounds, and the advantages of the novel macrocyclic compounds "pillar[*n*]arenes" compared with the four key classes of compounds.

1.2 Cyclodextrins

Cyclodextrins (CDs), a family of macrocyclic oligosaccharides, were discovered by A. Villiers in 1891 (Figure 1.1).[2,3] The CDs have been the most widely used in macrocyclic compounds, because CDs have a very long history of over 100 years. As shown in Figure 1.1a, CDs are cyclic oligosaccharides consisting of glucopyranose units attached by α-1,4-linkages. The most common and commercially available CDs consist of six, seven or eight glucopyranose units and are named α-, β-, and γ-CD, respectively. As shown in Figure 1.1b, CDs form bucket structures, therefore they possess a large but narrow opening. The CDs exhibit hydroxyl groups on both faces owing to the glucopyranose units. They are water-soluble compounds because the hydroxyl groups on both faces of CDs are arranged on the outside of the cavity, whereas the inside of the cavity is a hydrophobic microenvironment. The CDs can therefore capture suitable hydrophobic guests in this hydrophobic cavity in aqueous media. The cavity size of CDs increases with an increasing number of glucopyranose repeating units. The cavity diameter of CDs is about 4.9–8.0 Å (α-CD: 4.9 Å; β-CD: 6.2 Å; γ-CD: 8.0 Å). The host–guest property of CDs has been widely investigated, so numerous host–guest complexes have been reported.[4] The cavity size of α-CD is approximately 4.9 Å, as mentioned above, and therefore it is suitable to accommodate linear alkanes and simple aromatic compounds. In contrast, the cavity size of β-CD (*ca.* 6.2 Å) is larger than that of α-CD, therefore β-CD is able to accommodate various bulky hydrocarbons, such as adamantine and cyclohexane derivatives, and polyaromatic compounds, such as naphthalene and anthracene derivatives, which cannot be included in the α-CD cavity. Because the cavity size of γ-CD is larger than those of α- and β-CD, γ-CD can incorporate even larger guests. Two γ-CD molecules can sandwich a fullerene, whereas two aromatic guests or polymeric chains can be included in a single cavity of γ-CD. The CDs also possess many chiral centers, therefore they can be used as a chiral scaffold to afford moderate to high diastereomeric excesses (de's) for several diastereo-differentiating photoreactions.[5–9] Based on the host–guest property of CDs, various mechanically interlocked molecules (MIMs),

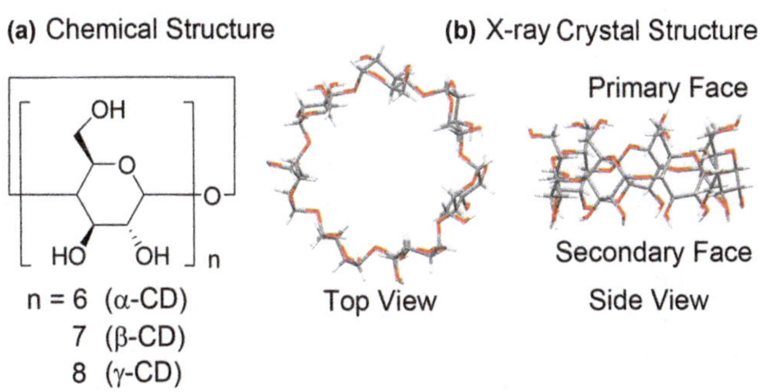

Figure 1.1 (a) Chemical and (b) X-ray crystal structures of cyclodextrins.

including rotaxanes, catenanes, molecular shuttles and polyrotaxanes, and topological gels have been constructed.[10–13]

Compared with other major macrocyclic compounds, the advantages of CD chemistry are as follows. (1) As native CDs are commercially available reagents, non-toxic and can form host–guest complexes with guests in aqueous media, the application of native CDs can be tested by anyone. The CDs can accommodate various hydrophobic guests including fats, coenzyme Q10, foods and medicines.[14] Therefore, their application in drug delivery systems (DDS), foods, beauty care and air fresheners has been investigated using the complexation of CDs with these guest molecules. This wide variety of applications resulted in the expansion of CD chemistry. (2) CDs have many highly reactive hydroxyl groups at both faces; therefore, functionalization of CDs can be achieved.[15,16]

The CDs are soluble in polar solvents, such as water, dimethyl fluoride (DMF) and dimethylsulfoxide (DMSO), but insoluble in nonpolar solvents. The low solubility of CDs in nonpolar solvents negates their use for various useful organic reactions. Well-established synthesis protocols for the selective modification of CDs in polar solvents have been extensively studied by many researchers, but yields of such functionalized CDs have been low. To overcome these disadvantages, organo-soluble CDs have been synthesized by introducing protecting moieties, such as methyl and benzyl groups, to the hydroxyl groups of CDs.[17] Various organic reactions can be enhanced using these protecting organo-soluble CDs.[18,19] Solubilization of CDs in nonpolar solvents allows host–guest complexation of CDs in organic solvents, which have been extensively studied by Akashi and Kida *et al.*[20–22] Organo-soluble CDs with protecting groups at the primary hydroxyl groups and the remaining secondary hydroxyl groups can form a chiral nano-capsule by intermolecular hydrogen bonds in nonpolar solvents (Figure 1.2). The chiral nano-capsule shows chiral recognition and kinetic resolution of the aromatic amine guests.

1.3 Crown Ethers and Their Derivatives

Historically, there have been a few cases where a mistake or an unexpected finding during an experiment has led to a huge breakthrough that has changed a scientific field. The invention of crown ethers by Pedersen in 1967 could be considered as one such case.[23,24] Simple crown ethers are cyclic compounds consisting of ethylene oxide repeating units. Pedersen accidentally discovered that these crown ethers can selectively accommodate alkali cations such as sodium and potassium cations. Oxygen atoms in crown ethers contribute to the binding because the cyclic alkanes do not form complexes with these cations. The driving force for the complexation is the coordination interactions between metal cations and oxygen atoms in crown ethers (Figure 1.3). Until the discovery of crown ethers, ion-recognition by synthetic compounds had not been reported, while the

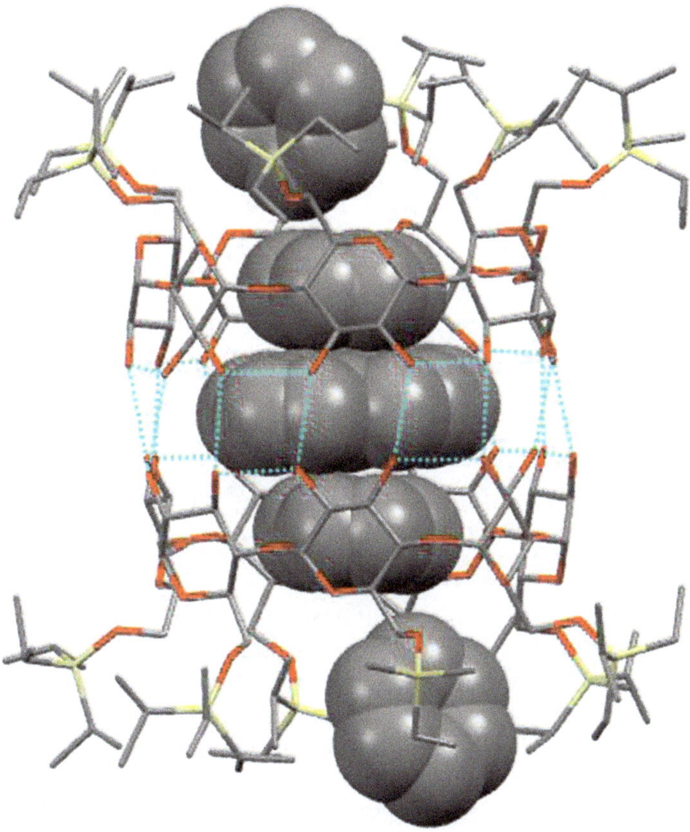

Figure 1.2 X-ray crystal structure of cyclodextrin capsule formed by intermolecular hydrogen bonds in nonpolar solvent.

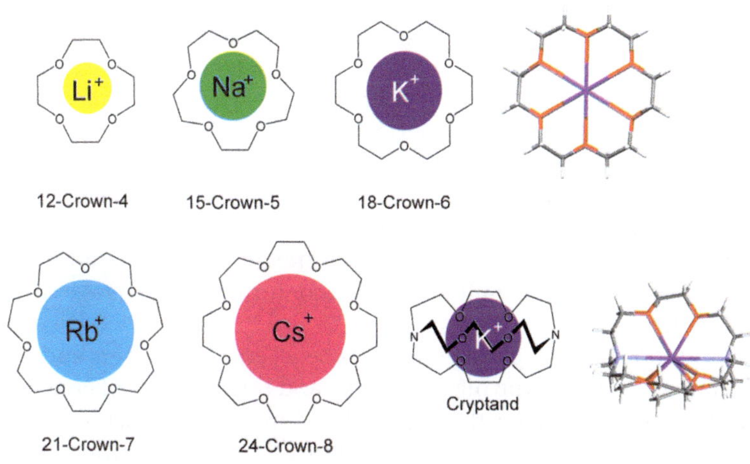

Figure 1.3 Chemical and X-ray crystal structures of simple crown ether⊃metal cation and cryptand⊃potassium cation complexes.

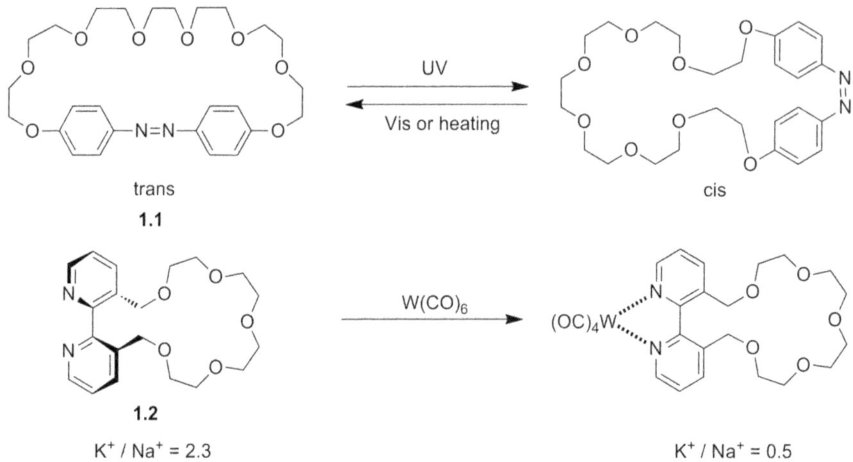

Figure 1.4 Photo- and metal ion-responsive structural changes of crown ethers with azobenzene and bipyridyl moieties.

selective recognition of ions is a fundamental function in a living system. Therefore, this discovery was a starting point not only in the fields of molecular recognition and supramolecular chemistry, but also in the field of biomimetic chemistry. The repeating units of crown ethers are water- and organo-soluble amphiphilic ethylene oxide moieties, therefore they can be used in both aqueous and organic media. The ion coordination ability of crown ethers in nonpolar solvents is stronger than that in polar solvents because an unpaired pair of electrons in oxygen atoms co-ordinates with metal cations, which locate in the inner cavity in nonpolar solvents.[25]

In contrast to CDs, functionalization of crown ethers can be achieved by various organic reactions owing to the high solubility of crown ethers. Therefore, various crown ether derivatives have been synthesized. One of the characteristic attributes of crown ethers is the conformational flexibility. Based on the conformational flexibility and high functionality, photo- (*e.g.*, **1.1**)[26] and metal ion-responsive crown ethers (*e.g.*, **1.2**)[27] have been synthesized. Structural changes of stimuli-responsive moieties, which are incorporated into the crown ether structure, are tools to facilitate conformational changes of crown ethers. The metal cation recognition ability can be increased by the stimuli-responsive conformational changes of crown ethers (Figure 1.4).

The formation of host–guest complexes between crown ethers and organic ammonium cations and the high solubility and functionality of crown ethers in organic solvents enable the synthesis of molecular machines such as molecular elevators and molecular shuttles.[28,29] Complexation of metal cations into the cavity of crown ethers contributes to the solubilization of metal cations in organic solvents and enhances the reactivity of counter anions because counter anions are bare when metal cations are included in the cavity of a crown ether. Therefore, crown ethers can be used as phase transfer catalysts. Crown ethers allow reactions to proceed through the active transport of polar reagents from polar solvents to nonpolar solvents.[30,31]

After the discovery of crown ethers, Lehn *et al.* synthesized cryptands, which contain a double-cyclic crown ether structure (see Figure 1.3).[32] Owing to the double-cyclic structure, cryptands can capture potassium ions selectively and strongly. This complexation behavior has received a great deal of attention because this system is very similar to the encapsulation of potassium ions by the antibacterial agent valinomycin.

1.4 Calix[*n*]arenes and Their Derivatives

Calix[*n*]arenes, which are phenolic macrocycles (Figure 1.5), were popularized by Gutsche in 1978,[33] although calixarenes date back to Baeyer's work on phenol-formaldehyde chemistry in the 1870s.[34] Calix[*n*]arenes are composed of phenolic units. The phenolic units are linked by methylene bridges at their 2- and 6-positions (*meta* positions). Gutsche *et al.* discovered that even-numbered calix[*n*]arenes ($n = 4, 6, 8$) were selectively obtained by tuning the reaction conditions for the phenolic derivatives and formaldehyde.

Odd-numbered calix[*n*]arenes ($n = 5, 7, 9$) and large calix[*n*]arene homologs were also formed by tuning reaction conditions, but owing to the low

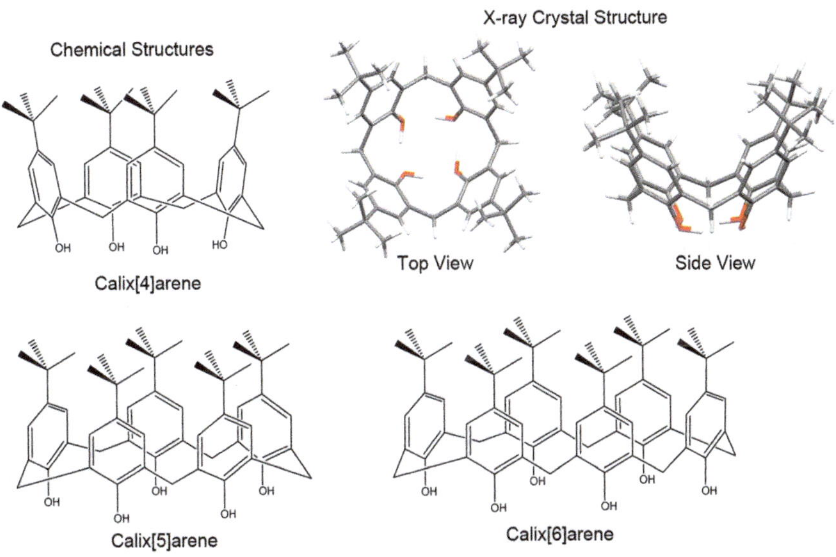

Figure 1.5 Chemical and X-ray crystal structures of *p-tert*-butyl calix[*n*]arenes ($n = 4, 5, 6$).

yields of the odd-numbered calix[*n*]arenes, their structures were only fully characterized after the discovery of even-numbered calix[*n*]arenes.[35–37] Owing to the intramolecular hydrogen bonds between phenolic moieties, they form unique calix-shaped structures, which is why they were named calix[*n*]arenes. The phenolic units are electron-donating moieties, therefore calix[*n*]arenes can form host–guest complexes with cationic moieties. Atwood *et al.* reported the selective complexation of C_{60} in the calix[8]arene cavity (**1.3**), and the application of this cavity for the purification of C_{60} from fullerene mixtures.[38] Shinkai *et al.* synthesized water-soluble calix[*n*]arenes (*e.g.*, **1.4**) by introducing water-soluble sulfonate moieties.[39,40] They noted that sulfonate calix[*n*]arenes can form host–guest complexes not only with cationic molecules but also with neutral molecules because hydrophilic–hydrophobic interactions are observed along with cation/π interactions in aqueous media. Their study expands on the host–guest chemistry of calix[*n*]arenes from organic to aqueous media.[41]

One of the characteristic attributes of calix[*n*]arenes is that they have various conformers.[42] Rotation of the phenolic units affords many conformers. For example, there are four conformers in calix[4]arenes: cone, partial cone, 1,2-alternate and 1,3-alternate (Figure 1.6). The conformational change is very useful when applying the calix[*n*]arene scaffold in sensors.[43,44]

Calix[*n*]arene derivatives have also been synthesized.[45] Thiacalix[*n*]arenes[46] and homooxacalix[*n*]arenes,[47] in which methylene bridges of the calix[*n*]arenes are replaced with sulfides and CH_2–O–CH_2 bonds,

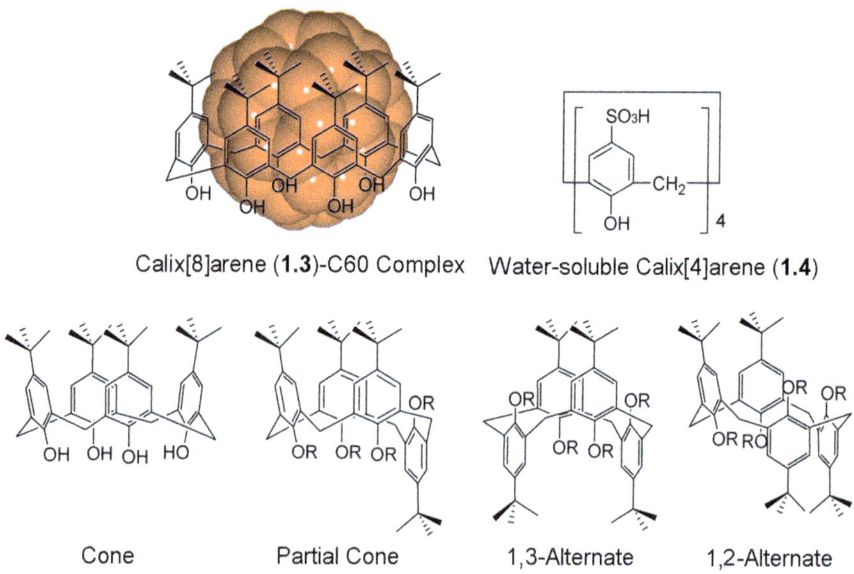

Figure 1.6 *t*-Butyl-calix[8]arene ⊃ C_{60} complex, water-soluble calix[4]arene with sulfuric acid moieties and conformation of calix[4]arenes.

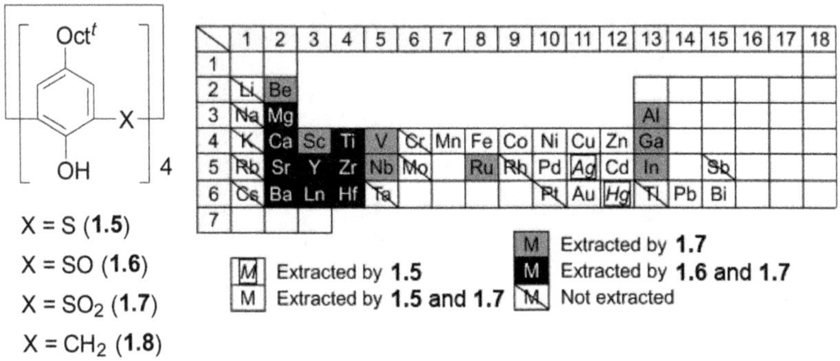

Figure 1.7 Periodic table of metal ions extracted by thiacalix[4]arenes. Reproduced with permission from ref. 46. Copyright 2006 American Chemical Society.

respectively, are new members of the calix[*n*]arene family. The sulfide bonds in thiacalix[*n*]arenes can be used as the coordination site for metal ions, therefore thiacalix[*n*]arenes can extract various metal ions. Thiacalix[*n*]arenes (**1.5**) can extract soft metal ions, and sulfinyl (**1.6**) and sulfonyl analogs (**1.7**) selectively extract hard and soft, and hard, metal ions, respectively (Figure 1.7).[46] Methylene-bridged calix[4]arene (**1.8**) hardly extract any metal ions, therefore the S bridges are very important for extracting metal ions. Application of host–guest complexes of thiacalix[*n*]arenes with lanthanide ions for fluorescence markers and MRI sensitizers has also been investigated.[48,49]

Resorcin[*n*]arenes[50] and calix[*n*]pyrroles[51] were synthesized using aromatic compounds such as resorcinol, or pyrogallol and pyrrole, as starting compounds, respectively. Resorcin[*n*]arenes have a wide upper rim and a narrow rim. The upper rim has eight phenolic moieties that can participate in hydrogen-bonding interactions. The bowl-shaped resorcin[*n*]arenes act as receptors for molecular recognition, as introduced by Aoyama *et al.*[50] They form host–guest complexes in organic solvents with small diacids, alcohols, sugars, and even steroids, featuring a 1 : 1 stoichiometry. Intermolecular hydrogen bonding between the eight phenolic moieties and these guests is the driving force holding the guest in the concavity of the resorcin[*n*]arene host.

Atwood *et al.* reported *C*-methylresorcin[4]arene (**1.9**) self-assembled to form a capsule consisting of six **1.9** molecules and eight water molecules.[50] *C*-Isobutylpyrogallol[4]arene (**1.10**) self-assembled to form a capsule consisting of two **1.10** molecules and eight zinc ions (Figure 1.8).[51]

Sessler *et al.* introduced the anion-recognition ability of calix[4]pyrrole (**1.11**).[52,53] **1.11** has the ability to bind halide anions efficiently through hydrogen-bonding interactions between pyrrolic NHs and the anions when the macrocyclic is fixed in the cone conformation. Counter cations are encapsulated at the same time in the opposite side of the bowl-shaped calix[4]pyrrole cavity because of the electron-rich pyrrole cavity (Figure 1.9).

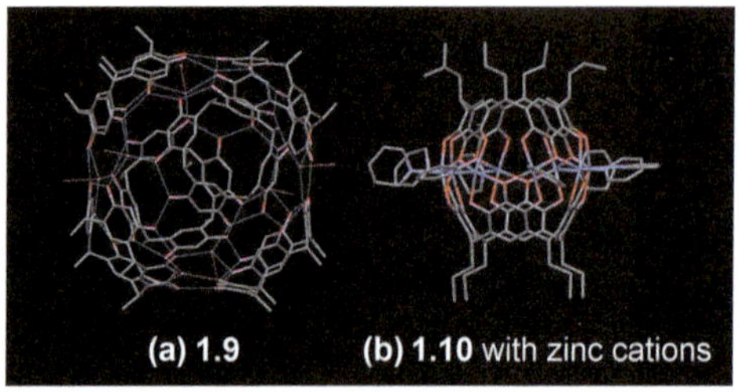

Figure 1.8 X-ray crystal structures of capsules from *C*-methylresorcin[4]arene (**1.9**) and *C*-isobutylpyrogallol[4]arene (**1.10**) with zinc cations.

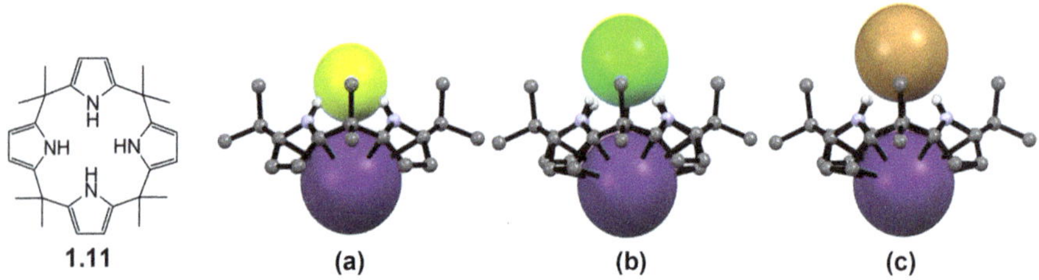

Figure 1.9 Chemical structure of calix[4]pyrrole **1.11** and single X-ray crystal structures of (a) the CsF, (b) CsCl, and (c) CsBr complexes of calix[4]pyrrole **1.11**. Reproduced with permission from ref. 52. Copyright 2014 American Chemical Society.

Cram *et al.* synthesized cavitands (*e.g.*, **1.12**), which are rigid tubular molecules, by bridging the phenolic moieties in pyrogallol[*n*]arene (Figure 1.10a).[54] By bridging cavitands through covalent bonds, rigid molecular capsules were realized. Water-soluble deep-cavitands, with the general structure shown in Figure 1.10b, were first reported by Gibb and Gibb in 2004.[55] Two bridging cavitands were covalently connected and thus formed deep-cavitands. Modification of octaacid (**1.13**) and hydrophilic dendritic moieties (**1.14**) into the cavitands contributes to the high water-solubility of deep-cavitands. Two cavitands assembled when guest molecules were encapsulated.

1.5 Cucurbit[*n*]urils CB[*n*] and Their Derivatives

Cucurbit[*n*]urils (CB[*n*], $n = 5$–10; Figure 1.11) are pumpkin-shaped macrocyclic compounds.[56–58] Although the synthesis of the cyclic hexamer CB[6] first appeared in the literature in 1905,[59] characterization and the host–guest property of CB[6] were reported by Mock *et al.* in 1981.[60] Unfortunately, the low solubility of CB[6] in aqueous media (0.018 mM) resulted in the slow development of CB[6] chemistry until 2000. In 2000, Kim *et al.* reported the synthesis of CB[*n*] homologs ($n = 5$–11).[61] The key to synthesizing CB[*n*] homologs is the reaction temperature. A lower reaction temperature

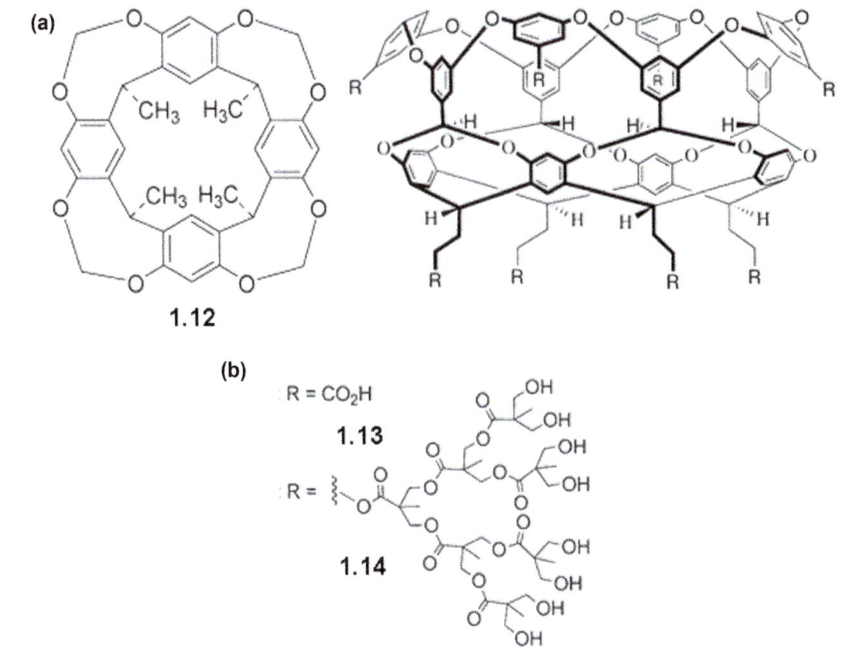

Figure 1.10 Chemical structure of cavitands (**1.12**) and deep-cavitands (**1.13, 1.14**).

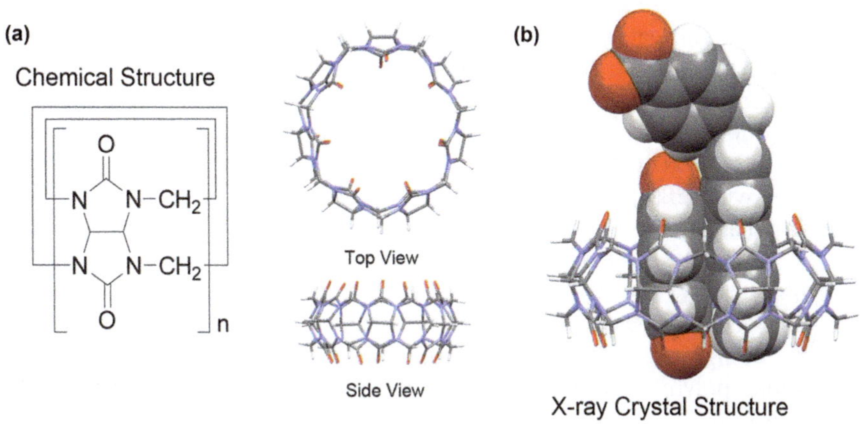

Figure 1.11 (a) Chemical and X-ray crystal structure of cucurbit[*n*]uril. (b) X-ray crystal structure of a charge-transfer complex stabilized by CB[8].

(75–90 °C) than that in the conventional synthesis of CB[6] (>110 °C) allowed for the formation of CB[*n*] homologs along with CB[6]. New CB[*n*] homologs, CB[5] and CB[7], which can be obtained in moderate yields, show moderate solubility in water (CB[5]: 420 mM; CB[7]: 370 mM), therefore CB[*n*] chemistry has been developed rapidly since Kim's discovery. They have a hydrophobic cavity and two carbonyl-laced portals, which allow them to form stable inclusion complexes with water-soluble guest molecules, such as cationic and neutral diamines, in aqueous media. Compared with the other macrocyclic compounds, such as CDs and calix[*n*]arenes, CB[*n*] has highly symmetrical entrances. Therefore, guest molecules can access both faces of CB[*n*]. Moreover, glycoluril units are connected by two methylene bridges, which increase the rigidity of the CB[*n*] structure. This rigid structure results in no conformational flexibility, therefore CB[*n*] can form host–guest complexes with high affinity and selectivity. In terms of cavity size, CB[6] (*ca.* 5.8 Å) is almost analogous to α-CD (*ca.* 4.7 Å). As with α-CD, the cavity size of CB[6] only accommodates positively charged linear alkanes and simple aromatic compounds. The cavity size of CB[7] (7.3 Å) is almost equal to that of β-CD (6.5 Å). As with β-CD, CB[7] binds a variety of positively charged polyaromatic compounds and bulky hydrocarbons, such as adamantanes and bicyclooctanes. Inoue and Kim *et al.* reported that CB[7] can form a very stable host–guest complex with a ferrocene derivative containing two ammonium cations ($K = 3.0 \times 10^{15}\ M^{-1}$), which is similar to the superior binding ability of enzyme–substrate compounds, such as that between avidin and biotin ($K > 10^{15}\ M^{-1}$).[62]

CB[8] (*ca.* 8.8 Å), which has a cavity comparable to that of γ-CD (*ca.* 8.3 Å), exhibits remarkable host–guest properties different from those of the smaller CB[6] and CB[7] homologs. CB[8] can encapsulate two guest compounds.[63] For example, two naphthalenes with cationic moieties can be included in the cavity of CB[8]. Methyl viologen cation radicals or tetrathiafulvalene cation radicals form stable π-dimers in the cavity of CB[8]. CB[8] also encapsulates two different molecules, electron-donating molecules and electron-accepting molecules, inside the cavity to form a stable 1 : 1 : 1 complex (see Figure 1.11). Electron-donors do not form a complex with CB[8] in the absence of an electron-acceptor. Therefore, the major driving force for the ternary 1 : 1 : 1 complex is the strong CT interaction between electron-donor and electron-acceptor molecules inside the CB[8] cavity. The ternary complexes can therefore be used as a method to connect an electron-acceptor and an electron-donor. Using the ternary complexation to connect molecules and polymers, various supramolecular polymers and gels have been reported by the Scherman and Zhang groups.[64]

Most cucurbit[*n*]uril homologs are practically insoluble in common solvents, which limit their use. Kim *et al.* successfully synthesized new soluble cucurbit[5]- and cucurbit[6]uril derivatives by the introduction of bulky cyclohexyl groups (Figure 1.12, **1.15** and **1.16**).[65] Interestingly, both are more soluble in water (*ca.* 2×10^{-1} M) than in methanol, DMF and DMSO (less than *ca.* 3×10^{-2} M), even though the cyclohexyl moieties are hydrophobic. The other long-standing problem in cucurbit[*n*]uril chemistry was their poor functionality. In 2003, Kim *et al.* reported the direct oxidation of CB[*n*] with $K_2S_2O_8$ in water to yield the perhydroxylated CB[*n*]

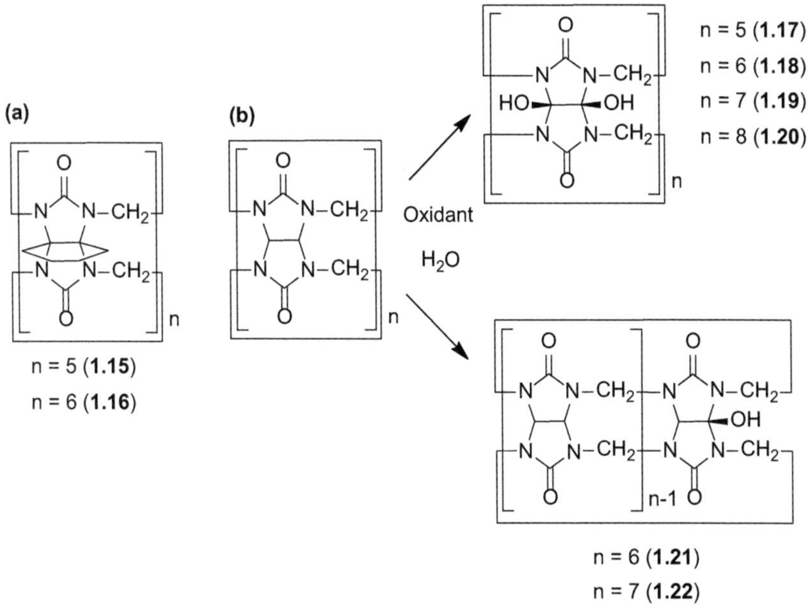

Figure 1.12 (a) Highly water-soluble CB[*n*] with cyclohexyl groups (**1.15** and **1.16**). (b) Mono- and per-hydroxylated CB[*n*]s (**1.17–1.22**) prepared by oxidation with APS.

derivatives (**1.17–1.20**).[66] Based on the synthetic protocol, Scherman *et al.* synthesized mono-hydroxylated CB[6] (**1.21**) by controlling the oxidation reaction.[67] Using the same protocol, mono-hydroxylated CB[7] (**1.22**) was also prepared by Kim *et al.*[68] Isaacs *et al.* reported the synthesis of mono-functionalized CB[*n*] derivatives by the cyclization method (Figure 1.13). Mono-functionalized CB[6] derivatives (**1.24**) were obtained by the condensation between linear glycoluril pentamer (**1.23**) and substituted phthalaldehydes.[69] Synthetic methods of mono-functionalized CB[7] derivatives were also developed by the same group. The condensation of linear glycoluril pentamer (**1.23**) with glycoluril bis(cyclic ethers) gave mono-functionalized CB[7]derivatives (**1.25**).[70] Overall, the functionalization of CB[*n*] can be achieved by oxidation of pre-formed cucurbit[*n*]urils and cyclization of linear glycoluril pentamer with aldehydes or glycoluril derivatives.

Various cucurbit[*n*]uril derivatives have been synthesized (Figure 1.14). Miyahara *et al.* reported half-cucurbit[*n*]urils, hemicucurbit[*n*]urils by condensation of ethyleneurea and paraformaldehyde in the presence of acid catalyst.[71] The reaction at 25 °C in 4 N HCl afforded the hexamer hemicucurbit[6]uril (**1.26**), in high yield (94%). In contrast, hemicucurbit[12]uril (**1.27**) was obtained in high yield (93%) when the reaction was conducted at 55 °C in 1 N HCl. Bambus[6]uril (**1.28**), which combines the structural features of both cucurbit[*n*]urils and hemicucurbit[*n*]urils, was introduced by

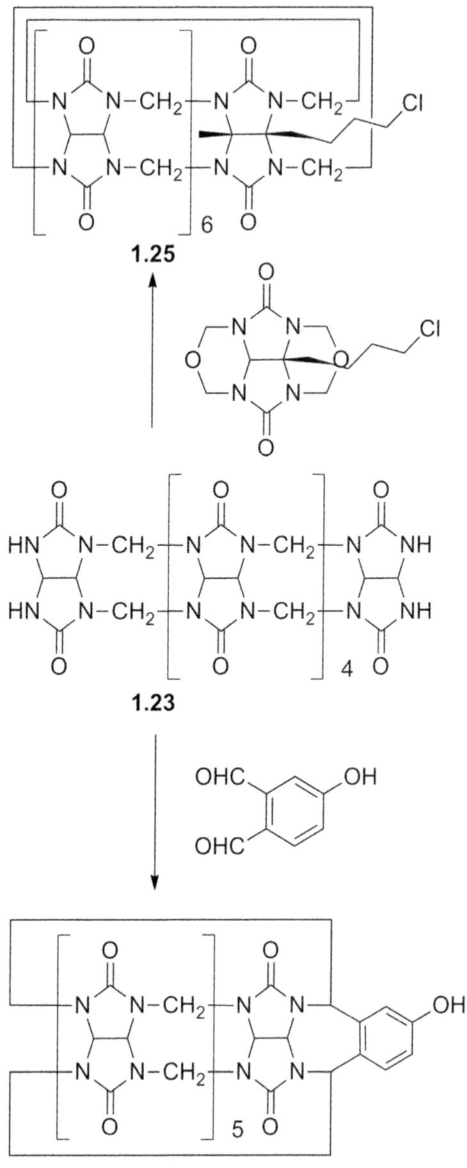

Figure 1.13 Mono-functionalization of CB[*n*]s by macrocyclization of the methylene-bridged glycoluril hexamer **1.23**.

Sindelar *et al.*[72] A key monomer is 2,4-dimethylglycoluril, in which substitutes protect the two nitrogen atoms. Condensation of 2,4-dimethylglycoluril and formaldehyde in HCl catalyst afforded bambus[6]uril, in which glycoluril units are connected by single methylene bridges (yield 30%). Tuning

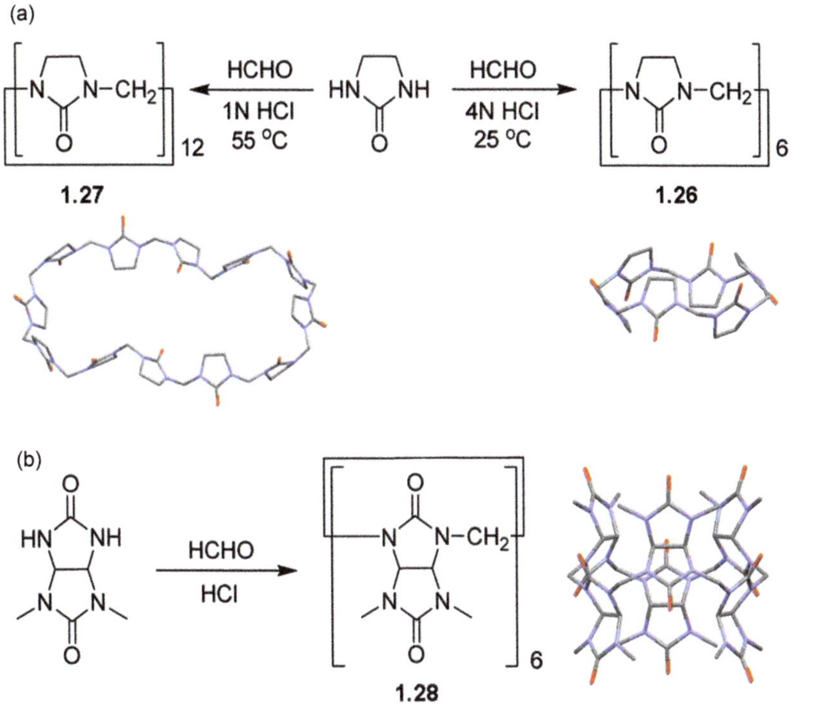

Figure 1.14 CB[*n*] derivatives: (a) hemicucurbit[*n*]urils (**1.26** and **1.27**) and (b) bambus[6]uril (**1.28**).

the reactive site and structure of the glycoluril monomers would result in new cucurbit[*n*]uril derivatives with fascinating structures and host–guest properties.

1.6 Other Synthetic Macrocyclic Compounds

The "blue box" [**1.29**, cyclobis(paraquat-*p*-phenylene) or CBPQT^{4+}],[73] developed by Stoddart *et al.*, is a useful building block to construct a wide range of interlocking molecules because the blue box forms effective charge-transfer complexes with a variety of electron-rich species (Figure 1.15). CBPQT^{4+} was synthesized by bipyridine and 1,4-bis-bromomethylbenzene. Stoddart *et al.* recently introduced an extended blue box, Ex^{n}Box^{4+} ($n=1$, **1.30**, $n=2$, **1.31**), by inserting a *p*-phenylene ring between the two pyridinium rings.[74] Stoddart *et al.* also synthesized a bicyclic hexacationic cyclophane, namely ExCage^{6+} (**1.32**), by changing the central 1,4-disubstituted benzenoid ring to a 1,3,5-trisubstituted ring.[75] Planar pyridinium units in conjugation with each other form linked face-to-face platforms with approximately 7 Å between the platforms. The six electron-deficient pyridinium binding units enable the formation of host–guest complexes

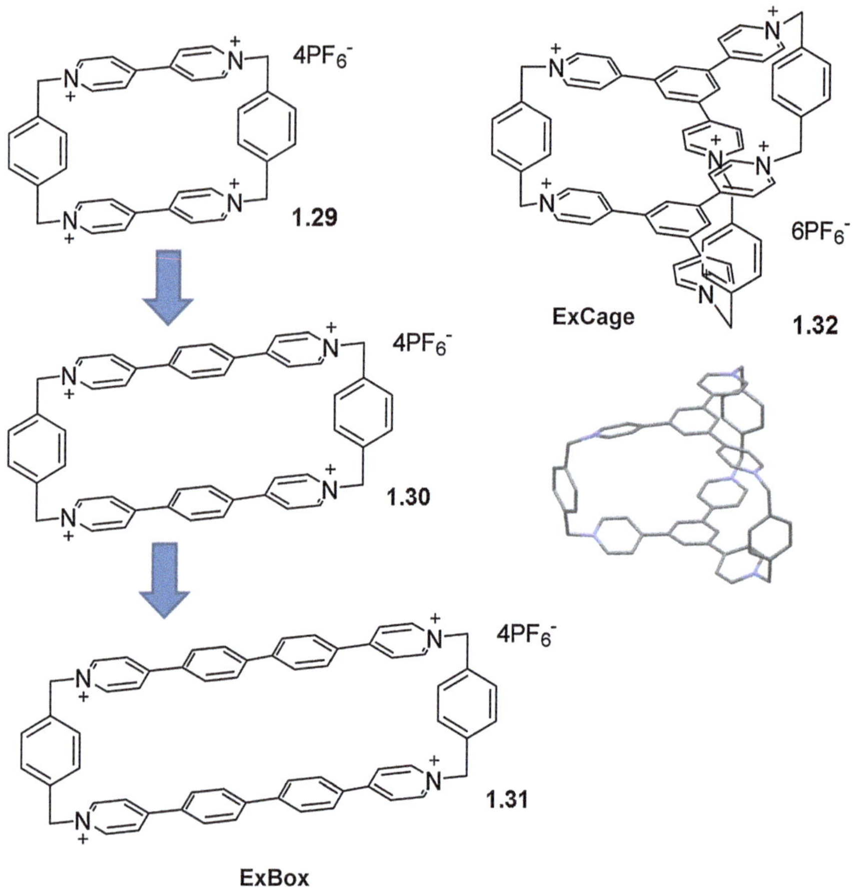

Figure 1.15 Blue box (**1.29**), Exboxs (**1.30** and **1.31**) and Excage (**1.32**).

with polycyclic aromatic hydrocarbons exhibiting high binding affinities. Sessler *et al.* synthesized a tetracationic macrocycle: a "Texas-sized" molecular box (**1.33**).[76,77] Tetracation **1.33** is flexible, therefore it has three crystallographically distinct structures, including boat-like, partial-chair and a more complete chair conformation (Figure 1.16).

A new discovery attracting interest in the field of macrocyclic chemistry is the "cyanostar" (**1.34**), which was introduced by Flood *et al.*[78,79] Cyanostars **1.34** are pentagonal star-shaped macrocyclic compounds with a one-pot synthesis in high yields that involves the cyclization of cyanostilbene repeat units. Their electropositive central cavity stabilizes anions because the CH hydrogen-bonding units are activated by the electron-withdrawing cyano groups.

Another interesting class of macrocyclic compounds are π-conjugated macrocycles.[80] Cyclic phenylenes are classified into *o*-, *m*-, and *p*-bridged phenylenes. Cyclic *o*-tetra- and hexaphenylenes have been investigated as

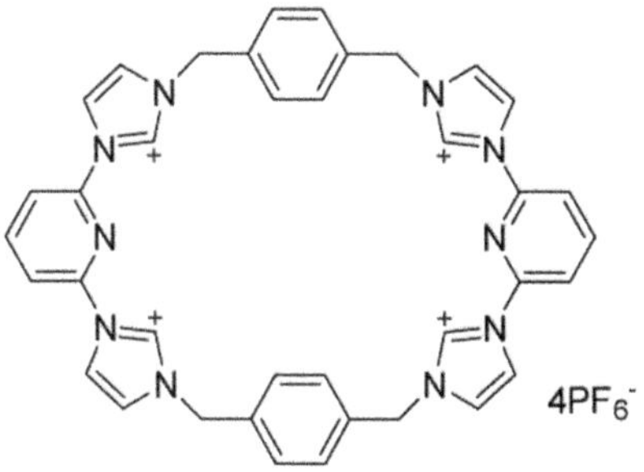

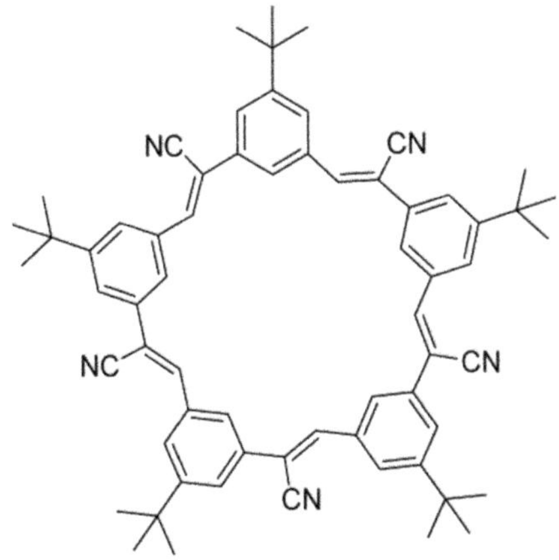

Figure 1.16 Texas-size box (**1.33**) and cyanostar (**1.34**).

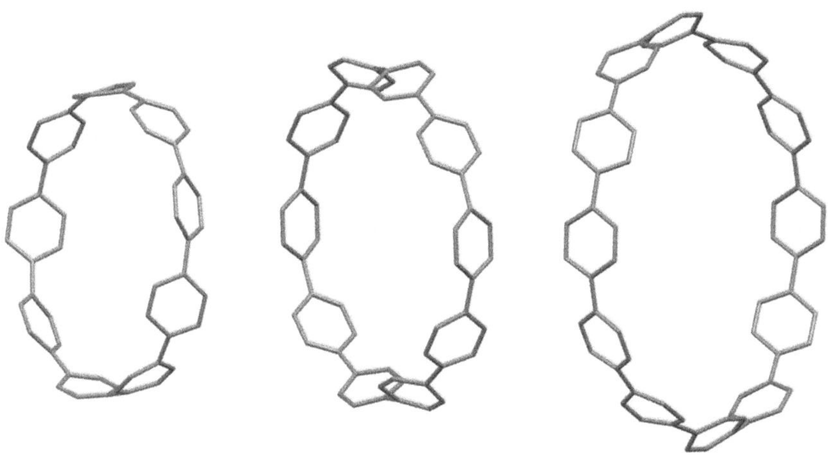

Figure 1.17 X-ray crystal structures of [9]-, [10]- and [12]-cycloparaphenylenes (CPP).

a consequence of their applications such as supramolecular scaffolds, chiral sources and liquid crystals.[81] Cycloparaphenylenes ([*n*]CPP)[82] were first introduced by Bertozzi *et al.* in 2008.[83] Following the original report of [9]CPP by Bertozzi in 2008,[83] various members of the CPP family were synthesized successively by the Itami[84] and Yamago[85] groups (Figure 1.17). An interesting characteristic of CPPs is that they are the shortest possible slices of armchair single-walled carbon nanotubes that retain aromatic rings.

1.7 Conclusion and Outlook

From the historical background of macrocyclic compounds, the factors that make macrocyclic compounds widely usable in a wide variety of applications are as follows:

(1) Easy to make: One or only a few reaction steps under normal conditions involving commercially available and cheap reagents, short reaction time, easy isolation and high yields are ideal.
(2) Versatile functionality: Easy derivatization by simple organic synthetic methods is ideal. This will enable wide-scale application in multiple fields.
(3) Unique host–guest property: Encapsulation of guest molecules that cannot be included in typical macrocyclic compounds and a high discrimination for a slight shape difference of the guest are ideal. Formation of host–guest complexes with guest molecules in various media, such as aqueous, organic, bulk liquid, solid and gas phases in some cases, is also one of the unique host–guest properties.
(4) Unknown structures: New structures, which have not been reported, are ideal for the discovery of new beneficial properties. New structures will contribute to providing insight into structural chirality, self-assembling properties, mobility and reactivity.

Table 1.1 presents the advantages and disadvantages of typical macrocyclic compounds in addition to pillar[*n*]arenes.

Table 1.1 Comparison between pillar[*n*]arenes and well-known host molecules.

Name / Features	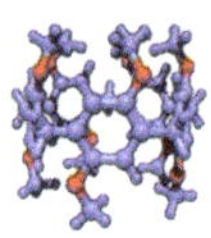 Pillar[n]arene	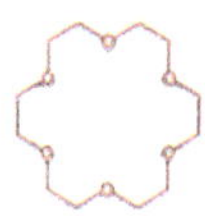 Crown Ether	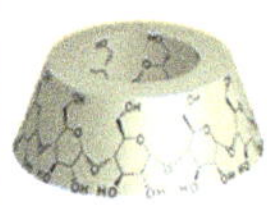 Cyclodextrin	 Calix[n]arene	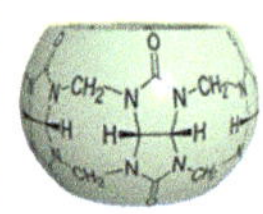 Cucurbit[n]uril
Synthesis/ Yield	◎	○	—	○	△
Functionality	◎	○	◎	○	△
Homologue	○ 5, 6 △(7-14)	◎12-21	○ α, β, γ-CD	○ 4, 6, 8 (5,7,9-20)	◎ 5-8 (10,14)
Shape	Pillar	Crown	Bucket	Calix	Pumpkin
Complexation (Aqueous)	○	△	○	○	◎
Complexation (Organic)	○	○	△	○	×
Original Property	Planar Chirality Symmetric	Flexible	Biocompatible Chirality Nonsymmetirc	Nonsymmetirc	Rigid Structure Symmetric

Synthesis and yields of pillar[*n*]arenes are superior to other typical host molecules (these points will be discussed in Chapter 2). Pillar[*n*]arenes show versatile functionality when compared with other host molecules, which are very similar to CDs (this point will be expounded in Chapter 3). Pillar[*n*]arenes show highly symmetrical cylindrical structures, large free volumes, conformational mobility and planar chirality, which are different from typical macrocyclic compounds (these points will be expanded on in Chapter 4). The host–guest properties of pillar[*n*]arenes are also unique. Owing to the cylindrical structure of pillar[*n*]arenes, guest molecules can access the cavity of pillar[*n*]arenes from both rims. Pillar[5]arenes show highly shape-selective host–guest complexation. The driving force for the complexation is the multiple physical interactions such as CH/π, cation/π, charge-transfer (CT) and CH/O interactions. A π-electron-rich cavity and crown ether-like arrangement of oxygen atoms at both rims contribute to the unique host–guest properties. Detailed host–guest properties of pillar[*n*]arenes will be provided in Chapter 5. In Chapter 6, we discuss the assembled structure of pillar[*n*]arenes in the bulk state. These chapters will provide the fundamental properties of pillar[*n*]arenes, which have been uncovered by pillar[*n*]arene chemists. We will discuss the application of pillar[*n*]arenes in detail in later chapters. Based on the various advantages, pillar[*n*]arenes have been applied in various fields of supramolecular materials, such as supramolecular polymers (Wang *et al.* in Chapter 7), interlocked molecules (Ogoshi *et al.* in Chapter 8), supramolecular assemblies (Zhao *et al.* in Chapter 9), hybrid materials (Yang *et al.* in Chapter 10), and biomedical applications (Hou *et al.* in Chapter 11). Furthermore, the introduction of pillar[*n*]arenes in host–guest chemistry has inspired many scientists to discover new and exciting macrocyclic compounds including asar[*n*]arenes, which were first introduced by the Stoddart group. In Chapter 12, Stoddart *et al.* review pillar[*n*]arene-related macrocyclic compounds including the asar[*n*]arenes.

References

1. J. M. Lehn, *Supramolecular Chemistry, Concepts and Perspectives*, Wiley-VCH, Weinheim, 1995.
2. A. Villiers, *C. R. Fr. Acad. Sci.*, 1891, 435–438.
3. G. Crini, *Chem. Rev.*, 2014, **114**, 10940–10975.
4. M. V. Rekharsky and Y. Inoue, *Chem. Rev.*, 1998, **98**, 1875–1918.
5. A. Nakamura and Y. Inoue, *J. Am. Chem. Soc.*, 2003, **125**, 966–972.
6. A. Nakamura and Y. Inoue, *J. Am. Chem. Soc.*, 2005, **127**, 5338–5339.
7. M. V. Rekharsky and Y. Inoue, *J. Am. Chem. Soc.*, 2002, **124**, 813–826.
8. C. Yang, T. Mori, Y. Origane, Y. H. Ko, N. Selvapalam, K. Kim and Y. Inoue, *J. Am. Chem. Soc.*, 2008, **130**, 8574–8575.
9. J. Yao, Z. Yan, J. Ji, W. Wu, C. Yang, M. Nishijima, G. Fukuhara, T. Mori and Y. Inoue, *J. Am. Chem. Soc.*, 2014, **136**, 6916–6919.

10. A. Harada, A. Hashidzume, H. Yamaguchi and Y. Takashima, *Chem. Rev.*, 2009, **109**, 5974–6023.
11. S. A. Nepogodiev and J. F. Stoddart, *Chem. Rev.*, 1998, **98**, 1959–1976.
12. G. Wenz, B.-H. Han and A. Müller, *Chem. Rev.*, 2006, **106**, 782–817.
13. Y. Okumura and K. Ito, *Adv. Mater.*, 2001, **13**, 485–487.
14. K. Uekama, F. Hirayama and T. Irie, *Chem. Rev.*, 1998, **98**, 2045–2076.
15. A. R. Khan, P. Forgo, K. J. Stine and V. T. D'Souza, *Chem. Rev.*, 1998, **98**, 1977–1996.
16. E. Engeldinger, D. Armspach and D. Matt, *Chem. Rev.*, 2003, **103**, 4147–4174.
17. A. J. Pearce and P. Sinaÿ, *Angew. Chem., Int. Ed.*, 2000, **39**, 3610–3612.
18. J. Terao, S. Tsuda, Y. Tanaka, K. Okoshi, T. Fujihara, Y. Tsuji and N. Kambe, *J. Am. Chem. Soc.*, 2009, **131**, 16004–16005.
19. J. Terao, *Chem. Rec.*, 2011, **11**, 269–283.
20. T. Kida, Y. Fujino, K. Miyawaki, E. Kato and M. Akashi, *Org. Lett.*, 2009, **11**, 5282–5285.
21. T. Kida, T. Iwamoto, Y. Fujino, N. Tohnai, M. Miyata and M. Akashi, *Org. Lett.*, 2011, **13**, 4570–4573.
22. T. Kida, T. Iwamoto, H. Asahara, T. Hinoue and M. Akashi, *J. Am. Chem. Soc.*, 2013, **135**, 3371–3374.
23. C. J. Pedersen, *J. Am. Chem. Soc.*, 1967, **89**, 7017–7036.
24. R. M. Izatt and J. J. Christensen, *Synthetuc Multidentate Macrocyclic Compounds*, Academic Press, New York, 1978.
25. B. O. Strasser and A. I. Popov, *J. Am. Chem. Soc.*, 1985, **107**, 7921–7924.
26. S. Shinkai, T. Minami, Y. Kusano and O. Manabe, *J. Am. Chem. Soc.*, 1983, **105**, 1851–1856.
27. J. Rebek, J. E. Trend, R. V. Wattley and S. Chakravorti, *J. Am. Chem. Soc.*, 1979, **101**, 4333–4337.
28. F. M. Raymo and J. F. Stoddart, *Chem. Rev.*, 1999, **99**, 1643–1664.
29. J. D. Badjić, V. Balzani, A. Credi, S. Silvi and J. F. Stoddart, *Science*, 2004, **303**, 1845–1849.
30. D. G. Lee and V. S. Chang, *J. Org. Chem.*, 1978, **43**, 1532–1536.
31. D. Landini, A. Maia, F. Montanari and P. Tundo, *J. Am. Chem. Soc.*, 1979, **101**, 2526–2530.
32. B. Dietrich, J. M. Lehn and J. P. Sauvage, *Tetrahedron Lett.*, 1969, **10**, 2885–2888.
33. C. D. Gutsche, *Calixarenes*, The Royal Society of Chemistry, Cambridge, 1989.
34. A. von Baeyer, *Ber. Dtsch. Chem. Ges.*, 1872, **5**, 25–26.
35. A. Ninagawa and H. Matsuda, *Makromol. Chem., Rapid Commun.*, 1982, **3**, 65–67.
36. Y. Nakamoto and S.-I. Ishida, *Makromol. Chem., Rapid Commun.*, 1982, **3**, 705–707.
37. D. R. Stewart and C. D. Gutsche, *J. Am. Chem. Soc.*, 1999, **121**, 4136–4146.
38. J. L. Atwood, G. A. Koutsantonis and C. L. Raston, *Nature*, 1994, **368**, 229–231.

39. S. Shinkai, S. Mori, T. Tsubaki, T. Sone and O. Manabe, *Tetrahedron Lett.*, 1984, **25**, 5315–5318.
40. S. Shinkai, S. Mori, H. Koreishi, T. Tsubaki and O. Manabe, *J. Am. Chem. Soc.*, 1986, **108**, 2409–2416.
41. D.-S. Guo and Y. Liu, *Acc. Chem. Res.*, 2014, **47**, 1925–1934.
42. A. Ikeda and S. Shinkai, *Chem. Rev.*, 1997, **97**, 1713–1734.
43. B. Schazmann, N. Alhashimy and D. Diamond, *J. Am. Chem. Soc.*, 2006, **128**, 8607–8614.
44. M. Takeuchi, M. Ikeda, A. Sugasaki and S. Shinkai, *Acc. Chem. Res.*, 2001, **34**, 865–873.
45. M. X. Wang, *Acc. Chem. Res.*, 2012, **45**, 182–195.
46. N. Morohashi, F. Narumi, N. Iki, T. Hattori and S. Miyano, *Chem. Rev.*, 2006, **106**, 5291–5316.
47. Z. Zhong, A. Ikeda and S. Shinkai, *J. Am. Chem. Soc.*, 1999, **121**, 11906–11907.
48. D. T. Schühle, J. Schatz, S. Laurent, L. Vander Elst, R. N. Muller, M. C. A. Stuart and J. A. Peters, *Chem. – Eur. J.*, 2009, **15**, 3290–3296.
49. D. T. Schühle, J. A. Peters and J. Schatz, *Coord. Chem. Rev.*, 2011, **255**, 2727–2745.
50. L. R. MacGillivray and J. L. Atwood, *Nature*, 1997, **389**, 469–472; Y. Aoyama, Y. Tanaka, H. Toi and H. Ogoshi, *J. Am. Chem. Soc.*, 1988, **110**, 634–635.
51. N. P. Power, S. J. Dalgarno and J. L. Atwood, *Angew. Chem., Int. Ed.*, 2007, **46**, 8601–8604.
52. S. K. Kim and J. L. Sessler, *Acc. Chem. Res.*, 2014, **47**, 2525–2536.
53. P. A. Gale, J. L. Sessler, V. Král and V. Lynch, *J. Am. Chem. Soc.*, 1996, **118**, 5140–5141.
54. D. J. Cram, S. Karbach, Y. H. Kim, L. Baczynskyj and G. W. Kallemeyn, *J. Am. Chem. Soc.*, 1985, **107**, 2575–2576.
55. Z. Laughrey and B. C. Gibb, *Chem. Soc. Rev.*, 2011, **40**, 363–386; C. L. D. Gibb and B. C. Gibb, *J. Am. Chem. Soc.*, 2004, **126**, 11408–11409.
56. J. W. Lee, S. Samal, N. Selvapalam, H.-J. Kim and K. Kim, *Acc. Chem. Res.*, 2003, **36**, 621–630.
57. J. Lagona, P. Mukhopadhyay, S. Chakrabarti and L. Isaacs, *Angew. Chem., Int. Ed.*, 2005, **44**, 4844–4870.
58. L. Isaacs, *Acc. Chem. Res.*, 2014, **47**, 2052–2062.
59. R. Behrend, E. Meyer and F. Rusche, *Justus Liebigs Ann. Chem.*, 1905, **339**, 1–37.
60. W. A. Freeman, W. L. Mock and N. Y. Shih, *J. Am. Chem. Soc.*, 1981, **103**, 7367–7368.
61. J. Kim, I. S. Jung, S. Y. Kim, E. Lee, J. K. Kang, S. Sakamoto, K. Yamaguchi and K. Kim, *J. Am. Chem. Soc.*, 2000, **122**, 540–541.
62. M. V. Rekharsky, T. Mori, C. Yang, Y. H. Ko, N. Selvapalam, H. Kim, D. Sobransingh, A. E. Kaifer, S. Liu, L. Isaacs, W. Chen, S. Moghaddam, M. K. Gilson, K. Kim and Y. Inoue, *Proc. Natl. Acad. Sci. U. S. A.*, 2007, **104**, 20737–20742.

63. Y. H. Ko, E. Kim, I. Hwang and K. Kim, *Chem. Commun.*, 2007, 1305–1315.
64. Y. Liu, H. Yang, Z. Wang and X. Zhang, *Chem. – Asian J.*, 2013, **8**, 1626–1632; U. Rauwald and O. A. Scherman, *Angew. Chem., Int. Ed.*, 2008, **47**, 3950–3953.
65. J. Zhao, H.-J. Kim, J. Oh, S.-Y. Kim, J. W. Lee, S. Sakamoto, K. Yamaguchi and K. Kim, *Angew. Chem., Int. Ed.*, 2001, **40**, 4233–4235.
66. S. Y. Jon, N. Selvapalam, D. H. Oh, J.-K. Kang, S.-Y. Kim, Y. J. Jeon, J. W. Lee and K. Kim, *J. Am. Chem. Soc.*, 2003, **125**, 10186–10187.
67. N. Zhao, G. O. Lloyd and O. A. Scherman, *Chem. Commun.*, 2012, **48**, 3070–3072.
68. Y. Ahn, Y. Jang, N. Selvapalam, G. Yun and K. Kim, *Angew. Chem., Int. Ed.*, 2013, **52**, 3140–3144.
69. D. Lucas, T. Minami, G. Iannuzzi, L. Cao, J. B. Wittenberg, P. Anzenbacher and L. Isaacs, *J. Am. Chem. Soc.*, 2011, **133**, 17966–17976.
70. B. Vinciguerra, L. Cao, J. R. Cannon, P. Y. Zavalij, C. Fenselau and L. Isaacs, *J. Am. Chem. Soc.*, 2012, **134**, 13133–13140.
71. Y. Miyahara, K. Goto, M. Oka and T. Inazu, *Angew. Chem., Int. Ed.*, 2004, **43**, 5019–5022.
72. J. Svec, M. Necas and V. Sindelar, *Angew. Chem., Int. Ed.*, 2010, **49**, 2378–2381.
73. W. R. Dichtel, O. Š. Miljanić, W. Zhang, J. M. Spruell, K. Patel, I. Aprahamian, J. R. Heath and J. F. Stoddart, *Acc. Chem. Res.*, 2008, **41**, 1750–1761.
74. M. Juricek, J. C. Barnes, E. J. Dale, W. G. Liu, N. L. Strutt, C. J. Bruns, N. A. Vermeulen, K. C. Ghooray, A. A. Sarjeant, C. L. Stern, Y. Y. Botros, W. A. Goddard, III and J. F. Stoddart, *J. Am. Chem. Soc.*, 2013, **135**, 12736–12746.
75. E. J. Dale, N. A. Vermeulen, A. A. Thomas, J. C. Barnes, M. Juricek, A. K. Blackburn, N. L. Strutt, A. A. Sarjeant, C. L. Stern, S. E. Denmark and J. F. Stoddart, *J. Am. Chem. Soc.*, 2014, **136**, 10669–10682.
76. B. M. Rambo, H.-Y. Gong, M. Oh and J. L. Sessler, *Acc. Chem. Res.*, 2012, **45**, 1390–1401.
77. H. Y. Gong, B. M. Rambo, E. Karnas, V. M. Lynch and J. L. Sessler, *Nat. Chem.*, 2010, **2**, 406–409.
78. S. Lee, C.-H. Chen and A. H. Flood, *Nat. Chem.*, 2013, **5**, 704–710.
79. B. E. Hirsch, S. Lee, B. Qiao, C.-H. Chen, K. P. McDonald, S. L. Tait and A. H. Flood, *Chem. Commun.*, 2014, **50**, 9827–9830.
80. M. Iyoda, J. Yamakawa and M. J. Rahman, *Angew. Chem., Int. Ed.*, 2011, **50**, 10522–10553.
81. A. Rajca and S. Rajca, *Angew. Chem., Int. Ed.*, 2010, **49**, 672–674.
82. H. Omachi, Y. Segawa and K. Itami, *Acc. Chem. Res.*, 2012, **45**, 1378–1389.
83. R. Jasti, J. Bhattacharjee, J. B. Neaton and C. R. Bertozzi, *J. Am. Chem. Soc.*, 2008, **130**, 17646–17647.
84. H. Takaba, H. Omachi, Y. Yamamoto, J. Bouffard and K. Itami, *Angew. Chem., Int. Ed.*, 2009, **48**, 6112–6116.
85. S. Yamago, Y. Watanabe and T. Iwamoto, *Angew. Chem., Int. Ed.*, 2010, **49**, 757–759.

CHAPTER 2

Synthesis of Pillar[n]arenes

TOMOKI OGOSHI* AND TADA-AKI YAMAGISHI

Graduate School of Natural Science and Technology, Kanazawa University, Kakuma-machi, Kanazawa 920-1192, Japan
*Email: ogoshi@se.kanazawa-u.ac.jp

2.1 Introduction

Historically, there have been a number of cases where the synthesis of new macrocyclic hosts has appeared as a product of chance. For example, crown ethers were first synthesized by Pedersen in 1967.[1] However, he obtained the crown ether as a minor product by chance when he synthesized another compound. He observed that etherification between bi-functional catechol and 1,2-bis(2-chloroethoxy)ethane afforded dibenzo-18-crown-6 as a minor product (Scheme 2.1).

However, Pederson focused on the minor products, and discovered a general synthesis method for crown ethers using metal cation templates. Hence, his discovery was a starting point for supramolecular chemistry using artificial macrocyclic compounds. Another topic with an interesting history is cucurbit[*n*]uril chemistry. Cucurbit[*n*]urils date back to 1905 when Behrend reported the condensation of glycoluril with paraformaldehyde.[2] The full characterization of the cyclic hexamer, cucurbit[6]uril, was reported by Mock *et al.* in 1981.[3] Unfortunately, its solubility in common solvents, except for strongly acidic aqueous media, is extremely low. This has been the main reason for the slow development of cucurbit[*n*]uril chemistry. In 2000, Kim *et al.*[4] observed that tuning the reaction conditions of cucurbit[6]uril

Monographs in Supramolecular Chemistry No. 18
Pillararenes
Edited by Tomoki Ogoshi
© The Royal Society of Chemistry 2016
Published by the Royal Society of Chemistry, www.rsc.org

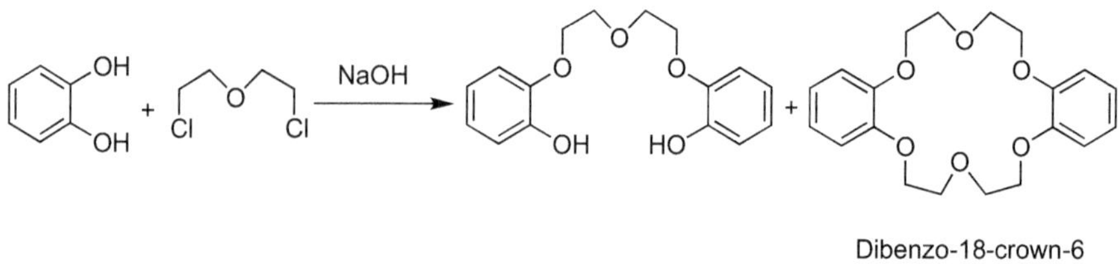

Scheme 2.1 Synthetic pathway of first crown ether by Pedersen.

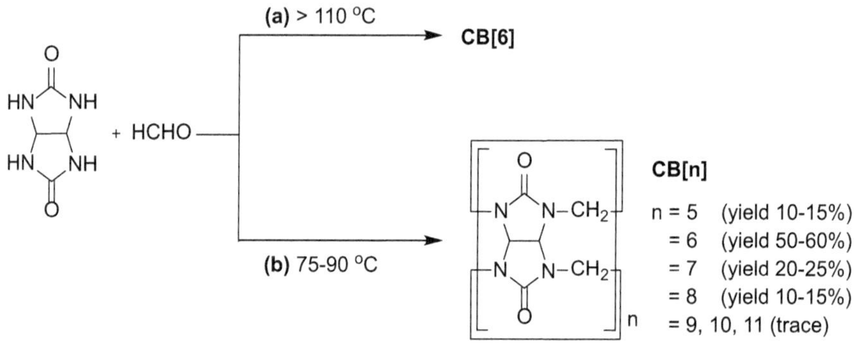

Scheme 2.2 (a) Conventional synthetic method of cucurbit[6]uril and (b) synthesis of cucurbit[*n*]uril homolog ($n = 5$–11) by Kim *et al.*[4]

synthesis resulted in the generation of smaller and larger cucurbit[*n*]uril homologs than cucurbit[6]uril. The key factor is the lower reaction temperature than that used in the conventional cucurbit[6]uril synthesis (Scheme 2.2).

Reaction at the lower temperature gave a family of cucurbit[*n*]urils from pentamer to 11-*mer*, although their yields, except for cucurbit[6]urils, were low (10–25%). Fortunately, the new synthetic receptors, smaller or larger than cucurbit[6]uril, show good solubility (CB[5]: *ca.* 420 mM; CB[7]: *ca.* 370 mM). Therefore, their discovery was a breakthrough in cucurbit[*n*]uril chemistry, and has led to a rapid increase in research in this area.[5,6]

The discovery of pillar[*n*]arenes was also by chance. In 2008, the synthesis of phenol-paraformaldehyde resins using a featureless phenolic derivative, 1,4-dimethoxybenzene, led to the report of the fascinating macrocycles, pillar[*n*]arenes, in the literature.[7] Initially, we could not understand why pillar[5]arenes could be obtained in high yields. However, extensive efforts by many pillar[*n*]arene chemists uncovered the mechanism for the synthesis of pillar[5]arenes, and led to the synthesis of larger pillar[*n*]arene homologs.[8,9] In this chapter, we provide a historical account of the synthesis of cyclic pentamers, pillar[5]arenes; cyclic hexamers, pillar[6]arenes; and cyclic oligomers, pillar[*n*]arenes larger than pillar[6]arenes.

2.2 Synthesis of Pillar[5]arenes

2.2.1 Synthesis of Pillar[5]arene from 1,4-Dimethoxybenzene as a Monomer

The first pillar[5]arene (**2.1**), which consists of five dimethoxybenzene units, was discovered by chance.[7] The discovery of this molecule was an important starting point for pillar[*n*]arene chemistry. From 2006, we investigated the reaction of 1,4-dimethoxybenzene with paraformaldehyde to synthesize new phenolic resins.

Generally, phenolic resins have been synthesized by reacting phenolic derivatives with aldehydes in acidic or basic conditions (Scheme 2.3).[10] *Ortho* and *meta* positions of phenolic derivatives are electron-rich; therefore, phenol derivatives react with aldehydes at the *ortho* and/or *para* sites of the phenolic derivatives.

The first reaction between a phenolic derivative and an aldehyde affords a hydroxymethyl phenol (Scheme 2.3b). The hydroxymethyl group is capable of reacting with the *ortho* or *para* sites of another phenolic derivative. Thus, the second reaction between the hydroxymethyl group and a phenolic derivative gives a phenolic dimer (Scheme 2.3c), which is connected by a methylene bridge. By repeating the reactions, phenolic polymers are obtained (Scheme 2.3d). Based on the reactive mechanism, we speculated that we could synthesize new phenolic polymers with controlled structures by designing phenolic monomers.[11] However, this idea was not new, because linear phenolic polymers had been synthesized using *para*-substituted phenolic derivatives. Linear phenolic polymers, in which phenolic units are connected by methylene bridges at the *ortho* position of the phenolic moieties, can be obtained by capping the reactive site at the *para* position of the phenolic monomer (Scheme 2.4a).

Under the controlled reaction conditions, macrocyclic compounds, calix[*n*]arenes, were prepared using *para*-substituted phenolic monomers.[12] When 1,2-dialkoxybenzenes are used as monomers, the reaction sites are the 4- and 5-positions of 1,2-dialkoxybenzenes, but not the 3- and 6-positions (Scheme 2.4b). This is because the electron density at the 4- and 5-positions is higher than that at the 3- and 6-positions. Steric hindrance of alkoxy

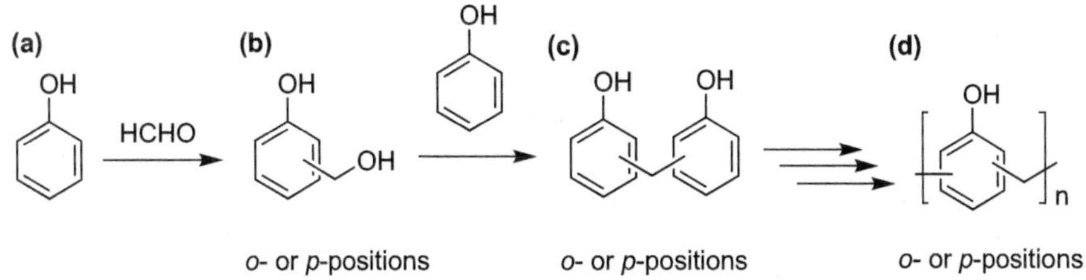

Scheme 2.3 Polymerization of phenol with paraformaldehyde.

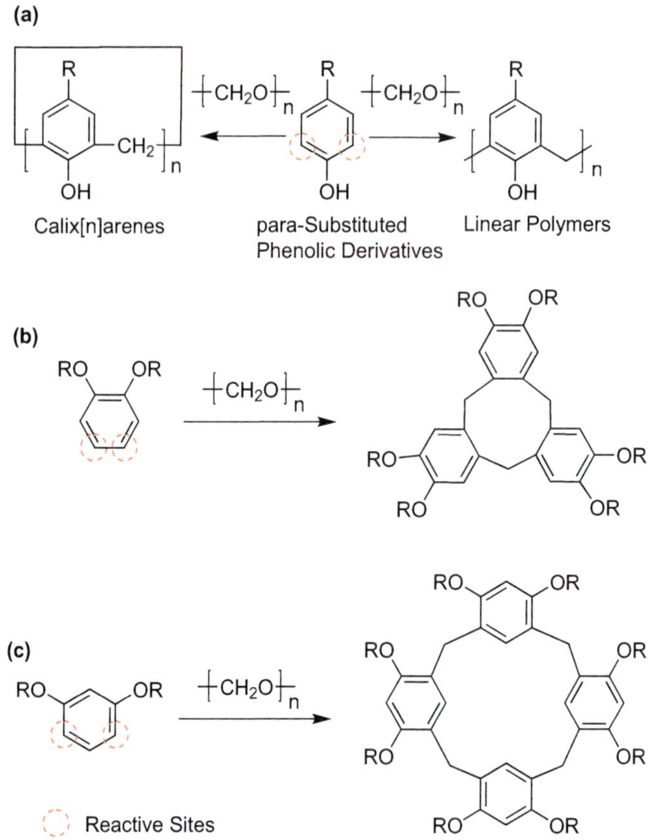

Scheme 2.4 Formation of macrocyclic compounds from phenolic monomers.

moieties reduces the reactivity at the 3- and 6-positions of 1,2-dialkoxybenzenes. By tuning the reaction conditions, cyclotriveratrylenes (CTVs),[13,14] which are macrocyclic trimers of 1,2-dialkoxybenzene (alias name, veratrole in 1,2-dimethoxybenzene) with a rigid bowl-shape, are obtained. When 1,3-dialkoxybenzenes are used as monomers, the reaction sites are the 4- and 6-positions (Scheme 2.4c). Reactivity of the 2- and 5-positions of 1,3-dialkoxybenzenes is low because of the steric hindrance of the alkoxy moieties and the lower electron density compared with the 4- and 6-positions. As with CTVs, by choosing appropriate aldehydes and tuning the reaction conditions, calix[4]resorcinarenes, which are rigid cyclic tetramers consisting of four resorcinols, were selectively obtained.[15,16] Based on these reactions, we chose 1,4-dimethoxybenzene as a monomer, which is a commercially available and cheap reagent (Scheme 2.5).

1,4-Dimethoxybenzene has four reaction sites that are electronically equal (2-, 3-, 5- and 6-positions). Therefore, one of the reaction sites should react with aldehyde, and afford the hydroxymethyl group. A subsequent reaction between the hydroxymethyl group and another 1,4-dimethoxybenzene yields

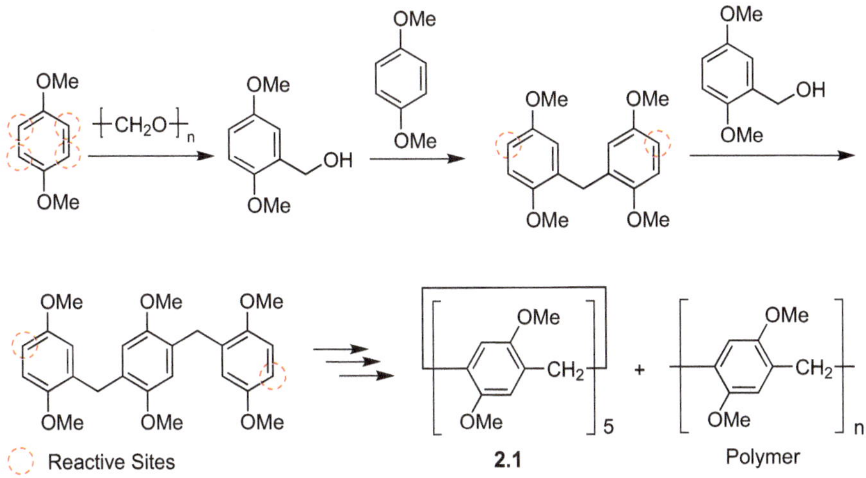

Scheme 2.5 Proposed pathway for the preparation of *para*-bridged pillar[5]arene and polymer.

a dimer structure. The dimer then reacts with another monomer with a hydroxymethyl group. The question arises as to which position of the dimer will react with the hydroxymethyl group on another monomer. There are three reactive sites in the dimer, the 3-, 5- and 6-positions. To clarify the reaction, we synthesized a phenolic polymer by reacting 1,4-dimethoxybenzne and paraformaldehyde using $FeCl_3$ as a Lewis acid catalyst. We then analyzed the polymer structure by ^1H and ^{13}C nuclear magnetic resonance (NMR) spectroscopy. A single peak from the methylene bridge was detected in the ^1H and ^{13}C spectra of the polymer, indicating that the polymerization reaction is position-selective and that the obtained polymer has a symmetrical structure. This indicates that the units are connected by the methylene bridge at the *para* positions (2- and 5-positions) of 1,4-dimethoxybenzenes and that the other reactive sites (3- and 6-positions) hardly reacted with paraformaldehyde. This arises from the steric hindrance of the methoxy moiety. The polymer obtained is therefore a phenolic polymer with structure control; therefore, we investigated the reaction conditions required to obtain a phenolic polymer with high molecular weight. We used Size-Exclusion Chromatography (SEC) measurements to monitor the polymerization using various Lewis acids (Figure 2.1). In $FeCl_3$, the product obtained was a polymer with a relatively high molecular weight (Figure 2.1a). However, when we used BF_3OEt_2 as a Lewis acid, surprisingly, the obtained product was not a polymer with a wide distribution but rather a particular oligomer (Figure 2.1b). To determine the reason for this observation, we first investigated the structure of the oligomer by ^1H NMR spectroscopy (Figure 2.2a).

Interestingly, the ^1H NMR spectrum of the oligomer was very simple: there were only three single peaks, from the benzene, methylene bridges and methyl moieties. This simple spectrum indicated a highly symmetrical cyclic

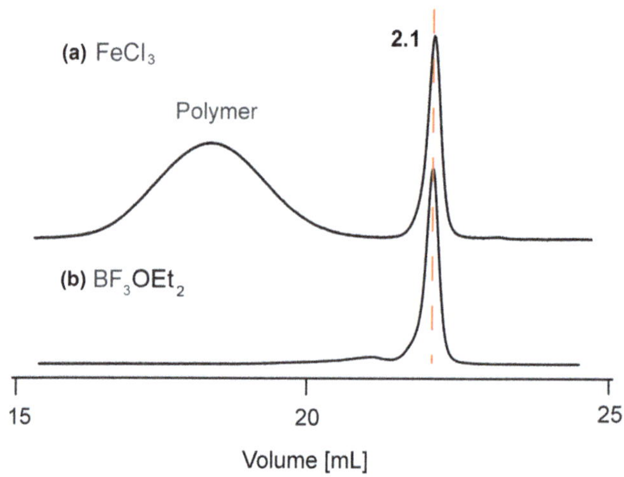

Figure 2.1 SEC traces of the products using Lewis acids.
Reproduced with permission from ref. 7. Copyright 2008 American Chemical Society.

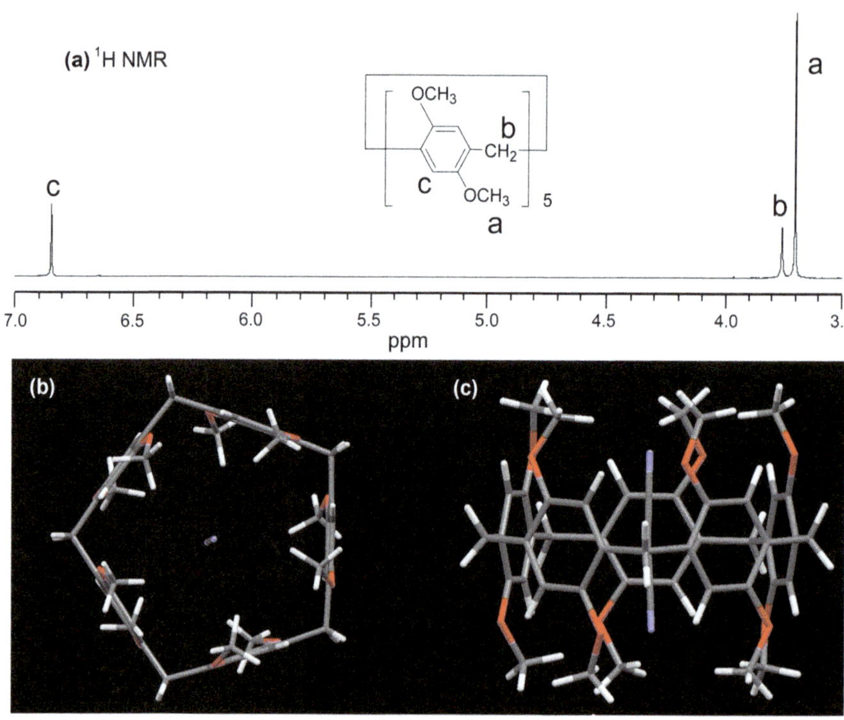

Figure 2.2 (a) 1H NMR spectrum and (b, c) X-ray crystal structure of per-methylated pillar[5]arene **2.1**.

structure of the oligomer. We also determined the mass of the oligomer by Fast Atom Bombardment Mass Spectrometry (FAB-MS) measurement. The mass of the product was 750, which corresponds to the cyclic pentamer. Therefore, the structure of the oligomer was proposed to be a highly symmetrical cyclic pentamer. Finally, we determined the key structural features of the oligomer by single X-ray crystalline analysis (Figure 2.2b and c). From the top view, its structure was a regular symmetrical pentagonal structure. Five 1,4-dimethoxybenzene units were connected by methylene bridges at the *para* position. From the side view, owing to the linkage at the *para* position, the structure had a highly symmetrical pillar-shape. Figure 2.2a shows the chemical structure of the macrocycle. The chemical structure is very similar to that of a calix[*n*]arene. However, one of the very important differences is the position of the methylene bridge. In calix[*n*]arenes, their units are connected by methylene bridges at the *meta* position (2,6-positions) of phenolic units. However, in this macrocycle, the units were connected by methylene bridges at the *para* position (2,5-positions). The different positions of the methylene bridges are very important and greatly affect their structure. In calix[*n*]arenes, owing to the *meta*-bridge linkage, their structures are open-ended and calix-shaped. This is why calix[*n*]arenes were named as such. However, because of the *para*-bridge linkage, the macrocycle obtained is not a calix-shaped structure but rather a pillar-shaped structure. We were reminded of the Parthenon at Athens in Greece in the motif of this macrocycle, therefore we named the cyclic pentamer pillar[5]arene.

To understand how the macrocycle, pillar[5]arene **2.1**, was selectively formed, we monitored the reaction by SEC (Figure 2.3).[17]

In the initial stage (after 10 seconds), various linear oligomers were mainly formed. However, we observed the peak from pillar[5]arene after 20 seconds. After 3 min, the major product was pillar[5]arene and there were small amounts of linear polymer contained in the product. The reaction is very quick, with 3 min being sufficient for completion of the reaction. The reaction proceeds under thermodynamic control, therefore pillar[5]arene can be obtained as a major product even within 3 min. However, the yield of pillar[5]arene was 28% in our first report on pillar[5]arene, and therefore, we searched for the optimized conditions for its synthesis. The feed ratio between 1,4-dimethoxybenzene and paraformaldehyde is very important to improve the yield of pillar[5]arene. Using an excess of 1,4-dimethoxybenzene to paraformaldehyde, the product was a mixture of oligomers and pillar[5]arene. Owing to the excess amount of 1,4-dimethoxybenzene, 1,4-dimethoxybenzene-end oligomers should be formed. The reaction of the 1,4-dimethoxybenzene-end oligomers does not take place because of the absence of paraformaldehyde. However, when we used excess amounts of paraformaldehyde to 1,4-dimethoxybenezene, the product was only pillar[5]arene. We therefore used excess paraformaldehyde, thus the end groups of the oligomers should be methylols. The methylol moieties are highly reactive, therefore they were converted to the thermally stable

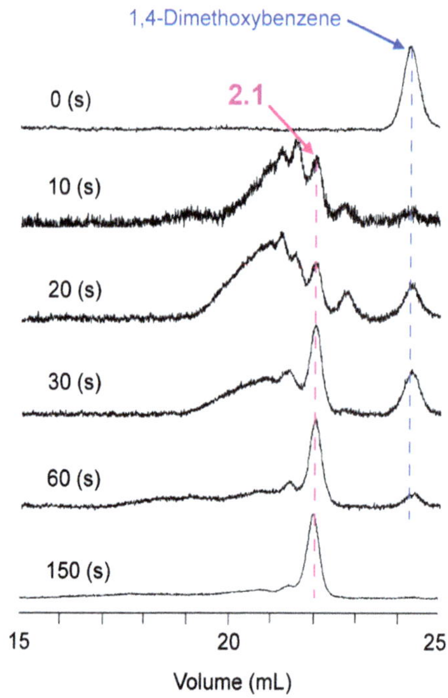

Figure 2.3 SEC traces of the products after quenching of the reaction with methanol. The feed ratio (1,4-dimethoxybenzene : paraformaldehyde) is 1 : 1. Reproduced with permission from ref. 17. Copyright 2011 American Chemical Society.

pillar[5]arene. The reaction using 3 equivalents of paraformaldehyde with 1,4-dimethoxybenzene over 3 min at 25 °C was observed to be the optimized conditions, which afforded pillar[5]arene in high yield (71%). The rapid high-yield synthesis of pillar[5]arene from commercially available reagents is outstanding compared with the synthesis of the other macrocyclic compounds, and has led to the widespread use of pillar[5]arenes in various fields of chemistry. Additionally, the first pillar[5]arene has now itself become a commercially available compound.[18]

2.2.2 Synthesis of Pillar[5]arenes from 1,4-Dialkoxybenzene Monomers

Introducing functional groups contributes to improving the host–guest property, changing the conformational properties and increasing the solubility of pillar[5]arenes. Therefore, first, we investigated the synthesis of pillar[5]arenes using 1,4-dialkoxybenzene monomers containing various alkoxy moieties in addition to methoxy moieties (Scheme 2.6).[19] In these cases, the products obtained were often a mixture of pillar[5]arene derivatives and linear oligomers. We successfully isolated pillar[5]arene derivatives

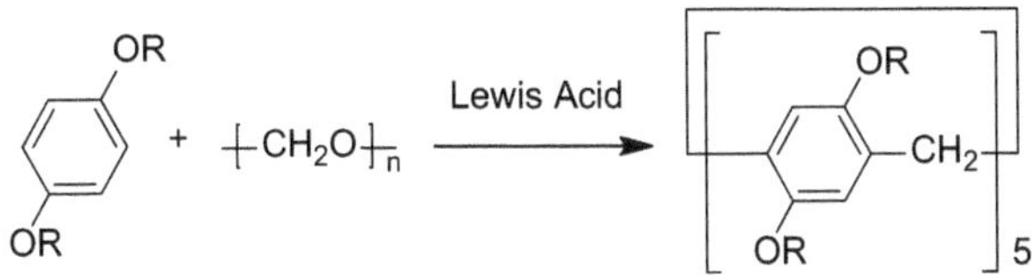

R = C_2H_5 (**2.2**) yield 23%
= C_3H_7 (**2.3**) yield 28%
= n-C_4H_9 (**2.4**) yield 8%
= n-C_5H_{11} (**2.5**) yield 14%
= n-C_6H_{13} (**2.6**) yield 14%
= n-$C_{12}H_{15}$ (**2.7**) yield 7%
= CH_2CH_2Br (**2.8**) yield 40%
= $CH_2CH_2CH_2Br$ (**2.9**) yield 44%
= $CH_2CH_2CH_2CH_2Br$ (**2.10**) yield 35%
= $CH_2C{\equiv}CH$ (**2.11**) yield 56%

Scheme 2.6 Preparation of simple per-alkylated pillar[5]arenes and pillar[5]arenes with ten reactive moieties from 1,4-dialkoxybenzenes.

carrying various alkyl chains including ethoxy (**2.2**), propoxy (**2.3**), butoxy (**2.4**), pentyloxy (**2.5**), hexyloxy (**2.6**) and dodecanoxy (**2.7**) moieties using silica gel chromatography.

The formation of polymer was hardly observed when 1,4-diethoxybenzene was used as a monomer. This result was the same as that observed in the synthesis of per-methylated pillar[5]arene (**2.1**) from 1,4-dimethoxybenzene. In contrast, the product was a mixture of polymer and pillar[5]arene in the case of monomers with long alkyl chains. Long alkyl substituents tended to suppress the cyclization, because of the steric hindrance of the long alkyl substituents. As a result of the formation of polymeric compounds, the yields of pillar[5]arenes decreased as the length of the alkyl substituents increased. The length and bulkiness of the alkyl chains largely resulted in the rotation of the constituent units (details are discussed in Chapter 4). The synthesis of pillar[5]arene derivatives containing reactive moieties is also very important to the synthesis of further functionalizing pillar[5]arenes. Pillar[5]arene derivatives containing reactive bromide[20,21] (**2.8**–**2.10**) and alkyne[22] (**2.11**) moieties could be synthesized by a cyclization reaction of 1,4-dialkoxybenzenes containing these functional groups. These pillar[5]arenes can be used for the synthesis of further functionalizing pillar[5]arenes owing to the high reactivity of these functional moieties (details are discussed in Chapter 3).

The reaction of non-symmetrical monomers, such as 1-ethoxy-4-methoxybenzene and 1-butoxy-4-methoxybenzene, affords the non-symmetrical pillar[5]arenes (Scheme 2.7) ethoxymethoxypillar[5]arene (**2.12**) and butoxymethoxypillar[5]arene (**2.13**), respectively.[23–25]

However, their detailed structures were not clarified. The Huang group and the Cao and Meier group reported the detailed structures of non-symmetric pillar[5]arenes. There are three possible types (A–C) of

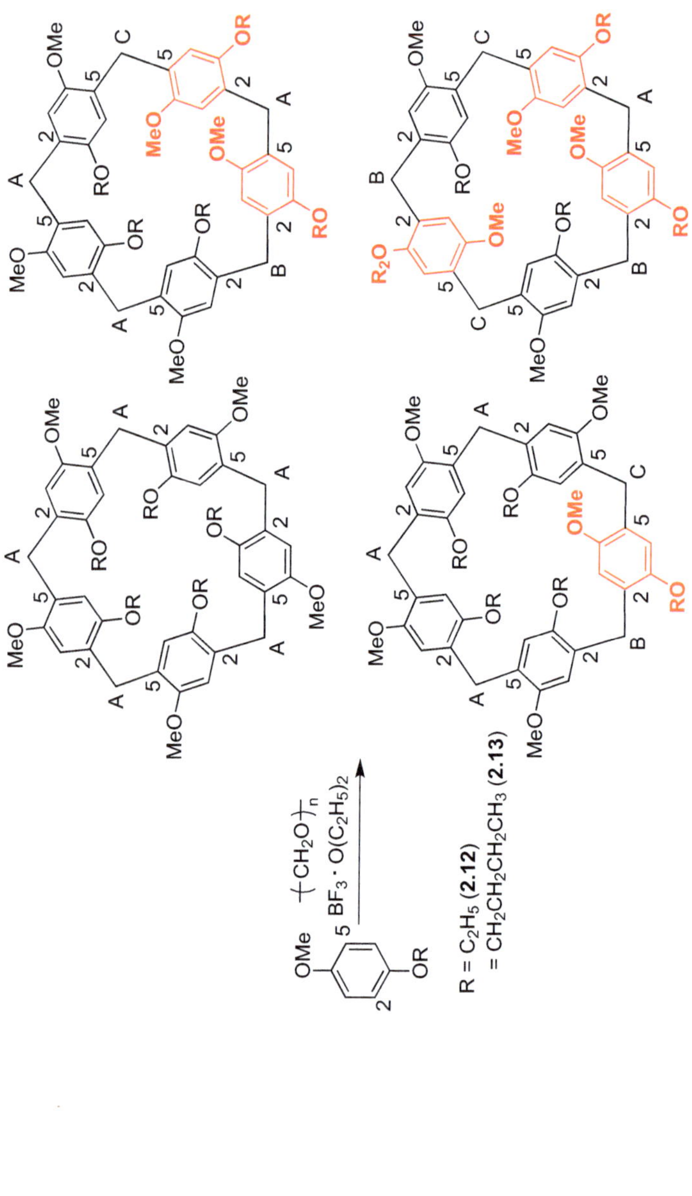

Scheme 2.7 Formation of four constitutional isomers from 1,4-dialkoxybenzene with different alkyl chains.

condensation reaction for the methylene bridge formation between the benzene rings. Type A is generated by reacting the 2- and 5-positions of non-symmetrical 1,4-dialkoxybenzene. In this case, same-length substituents orient in the same directions. Formation of the methylene bridge between the 2- and 2- or the 5- and 5-positions of non-symmetrical 1,4-dialkoxybenzene monomers results in Type B or C, respectively. In both cases, the same alkoxy substituents position on alternating sides. The three types of methylene bridge link, A, B, and C, can generate four constitutional isomers. The experimental distribution of the four isomers is 4.1 : 4.4 : 5.0 : 1.0, which is almost the same as the theoretical prediction (5 : 5 : 5 : 1). The lack of regioselectivity means that there is no electronic or steric preference for the reaction. The selective synthesis of one constitutional isomer from the four constitutional isomers remains a challenge and would contribute to revealing the mechanism for the synthesis of pillar[5]arenes.

1,4-Dialkoxybenzene monomers with tri(ethylene oxide) and ester moieties did not afford pillar[5]arene derivatives. In the case of monomers with tri(ethylene oxide) and ester moieties, these moieties should form complexes with Lewis acids. 1,4-Dialkoxybenzene monomers containing benzyl substituents also could not yield cyclic compounds. A Friedel–Crafts reaction took place between the benzyl moieties and paraformaldehyde in the case of 1,4-dialkoxybenzene with benzyl moieties, which led to the formation of insoluble gels. When 1,4-xylene and 1,4-diethylbenzene were used as the monomers, the products obtained were insoluble in all solvents. As described in Chapter 12, when 1,2,4,5-tetramethoxybenzene is used as a monomer, the product obtained is a highly symmetrical hexagonal macrocyclic compound.[26] In this case, to overcome the steric hindrance of the methoxy substituents, heating is necessary to obtain the cyclic compounds (details in Chapter 12). Pursuing further knowledge regarding the substituent effects on the cyclization reaction would enable the synthesis of new macrocyclic compounds even when simple monomers are used.

2.2.3 Mechanism of the Formation of Pillar[5]arenes – Thermodynamic Control

New macrocyclic compounds have been synthesized by chemists. However, the synthesis of new macrocyclic compounds has many problems, such as low yields, difficult purification and multi-step reactions. These drawbacks inhibit the wide use of the new macrocyclic compounds. In contrast, pillar[5]arenes can be synthesized in air by mixing commercially available and cheap reagents, and purified by simple re-crystallization. In addition to the easy synthesis of pillar[5]arenes, they are obtained in high yields. These advantages are the reasons why pillar[*n*]arene chemistry has progressed. However, the question remains as to why the cyclic pentamers, pillar[5]arenes, are selectively obtained in high yields. As previously stated, the feed ratio between 1,4-dimethoxybenzene and paraformaldehyde plays an important role in the yields. The other factors are the Lewis acids, solvents and reaction temperatures. From the comprehensive studies by pillar[*n*]arene chemists, the solvent used for the reaction is a key factor in the formation of cyclic pentamers, pillar[5]arenes.

From our first investigation into the synthesis of pillar[5]arenes, the Lewis acid was shown to affect the cyclization reaction (see Figure 2.1).[7] Using BF_3OEt_2 gives the macrocyclic pentamers, pillar[5]arenes, in high yields, whereas use of other Lewis acids decreased their yields. Therefore, BF_3OEt_2 was predicted to act as a template to form cyclic pentamers. However, this prediction was incorrect. Pillar[5]arenes can be obtained by choosing an appropriate organic acid. The Szumna group[27] applied organic acids instead of Lewis acids. Acetic acid and HCl did not promote the cyclization reaction. *p*-Toluenesulfonic acid promoted the formation of pillar[5]arene, but the yield was low (38%). Similar results were reported by the Zhang and Yang groups.[28] When using 15 mol% of *p*-toluenesulfonic acid in CH_2Cl_2, the cyclization occurred very slowly, and yielded only a trace amount of pillar[5]arene. A long reaction time and reflux conditions were required to obtain pillar[5]arene using *p*-toluenesulfonic acid in high yield (65%). When using trifluoroacetic acid (TFA), which is generally used as a catalyst for promoting various organic reactions, pillar[5]arene was obtained in high yield (81%). It is important to note that the reaction is not moisture-sensitive. Even in the presence of a small amount of water, the yield of pillar[5]arene was high (77%).

Lewis acid- or organic acid-catalyzed synthesis of pillar[*n*]arenes with different monomers has been reported by Cao and Meier *et al.*,[29] Huang *et al.*[30] and Nierengarten *et al.* (Scheme 2.8).[31]

Cao and Meier synthesized pillar[5]arene derivatives using 2,5-bis(alkoxymethyl)-1,4-diethoxybenzenes as monomers.[29] For these monomers, *p*-toluenesulfonic acid in CH_2Cl_2 was the best catalyst for the cyclization reaction. In their system, apart from cyclic pentamers (pillar[5]arenes) as major products, cyclic hexamers (pillar[6]arenes) were synthesized as minor products (*ca.* 10%). Huang *et al.* synthesized pillar[5]arenes from 2,5-alkoxybenzyl alcohols and 2,5-alkoxybenzyl bromides.[30] When the alcohols were used as monomers, the appropriate catalyst was $FeCl_3$, $AlCl_3$, $SnCl_4$ or BF_3OEt_2. However, when the bromides were used, the appropriate catalyst was $FeCl_3$, $AlCl_3$, $ZnCl_2$ or $SnCl_4$. These studies clarified that Friedel–Crafts alkylation is the main reaction mechanism for the formation of pillar[5]arenes. However, from the viewpoint of the commercial availability of the monomers and the number of reaction steps, the reaction system between 1,4-dialkoxybenzenes and paraformaldehyde is convenient for synthesis of pillar[5]arene derivatives and has been the main reaction method for its synthesis.

Nierengarten *et al.* revealed the reaction mechanism of pillar[5]arenes using 1,4-dialkoxybenzene derivatives (Scheme 2.8d).[31] To synthesize pillar[6]arenes, a co-cyclization of 2,5-bis(bromomethyl)-1,4-diethoxybenzene with 1,4-diethoxybenzene was performed. However, the product obtained was not cyclic hexamers (pillar[6]arenes), but cyclic pentamers (pillar[5]arenes). This

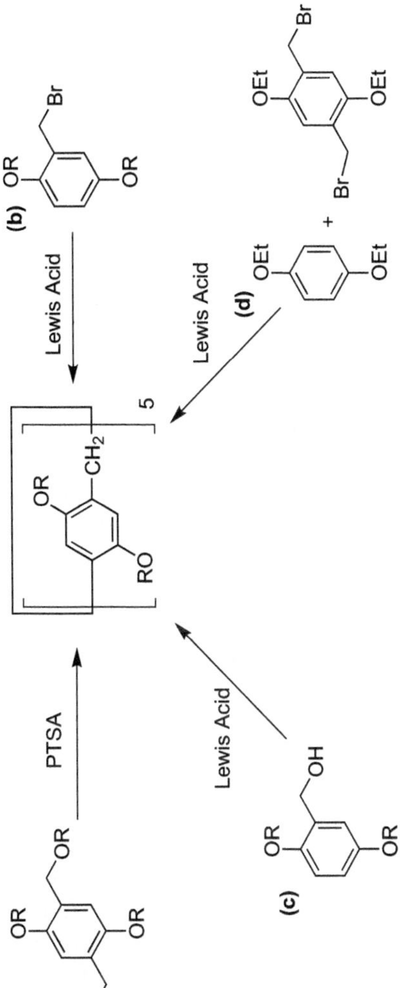

Scheme 2.8 Synthesis of pillar[5]arenes from 1,4-dialkoxybenzene derivatives as a monomer.

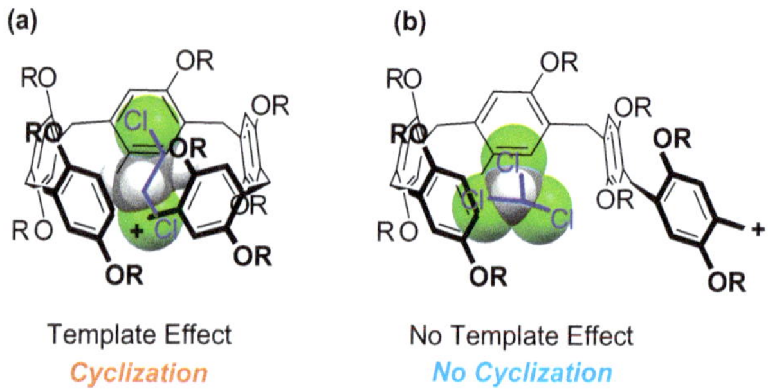

Figure 2.4 (a) Linear shaped dichloroethane acts as template for cyclic pentamers, pillar[5]arenes. (b) Branched shaped chloroform cannot act as a template for particular pillar[*n*]arene homologs.

result indicates that the dynamic covalent bond formation, methylene exchange reactions, occur under the Friedel–Crafts conditions.

We used 1,2-dichloroethane as a solvent for the synthesis of pillar[5]arenes because of the high yield.[7] However, investigation of the solvent effects has clarified that the solvents are very important for the formation of pillar[5]arenes.

Szumna and co-workers investigated the effect of solvent on the synthesis of pillar[5]arenes.[27] When dichloroethane was used as a solvent, the yields of pillar[5]arenes were relatively high. In contrast, the yield of pillar[5]arene was low when using chloroform as a solvent. They investigated host–guest complexation between pillar[5]arene and dichloroethane in chloroform. The association constant of the complex between pillar[5]arene and dichloroethane in deuterium chloroform, determined by ^{1}H NMR titration, was 650 M^{-1}. The chemical structure and properties of the solvent (chloroform) and the guest (dichloroethane) are very similar, therefore the complexation is derived from the size-fitting between host and guest.

Dichloroethane is a linear-shaped molecule, while chloroform has a branch-shaped structure with three chlorine moieties. They concluded that linear dichloroethane acts as a template for the synthesis of pillar[5]arenes (Figure 2.4).

Hence, the solvent is an important factor for the selective synthesis of pillar[5]arene. However, the effect of the reaction temperature on the cyclization reaction has not been investigated. Reaction temperature is a very important factor for the synthesis of cucurbit[*n*]uril and calix[*n*]arene homologs. This should therefore be the subject of further investigations.

2.3 Synthesis of Pillar[6]arenes under Kinetic and Thermodynamic Control

The type of solvent is a key factor for the selective synthesis of pillar[5]arenes. This observation is a clue to obtaining the high-yielding synthesis of larger pillar[*n*]arene homologs ($n>6$). Cyclic hexamers, pillar[6]arenes,

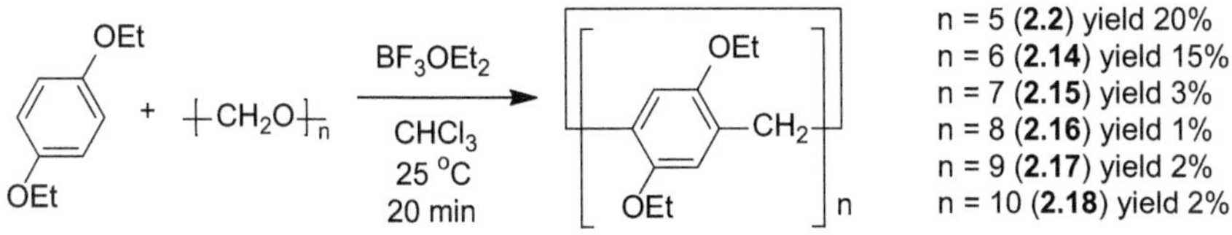

Scheme 2.9 Synthesis of pillar[*n*]arene homologs ($n=5$–10) from 1,4-diethoxybenzene by Hou *et al.*[8]

which were first prepared by Cao, Meier and co-workers, were synthesized using chloroform as a solvent in many cases, but with low yields even under optimized conditions.[29] Li *et al.* reported that the reaction of 1,4-diethoxybenzene with paraformaldehyde and BF_3OEt_2 in chloroform afforded a mixture of various pillar[*n*]arene homologs [**2.2** ($n=5$), **2.14** ($n=6$), **2.15** ($n=7$), **2.16** ($n=8$), **2.17** ($n=9$) and **2.18** ($n=10$)] under the optimized reaction time (Scheme 2.9).[8]

Therefore, chloroform does not act as a template for particular pillar[*n*]arene homologs, and pillar[6]arene is obtained under kinetic control. Yields of the pillar[*n*]arene homologs under kinetic control were much lower than those obtained under thermodynamic control. Therefore, we investigated several reaction solvents that fit into the pillar[6]arene cavity to act as a template solvent for the synthesis of pillar[6]arenes under thermodynamic control.

Pillar[6]arenes have larger cavities than pillar[5]arenes, and therefore can capture various bulky hydrocarbons (the host–guest properties of pillar[6]arene are mentioned in Chapter 5). We speculated that bulky hydrocarbons act as template solvents for the synthesis of pillar[6]arenes, therefore we chose bulky hydrocarbons, such as cyclohexane, chlorocyclohexane, cyclooctane and decalin, as solvents.[32] The limitation for the use of these hydrocarbons is the poor solubility of 1,4-dialkoxybenzene monomers, therefore we used 1,4-bis(methylcyclohexyl ether)benzene as a monomer (**2.19**), which is soluble in various hydrocarbon solvents because the cyclohexyl moiety is bulky and lipophilic. Pillar[6]arene (**2.21**) was obtained as a major product in high yield (87%) and pillar[5]arene (**2.20**) was obtained as a minor product (yield: 3%) when chlorocyclohexane was used as a solvent. Interestingly, pillar[*n*]arene homologs were not obtained using the other hydrocarbons as solvent. This is because pillar[*n*]arenes form more stable host–guest complexes with haloalkanes than with alkanes (details in Chapter 5). We monitored the synthesis of pillar[6]arene in chlorocyclohexane by ^{1}H NMR measurement. At the beginning of the reaction, the major product was linear oligomers, but these oligomers converted to pillar[6]arenes as the reaction time increased, indicating that pillar[6]arene is a thermodynamically stable compound in the reaction system (Figure 2.5).

This trend is the same as for the synthesis of pillar[5]arene in the template solvent of dichloroethane. Another interesting aspect is that the reaction is reversible between oligomers and the macrocyclic compounds. Nierengarten *et al.* investigated the dynamic covalent bond formation using a linear trimer

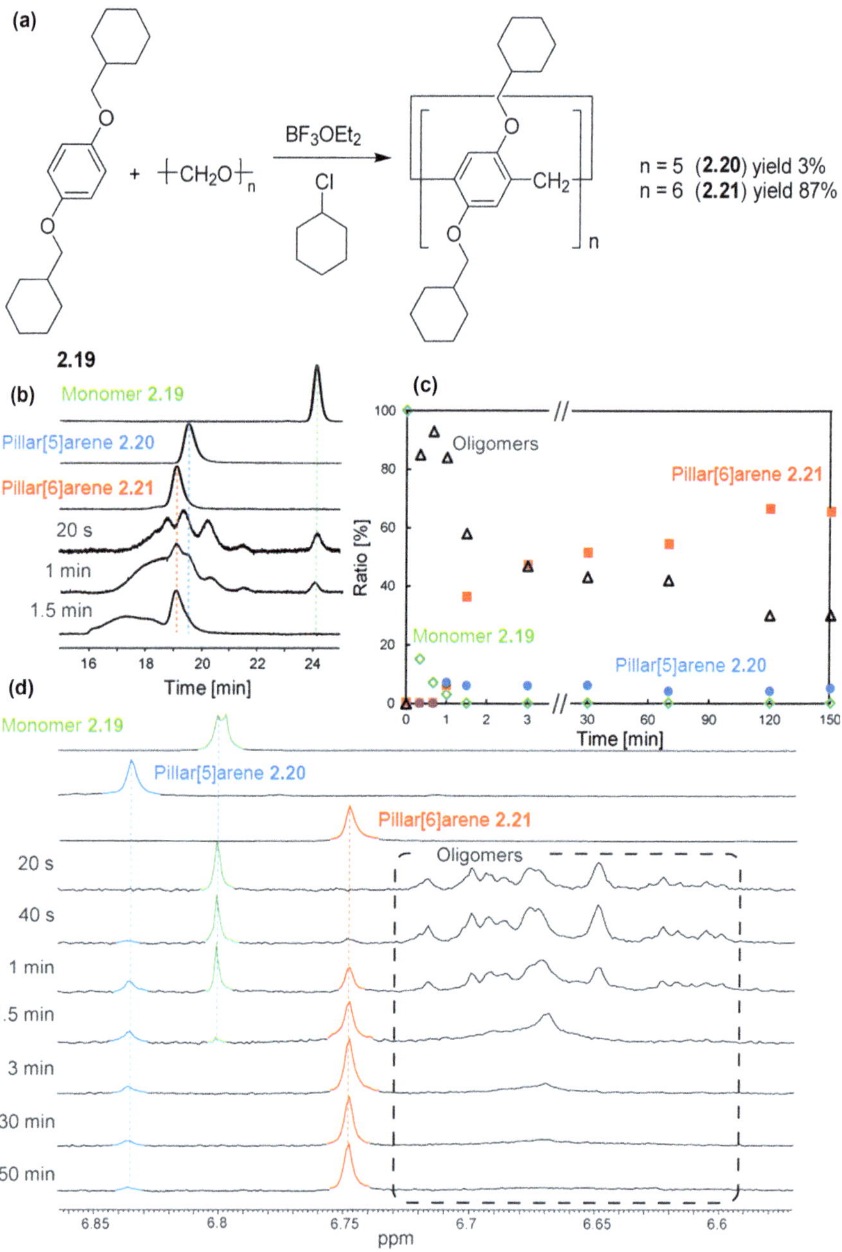

Figure 2.5 Selective synthesis of pillar[6]arene using chlorocyclohexane as template solvent. (a) Synthetic scheme of pillar[6]arene. Monitoring the reaction by (b) SEC and (d) partial ^{1}H NMR spectra ($CDCl_3$, 25 °C) and (c) ratios of monomer **2.19** (green diamonds), pillar[5]arene **2.20** (blue circles), pillar[6]arene **2.21** (red squares) and oligomers (black triangles) determined by ^{1}H NMR of the reaction mixtures over time.
Reproduced from ref. 32 with permission from The Royal Society of Chemistry.

as a monomer. Reaction of the trimer with paraformaldehyde and BF_3OEt_2 afforded the cyclic pentamer, but did not afford the expected cyclic hexamer (Scheme 2.10).[31]

On the basis of the dynamic covalent bond formation, we investigated the dynamic interconversion between pillar[5]arene **2.20** and pillar[6]arene **2.21** (Figure 2.6).[32] Pillar[6]arene converted to pillar[5]arene with BF_3OEt_2 in

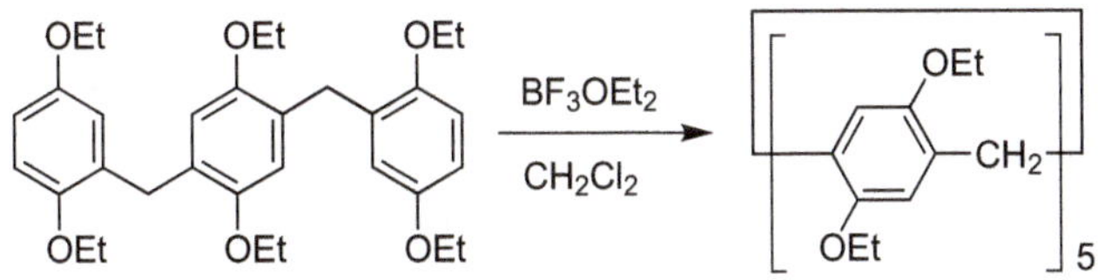

Scheme 2.10 Preparation of pillar[5]arene from trimer.

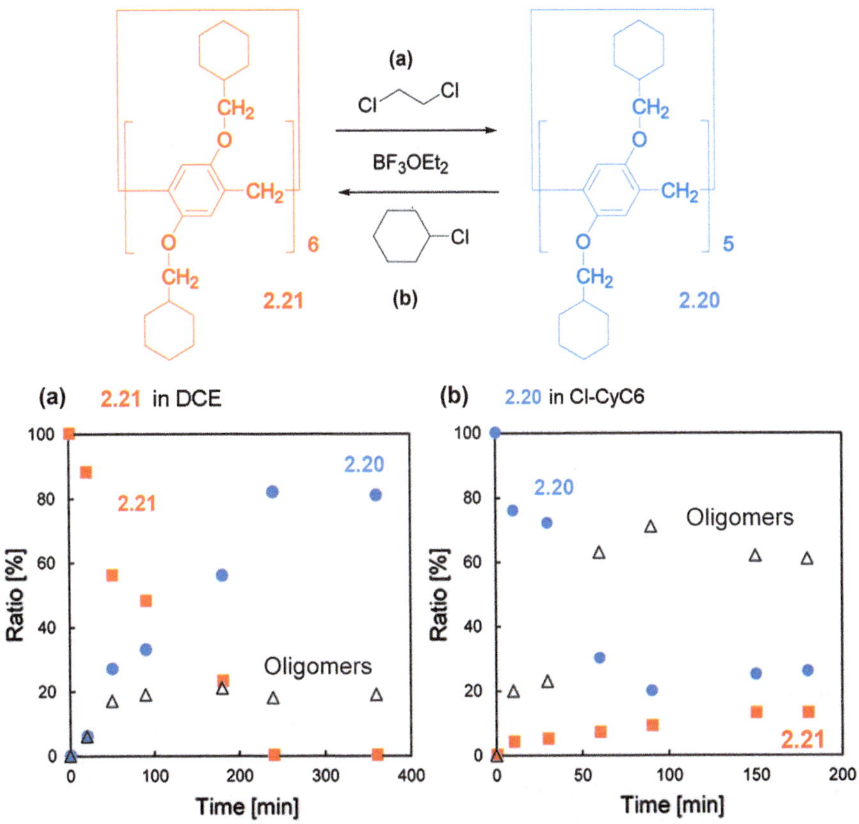

Figure 2.6 Interconversion between pillar[5]arene **2.20** and pillar[6]arene **2.21**. Ratios of **2.20** (blue circles), **2.21** (red squares) and the oligomers (black triangles) by cleaving (a) **2.21** in dichloroethane and (b) **2.20** in chlorocyclohexane at 80 °C, which were monitored by ^{1}H NMR analysis of the reaction mixtures over time.
Reproduced from ref. 32 with permission from The Royal Society of Chemistry.

dichloroethane, but did not convert in chlorocyclohexane. In contrast, pillar[5]arene changed to pillar[6]arene with BF_3OEt_2 in chlorocyclohexane, but did not change in dichloroethane. These experiments clearly showed the cleavage of Ar–CH_2 bonds in the presence of Lewis acids, and demonstrated that obtaining thermally stable macrocyclic compounds depends on the solvent.

2.4 Synthesis of Larger Pillar[*n*]arene Homologs under Kinetic Control

Using template solvents, pillar[5]- and pillar[6]arenes can be obtained in high yields.[17,32] Therefore, to date, pillar[5]- and pillar[6]arenes have been widely used. However, larger pillar[*n*]arene homologs ($n>7$) have larger cavity sizes than pillar[5]- and pillar[6]arenes, so they are able to capture larger guest molecules that are unable to be included in the cavities of pillar[5]- and pillar[6]arenes. Therefore, larger pillar[*n*]arene homologs have far more potential than pillar[5]- and pillar[6]arenes to further pillar[*n*]arene chemistry. Pillar[7–10]arenes were first reported by Hou and co-workers (see Scheme 2.9).[8] They used 1,4-diethoxybenzene as a monomer and chloroform as a solvent. Chloroform does not act as a template solvent for particular pillar[*n*]arene homologs, therefore the product was a kinetically stable mixture of pillar[*n*]arene homologs under the optimized reaction conditions. Silica gel chromatography enabled isolation of the new larger pillar[*n*]arene homologs ($n=7$–10). This method has advantages and disadvantages. The starting compound, 1,4-diethoxybenzne, is commercially available and cheap. However, a disadvantage is that the mixture after the reaction contained both pillar[5–15]arenes and soluble linear polymers and oligomers. The formation of soluble linear polymers and oligomers makes it difficult to isolate pillar[*n*]arene homologs because the solubility, polarity and molecular weight of the linear polymers and oligomers are very similar to those of the large pillar[*n*]arene homologs. To avoid this problem, we investigated the synthesis of larger pillar[*n*]arene homologs by a ring-opening reaction of cyclic pentamer, pillar[5]arene **2.2** (Figure 2.7).[9] The reaction of pillar[5]arene with BF_3OEt_2 in chloroform afforded a kinetically stable mixture of pillar[*n*]arene homologs ($n=5$–15) under the optimized reaction conditions. New pillar[*n*]arene homologs ($n=11$–13) were successfully isolated with known pillar[*n*]arene homologs ($n=6$–10) by silica gel chromatography. Disadvantages of this method are as follows: (1) we need to synthesize the starting compound pillar[5]arene **2.2** in advance, and (2) the yields of pillar[5–13]arenes are poor compared with Hou's method. However, the advantage of this method is that the formation of soluble byproducts, such as linear polymers and oligomers, is avoided. The lack of soluble byproducts makes the separation of pillar[*n*]arene homologs easy, thus leading to the isolation of new larger pillar[11–13]arene homologs.

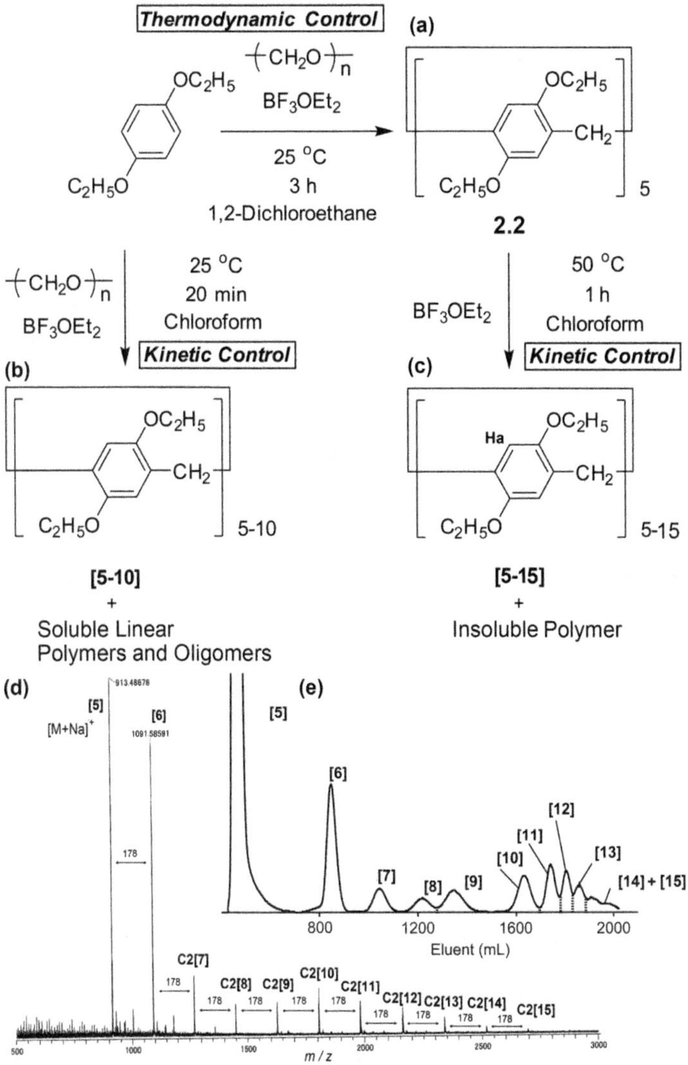

Figure 2.7 Synthesis of (a) pillar[5]arene **2.2** and (b) pillar[5–10]arenes from 1,4-diethoxybenzene and (c) pillar[5–15]arenes prepared from ring-opening reaction of pillar[5]arene **2.2**. (d) Electrospray ionization Mass Spectrometry (ESI-MS) spectrum of the mixture after the conversion reaction for 5 min. (e) Chromatogram of the mixture after the conversion reaction for 1 h. Reproduced with permission from ref. 9. Copyright 2014 American Chemical Society.

2.5 Conclusion and Outlook

This chapter describes the history of the first synthesis of pillar[5]arene, the cyclization mechanism of the formation of pillar[*n*]arenes, and the synthesis of larger pillar[*n*]arene homologs. Pillar[5]- and pillar[6]arenes have been obtained under thermodynamic control using template solvents, therefore yields of pillar[5]- and pillar[6]arenes are high. These high-yield syntheses have led to their use in various fields of chemistry. However, synthesis of larger pillar[*n*]arene homologs ($n>7$) has been performed under kinetic control, thus their yields are quite low. One of the challenges is, therefore, the synthesis of larger pillar[*n*]arene homologs under thermodynamic control using appropriate template solvents. To achieve this, it is necessary to determine the host–guest properties of larger pillar[*n*]arene homologs. However, guest molecules for larger pillar[*n*]arene homologs should have large volumes with high molecular weights, hence the guests should be solids. Therefore, the solvent–template method applied for the synthesis of pillar[5]- and pillar[6]arenes would have limitations for obtaining larger pillar[*n*]arenes. Hence, to obtain larger pillar[*n*]arene homologs, new synthetic methods are necessary. As was shown by the historical account of pillar[5]arene, the serendipitous discovery of a new selective synthesis method of larger pillar[*n*]arene homologs would be welcomed.

References

1. C. J. Pedersen, *J. Am. Chem. Soc.*, 1967, **89**, 7017–7036.
2. R. Behrend, E. Meyer and F. Rusche, *Justus Liebigs Ann. Chem.*, 1905, **339**, 1–37.
3. W. A. Freeman, W. L. Mock and N. Y. Shih, *J. Am. Chem. Soc.*, 1981, **103**, 7367–7368.
4. J. Kim, I. S. Jung, S. Y. Kim, E. Lee, J. K. Kang, S. Sakamoto, K. Yamaguchi and K. Kim, *J. Am. Chem. Soc.*, 2000, **122**, 540–541.
5. J. W. Lee, S. Samal, N. Selvapalam, H. J. Kim and K. Kim, *Acc. Chem. Res.*, 2003, **36**, 621–630.
6. J. Lagona, P. Mukhopadhyay, S. Chakrabarti and L. Isaacs, *Angew. Chem., Int. Ed.*, 2005, **44**, 4844–4870.
7. T. Ogoshi, S. Kanai, S. Fujinami, T. Yamagishi and Y. Nakamoto, *J. Am. Chem. Soc.*, 2008, **130**, 5022–5023.
8. X. B. Hu, Z. X. Chen, L. Chen, L. Zhang, J. L. Hou and Z. T. Li, *Chem. Commun.*, 2012, **48**, 10999–11001.
9. T. Ogoshi, N. Ueshima, F. Sakakibara, T. Yamagishi and T. Haino, *Org. Lett.*, 2014, **16**, 2896–2899.
10. W. Hesse, *Ullmann's Encyclopedia of Industrial Chemistry*, Wiley-VCH Verlag GmbH & Co. KGaA, 2000.
11. T. Ogoshi, K. Kitajima, K. Umeda, S. Hiramitsu, S. Kanai, S. Fujinami, T. Yamagishi and Y. Nakamoto, *Tetrahedron*, 2009, **65**, 10644–10649.
12. C. D. Gutsche, *Calixarenes*, Royal Society of Chemistry, Cambridge, 1989.
13. M. J. Hardie, *Chem. Soc. Rev.*, 2010, **39**, 516–527.
14. T. Brotin and J. P. Dutasta, *Chem. Rev.*, 2009, **109**, 88–130.
15. A. G. S. Hoegberg, *J. Am. Chem. Soc.*, 1980, **102**, 6046–6050.
16. Y. Aoyama, Y. Tanaka, H. Toi and H. Ogoshi, *J. Am. Chem. Soc.*, 1988, **110**, 634–635.

17. T. Ogoshi, T. Aoki, K. Kitajima, S. Fujinami, T. Yamagishi and Y. Nakamoto, *J. Org. Chem.*, 2011, **76**, 328–331.
18. Dimethoxypillar[5]arene, CAS Number: 1188423-16-6 Product Number: D4471 from TCI.
19. T. Ogoshi, K. Kitajima, T. Aoki, S. Fujinami, T. Yamagishi and Y. Nakamoto, *J. Org. Chem.*, 2010, **75**, 3268–3273.
20. T. Ogoshi, N. Ueshima, T. Yamagishi, Y. Toyota and N. Matsumi, *Chem. Commun.*, 2012, **48**, 3536–3538.
21. Y. Yao, M. Xue, X. D. Chi, Y. J. Ma, J. M. He, Z. Abliz and F. H. Huang, *Chem. Commun.*, 2012, **48**, 6505–6507; Y. Chang, K. Yang, P. Wie, S. Huang, Y. Pei, W. Zhao and Z. Pei, *Angew. Chem., Int. Ed.*, 2005, **44**, 4844–4870.
22. H. M. Deng, X. Y. Shu, X. S. Hu, J. Li, X. S. Jia and C. J. Li, *Tetrahedron Lett.*, 2012, **53**, 4609–4612.
23. T. Ogoshi, K. Kitajima, T. Yamagishi and Y. Nakamoto, *Org. Lett.*, 2010, **12**, 636–638.
24. Z. Zhang, Y. Luo, B. Xia, C. Han, Y. Yu, X. Chen and F. Huang, *Chem. Commun.*, 2011, **47**, 2417–2419.
25. Y. H. Kou, H. Q. Tao, D. R. Cao, Z. Y. Fu, D. Schollmeyer and H. Meier, *Eur. J. Org. Chem.*, 2010, **2010**, 6464–6470.
26. S. T. Schneebeli, C. Cheng, K. J. Hartlieb, N. L. Strutt, A. A. Sarjeant, C. L. Stern and J. F. Stoddart, *Chem. – Eur. J.*, 2013, **19**, 3860–3868.
27. T. Boinski and A. Szumna, *Tetrahedron*, 2012, **68**, 9419–9422.
28. K. Wang, L. L. Tan, D. X. Chen, N. Song, G. Xi, S. X. A. Zhang, C. Li and Y. W. Yang, *Org. Biomol. Chem.*, 2012, **10**, 9405–9409.
29. D. Cao, Y. Kou, J. Liang, Z. Chen, L. Wang and H. Meier, *Angew. Chem., Int. Ed.*, 2009, **48**, 9721–9723.
30. Y. J. Ma, Z. B. Zhang, X. F. Ji, C. Y. Han, J. M. He, Z. Abliz, W. X. Chen and F. H. Huang, *Eur. J. Org. Chem.*, 2011, 5331–5335.
31. M. Holler, N. Allenbach, J. Sonet and J. F. Nierengarten, *Chem. Commun.*, 2012, **48**, 2576–2578.
32. T. Ogoshi, N. Ueshima, T. Akutsu, D. Yamafuji, T. Furuta, F. Sakakibara and T. A. Yamagishi, *Chem. Commun.*, 2014, **50**, 5774–5777.

CHAPTER 3

Functionalization of Pillararenes

TOMOKI OGOSHI* AND TADA-AKI YAMAGISHI

Graduate School of Natural Science and Technology, Kanazawa University, Kakuma-machi, Kanazawa 920-1192, Japan
*Email: ogoshi@se.kanazawa-u.ac.jp

3.1 Introduction

Functionalization of macrocyclic compounds is very important because it can change the solubility and physical properties of the macrocyclic compounds.[1–3] For example, the solubility of cyclodextrins can be changed by capping the OH moieties.[1,2] Native cyclodextrins with many OH moieties are soluble in polar solvents, such as water, DMF and DMSO, but not soluble in nonpolar solvents, such as halogenated solvents. In contrast, per-methylated cyclodextrins, in which OH groups are protected by methyl moieties, are soluble in both polar and nonpolar solvents. Moreover, position-selective functionalization of macrocyclic compounds has expanded the possibility for material applications.[1,2,4–12] Mono-functionalized macrocyclic compounds are very important as building blocks for constructing supramolecular polymers and sensors, which detect substrates using color or fluorescence changes. Based on a di-functional monomer, 1D linear polymers, consisting of macrocyclic compounds as repeating units, can be produced. Crown ether-based 1D linear polymers, which were synthesized by Takata *et al.*, are good inter-molecular cross-linking agents between polymers.[13–15] However, the selective functionalization of macrocyclic compounds is still a challenging topic because numerous constitutional isomers

Monographs in Supramolecular Chemistry No. 18
Pillararenes
Edited by Tomoki Ogoshi
© The Royal Society of Chemistry 2016
Published by the Royal Society of Chemistry, www.rsc.org

are generated during the functionalization process. This makes isolation difficult and provides low yields of the particular isomers.[16–18]

Compared with other typical host molecules, one of the important advantages of pillar[*n*]arene chemistry is the versatile functionality.[19,20] A wide variety of multi-functionalized pillar[*n*]arene derivatives, which are difficult to obtain using other typical macrocyclic platforms, can be synthesized in moderate yields. To date, pillar[5]- and pillar[6]arenes, containing five and six repeat units, respectively, have mainly been used because they can be obtained in relatively high yields by a one-step reaction (details in Chapter 2). Pillar[5]- and pillar[6]arenes have 10 and 12 substituents, respectively. Figure 3.1 shows the assignment of positions in pillar[5]- and pillar[6]arenes.[21,22]

We have introduced a nomenclature system for identifying the functionalizing position in pillar[*n*]arenes.[21] The numbers and letters refer to the upper/lower rims and unit positions of the pillar[*n*]arenes, respectively. Mono-, di-, tetra-, penta- and per-functionalized pillar[5]arenes have been obtained to date. In the case of pillar[6]arenes, mono-, di- and per-functionalization have been achieved. Mono- and per-functionalized pillar[*n*]arenes have only one constitutional isomer. Therefore, synthesis of mono- and per-functionalized pillar[*n*]arenes is relatively easy. However, multi-functionalized pillar[*n*]arenes have many conformers.[22] For example, di-functionalized pillar[5]arenes have five possible conformers (A1/B2, A1/C1, A1/A2, A1/B1 and A1/C2). The number of possible conformers increased as the number of the functional groups increased: the tri-functionalized pillar[5]arene has 10 constitutional isomers (A1/B1/C1, A1/B1/D1, A1/B1/C2, A1/B1/D2, A1/A2/B1, A1/A2/B2, A1/A2/C1, A1/A2/C2, A1/B2/C1 and A1/B2/D1). As the number of rings increased, the number of conformers also increased: di-functionalized pillar[6]arenes have seven isomers (A1/B1, A1/C1, A1/D1, A1/A2, A1/B2, A1/C2 and A1/D2), while di-functionalized pillar[5]arenes have five. Isolation of a particular constitutional isomer from a mixture is quite difficult, therefore position-selective functionalization based on the unique

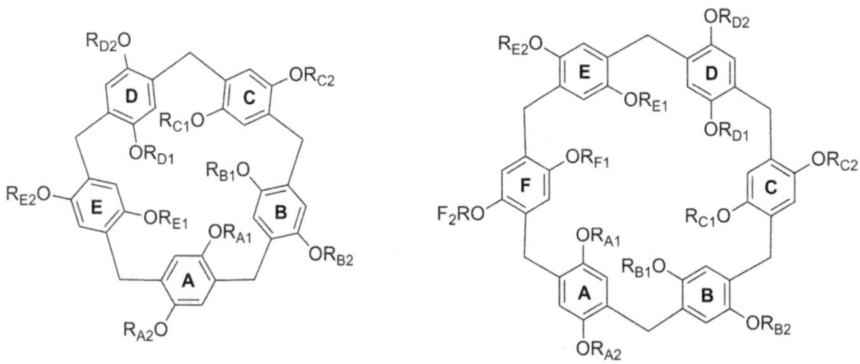

Figure 3.1 Assignment of positions in pillar[5]- and pillar[6]arenes. Stereoisomers generated by rotation of constituent units are not included.

organic synthesis is necessary to synthesize multi-functionalized pillar[*n*]-arenes. Position-selective functionalization has been achieved by two methods to date. One is the post-functionalization of pre-formed pillar[*n*]arenes, which was first developed by our groups.[23] The other is the synthesis of pillar[*n*]arenes by co-cyclization of two or more different monomers, which was first reported by Huang *et al.*[24] In this chapter, we discuss the functionalization of pillar[*n*]arenes by two methods for the synthesis of functionalized pillar[*n*]arene derivatives.

3.2 Per-functionalization of Pillar[*n*]arenes by Pre-formed Pillar[*n*]arenes

Per-functionalized pillar[*n*]arenes have only one isomer, therefore their synthesis is relatively easy. One of the reliable routes for per-functionalized pillar[*n*]arenes is by using pre-formed reactive pillar[*n*]arenes. Per-hydroxylated pillar[*n*]arenes having phenolic moieties at both rims are one of the useful key compounds for the synthesis of various functionalized pillar[*n*]arene derivatives because phenolic moieties are highly reactive. Removing the methoxy groups on the rim of pillar[5]arene (**3.1**) with BBr_3 afforded a completely de-protected pillar[5]arene with 10 OH moieties (**3.6**) in quantitative yield (Scheme 3.1).[25]

Scheme 3.2 shows per-functionalized pillar[5]arene derivatives prepared from per-hydroxylated pillar[5]arene **3.6** as a starting compound. Applying an efficient organic reaction is required for the perfect introduction of functional groups at all 10 reaction sites. A straightforward and efficient functionalization process for per-hydroxylated pillar[5]arene is etherification, which has been used for functionalization of the lower rims of phenolic moieties in calix[*n*]arene derivatives. Various functionalized pillar[5]arene derivatives can be obtained by etherification of per-hydroxylated pillar[5]-arene with an alkyl-halide in the presence of appropriate bases, such as NaH and K_2CO_3. Introduction of the substituents can change various physical

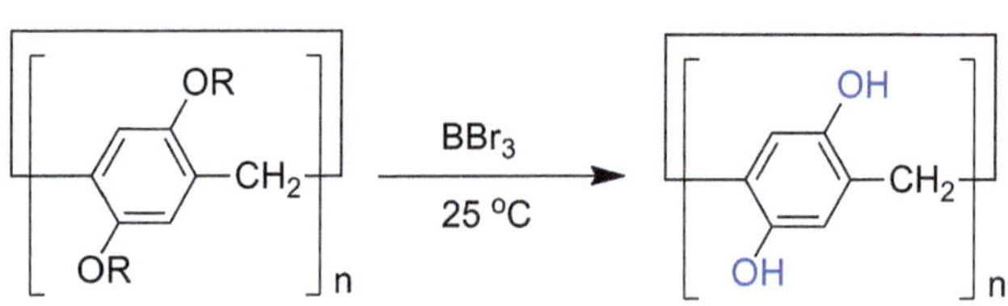

R = CH_3 (n = 5) **3.1**
C_2H_5 (n = 6) **3.2**
(n = 7) **3.3**
(n = 9) **3.4**
(n = 10) **3.5**

(n = 5) **3.6**
(n = 6) **3.7**
(n = 9) **3.8**
(n = 10) **3.9**

Scheme 3.1 Synthesis of per-hydroxylated pillar[*n*]arenes (*n* = 5, 6, 7, 9 and 10) by de-protection of alkoxy moieties.

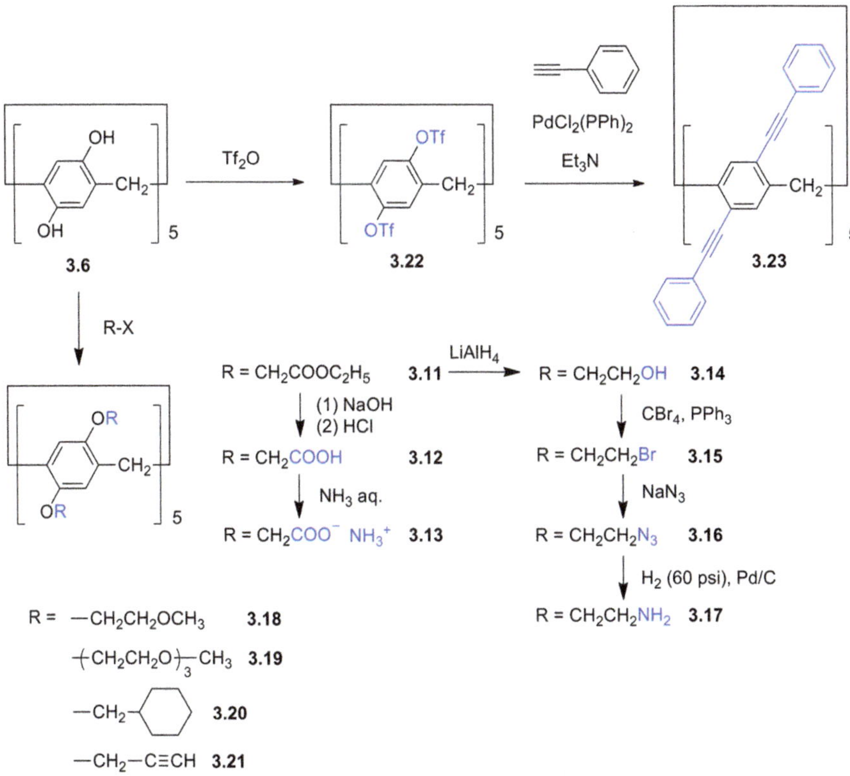

Scheme 3.2 Synthesis of per-functionalized pillar[5]arene derivatives from pre-formed per-hydroxylated pillar[5]arene.

properties of pillar[5]arenes, add new functionality and open the possibility of material applications of pillar[5]arenes. The pillar[5]arene core is surrounded by 10 functional groups; the type of functional group determines the physical properties of pillar[5]arene derivatives, such as dendrimers. For example, the substituents largely affect the solubility of pillar[5]arenes. The starting compound, per-hydroxylated pillar[5]arene **3.6**, can be soluble in polar solvents, such as methanol, acetone, tetrahydrofuran (THF), acetonitrile, DMSO and DMF, but is not soluble in halogenated solvents, such as chloroform and dichloromethane, toluene and water. In contrast, the introduction of water-soluble moieties, such as carboxylate anions (**3.13**)[23] and amine moieties (**3.17**),[26] led to water-soluble pillar[5]arenes.

Pillar[5]arene with 10 carboxylate anions (**3.13**) can be used as nanovalves[27] (details in Chapter 10) and ^{128}Xe-based biosensors.[28] Pillar[5]arenes with 10 ethylene oxide (**3.18**) and tri(ethylene oxide) moieties (**3.19**) were also synthesized by etherification.[29] Pillar[5]arene with 10 tri(ethylene oxide) moieties **3.19** was completely soluble in aqueous media, while pillar[5]arene with 10 ethylene oxide moieties **3.18** was insoluble. The introduction of longer water-soluble parts is required to increase the water solubility.

Interestingly, pillar[5]arene with 10 tri(ethylene oxide) moieties **3.19** was soluble in aqueous media at 25 °C, but not soluble with heating (45 °C), indicating that **3.19** showed a lower critical solution temperature (LCST). This is because pillar[5]arene with 10 tri(ethylene oxide) is an amphiphilic molecule: the core of pillar[5]arene consists of water-insoluble hydrophobic benzene and methylene bridge moieties and the 10 tri(ethylene oxide) arms are water-soluble amphiphilic moieties.

The introduction of substituents can change the conformational flexibility of pillar[*n*]arenes: modification of the bulky substituents of the cyclohexylmethyl moieties by etherification (**3.20**) leads to inhibition of the rotation units, and enables the separation of pillar[5]arene enantiomers (see Chapter 4 for details).[30] Various functional pillar[5]arenes can be produced from per-hydroxylated pillar[5]arene **3.6** by modification of various functional groups.

The other useful functionalization method of pillar[5]arenes from per-hydroxylated pillar[5]arene **3.6** is Pd-catalyzed cross-coupling to afford highly conjugated pillar[5]arenes, which show an additional photoluminescence property. Sonogashira coupling between a reactive intermediate triflate group-modified pillar[5]arene (**3.22**) and ethynylbenzene afforded phenylethynyl group-modified pillar[5]arene (**3.23**).[31] The highly conjugated pillar[5]arene showed a new emission band derived from the spatial π-delocalization of the repeating units within the cavity. The fluorescence property was largely affected by the measurement temperature because swinging movements of the π-conjugated units increased as the measurement temperatures increased. The fluorescent pillar[5]arene **3.23** can be used as a sensor where a fluorescence change is induced by the inclusion of guests.

The synthetic protocol for per-functionalized pillar[5]arene derivatives from per-hydroxylated pillar[5]arene can be applied to the synthesis of larger per-functionalized pillar[*n*]arene derivatives. A per-hydroxylated pillar[6]arene with 12 phenolic moieties (**3.7**) was synthesized by Huang and Hou *et al.* (Scheme 3.3).[32] **3.7** can be synthesized by de-protection of ethoxy moieties (**3.2**) with BBr_3, which is the same method as used for the synthesis of per-hydroxylated pillar[5]arene.

Water-soluble pillar[6]arenes with 12 carboxylate anions (**3.24**) have been synthesized by etherification of alkyl-halide with per-hydroxylated pillar[6]-arene (Scheme 3.3).[33] Pillar[6]arenes with 12 mono- (**3.25**) or tri(ethylene oxide) moieties (**3.26**) were also synthesized by etherification.[34,35] As with the pillar[5]arene with 10 tri(ethylene oxide) chains, **3.19**, tri(ethylene oxide)-substituted pillar[6]arene **3.26** also showed LCST behavior. The *trans* form of an azobenzene-guest derivative formed a host–guest complex with **3.26**, but the *cis* form of the azobenzene guest did not. Based on the photo-responsive host–guest system between **3.26** and the azobenzene guest, photo-switching of LCST behavior was demonstrated.

As with **3.6** and **3.7**, per-hydroxylated pillar[*n*]arene derivatives ($n = 7$, **3.8**; $n = 9$, **3.9**; $n = 10$, **3.10**) have been synthesized by the complete de-protection of ethoxy moieties on pillar[*n*]arenes ($n = 7$, **3.3**; $n = 9$, **3.4**; $n = 10$, **3.5**).[36–38]

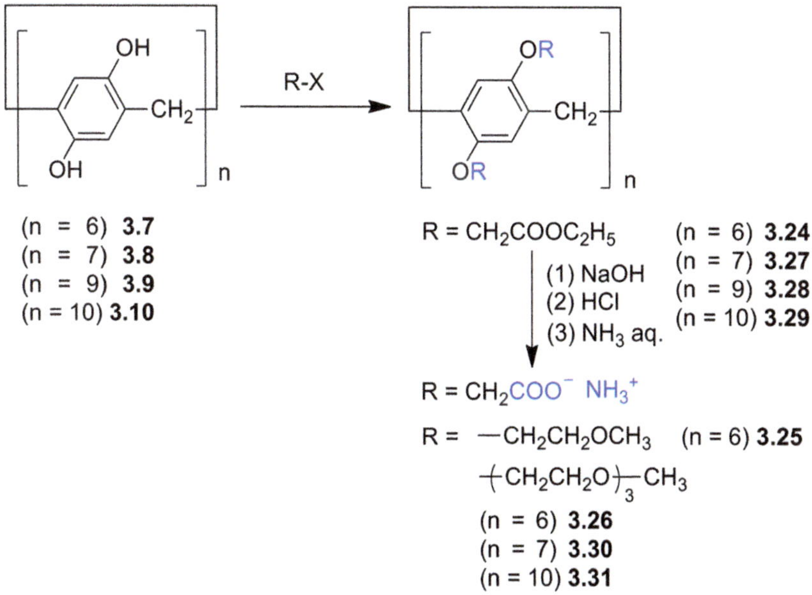

Scheme 3.3 Synthesis of per-functionalized pillar[*n*]arene homologs ($n \geqq 6$) from pre-formed per-hydroxylated pillar[*n*]arenes.

However, because of the low yields of the larger pillar[*n*]arenes, as these compounds were isolated as intermediates, they have not been completely characterized. Water-soluble pillar[7]-, pillar[9]- and pillar[10]arenes were also prepared by modification of the carboxylate moieties on the rims of per-hydroxylated pillar[7]- (**3.27**),[36] pillar[9]- (**3.28**)[37] and pillar[10]arenes (**3.29**).[39] Tri(ethylene oxide)-substituted pillar[7]- (**3.30**)[40] and pillar[10]arenes (**3.31**)[38] were synthesized. As was observed with the tri(ethylene oxide)-substituted pillar[5]- (**3.19**) and pillar[6]arenes (**3.26**), **3.30** and **3.31** also showed tunable LCST behavior based on their host–guest systems.

3.3 Per-functionalization of Pillar[*n*]arenes by Cyclization of Dialkoxybenzene Derivatives with Reactive Moieties

The other useful key compounds used to synthesize per-functionalized pillar[5,6]arenes are pillar[5,6]arenes having alkynes and azide moieties, because the reaction between alkynes and azide moieties in the presence of copper (I), the copper (I)-catalyzed Huisgen alkyne–azide 1,3-dipolar cycloaddition reaction (CuAAC reaction), is high yielding, very quick and unlikely to produce byproducts. Pillar[5]arene with 10 alkynes can be accessed by cyclization of a dialkoxybenzene monomer with two alkyne moieties (Chapter 2, Scheme 2.6, **2.11**)[41] or by reaction of per-hydroxylated pillar[5]arene with propargyl bromide (Scheme 3.2, **3.21**).[42]

We synthesized various functionalized pillar[5]arenes by the CuAAC reaction between decaalkynes **3.21** and azides (Scheme 3.4).[41] Reactions of the "clickable" decaalkyne **3.21** with hexylazide, benzylazide, phenylazide and pyrenylazide afforded functionalized pillar[5]arenes with 10 hexyl (**3.32**), benzyl (**3.33**), phenyl[42] (**3.34**) and pyrenyl moieties (**3.35**), respectively. Zhao *et al.* reported biocompatible pillar[5]arenes with amphiphilic oligoethylene glycol chains (**3.36** and **3.37**) by a CuAAC reaction between **3.21** and

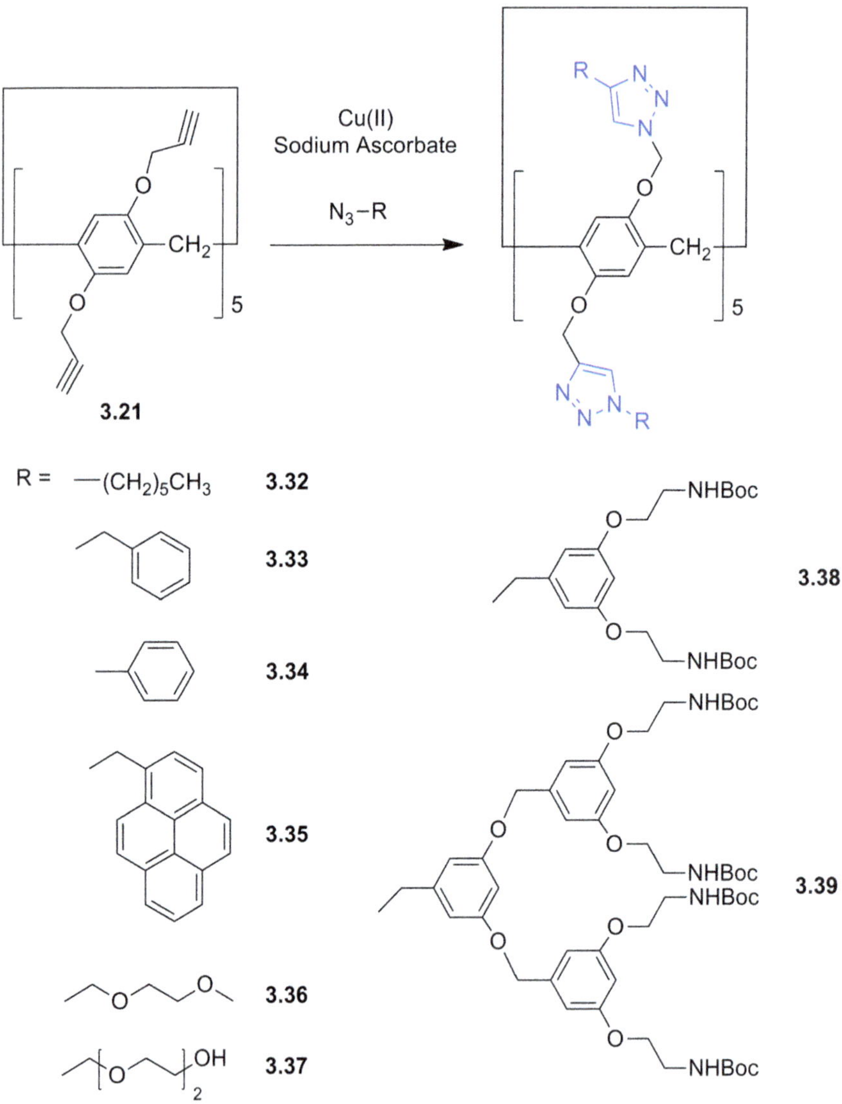

Scheme 3.4 Synthesis of per-functionalized pillar[5]arenes from per-ethylated pillar[5]arene *via* CuAAC reaction.

oligoethylene glycols with an azide moiety.[43] Nierengarten *et al.* synthesized dendritic pillar[5]arenes (**3.38** and **3.39**) by CuAAC reaction between the dodecaalkyne **3.21** and cationic dendrons with an azide moiety.[44] The dendritic pillar[5]arenes can efficiently interact with dense plasmid DNA, and good transfection efficiencies were found for both dendritic pillar[5]-arenes. Nierengarten *et al.* have synthesized pillar[5]- and pillar[6]arenes with 10 (**3.44**) and 12 azide moieties (**3.46**) by a two-step reaction (Scheme 3.5).[45]

First, cyclization of dialkoxybenzene monomer with two bromide moieties with paraformaldehyde also afforded pillar[5]- and pillar[6]arenes with 10 (**3.40–3.42**) and 12 bromide moieties (**3.43**). The cyclization process for synthesis of **3.43** is a kinetically controlled system, therefore the reaction temperature and time were tuned to obtain 10 and 12 bromides in moderate yields. Then, clickable pillar[5]- (**3.44** and **3.45**) and pillar[6]arenes **3.46** with 10 and 12 azides were obtained by reaction of pillar[5]- (**3.40** and **3.42**) and pillar[6]arenes **3.43** having 10 and 12 bromide moieties, respectively, with

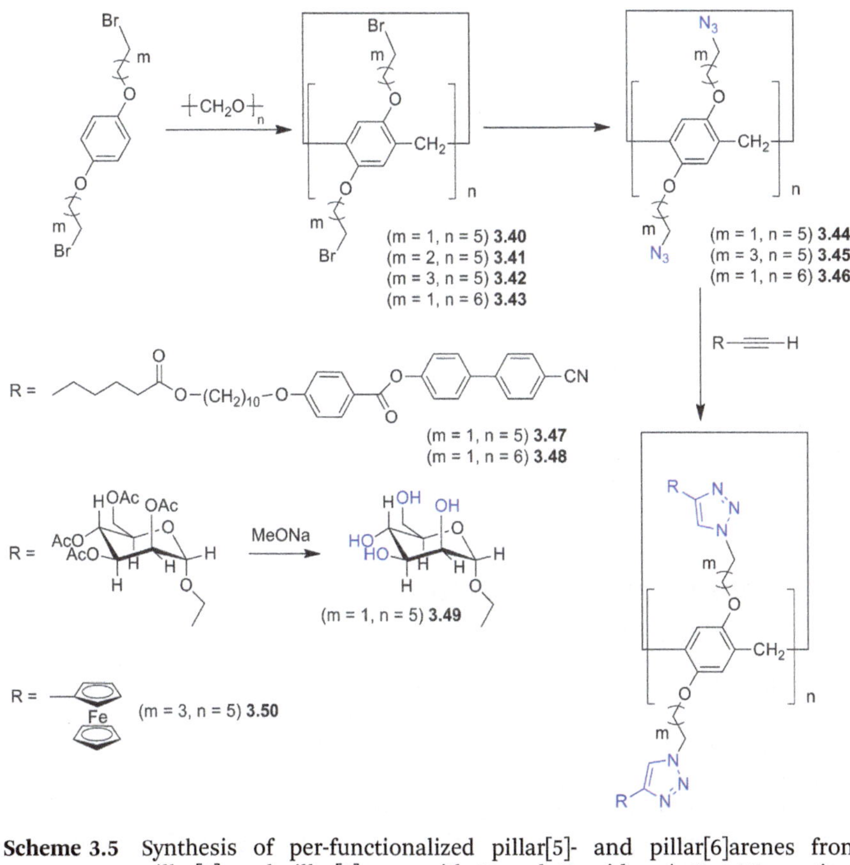

Scheme 3.5 Synthesis of per-functionalized pillar[5]- and pillar[6]arenes from pillar[5]- and pillar[6]arene with 10 and 12 azides *via* CuAAC reaction.

sodium azide in DMF. The CuAAC reactions of the decaazides **3.44** and dodecaazide **3.46** with the mesomorphic subunit with an alkyne moiety afforded pillar[5]- and pillar[6]arenes with 10 (**3.47**) and 12 mesomorphic subunits (**3.48**), respectively. These mesogenic pillar[5]- and pillar[6]arenes showed a broad enantiotropic mesophase, while only a monotropic mesophase was observed in the unit model with two mesomorphic units. Nierengarten *et al.* synthesized a pillar[5]arene with 10 mannose moieties (**3.49**) from the decaazide **3.44**. The glycoconjugate showed promise as an inhibitor of the adhesion of a uropathogenic *Escherichia coli* to red blood cells.[46] Pei *et al.* synthesized an amphiphilic pillar[5]arene with 10 ferrocenium moieties (**3.50**) by a CuAAC reaction between the decaazide **3.45** and ferrocene with an alkyne moiety.[47] The amphiphilic pillar[5]arenes with 10 ferrocenium moieties formed vesicles, but the vesicles collapsed by reduction from ferrocenium to ferrocene.

The pillar[5]- and pillar[6]arenes containing 10 and 12 bromo groups (**3.40**–**3.43**, **3.51**), which are intermediates for the synthesis of decaazide and dodecaazide, are also good key compounds for obtaining cationic pillar[5]- and pillar[6]arenes (Scheme 3.6).

Reaction of **3.40**, **3.41**, **3.43** or **3.51** with 1-methylimidazole,[48,49] pyridine,[50] amine[51–54] and tributylphosphine[55] afforded cationic pillar[5]- and pillar[6]arenes (**3.52**–**3.56**). A pillar[5]arene with 10 ionic liquids (imidazolium cations; **3.53**) was produced in the liquid state at 25 °C by choosing an appropriate counter anion.[48] A pillar[5]arene with 10 phosphonium cations (**3.56**) is amphiphilic and acts as a substrate-selective phase-transfer catalyst. Oxidation of linear alkane 1-hexene to 1-pentanal by $KMnO_4$ was >99%, whereas that of the branched alkene, 4-methyl-1-hexene, was only 31%, even under ideal conditions.[55] An etherification reaction between an alkyl-halide and phenolic compounds is a good method to synthesize functionalized pillar[5]arene. Etherification of pillar[5]arene containing 10 bromide with a coumarin derivative with a phenolic moiety gave a pillar[5]arene carrying 10 coumarin moieties (**3.57**)[56]; **3.57** acted as a fluorescence sensor for methyl parathion. Pillar[5]- and pillar[6]arenes with 10 and 12 bromide moieties should be good key compounds for the synthesis of various functionalized pillar[5]- and pillar[6]arenes.

3.4 Mono-functionalization of Pillar[5]- and Pillar[6]arenes

There are two different approaches to synthesizing mono-functionalized pillar[*n*]arenes; co-cyclization of different monomers, and mono-deprotection and functionalization of pre-formed pillar[*n*]arenes. Stoddart *et al.* reported formation of a pillar[5]arene with one bromide moiety (**3.51**) by co-cyclization of dialkoxybenzene derivatives (Scheme 3.7).[57]

In this approach, the products obtained are a mixture of pillar[5]arenes with no bromide, one bromide and more than two bromide groups. By silica

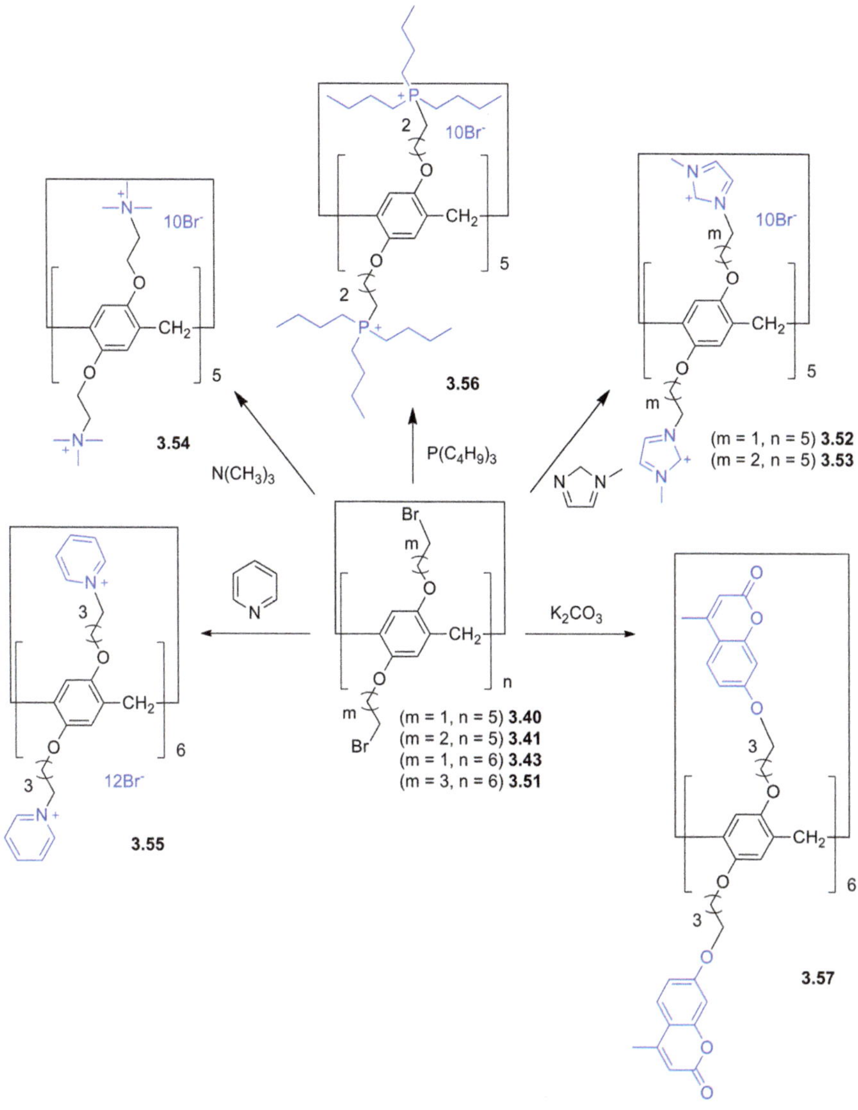

Scheme 3.6 Synthesis of per-functionalized pillar[5]- and pillar[6]arenes from pillar[5]- and pillar[6]arene with 10 and 12 bromides *via* etherification and cationation.

gel chromatography, pillar[5]arene with one bromide moiety (**3.58**) can be obtained (yield 10%). A clickable mono-azide pillar[5]arene (**3.59**) can be obtained by reaction of the mono-bromide **3.58** with sodium azide. The mono-azide **3.59** is a useful key compound to synthesize mono-functionalized pillar[5]arenes with various functional groups through the CuAAC reaction. The CuAAC reaction between **3.59** and pyrene with an alkyne moiety afforded a pillar[5]arene with one pyrene moiety (**3.60**);[57] **3.60** can be used as

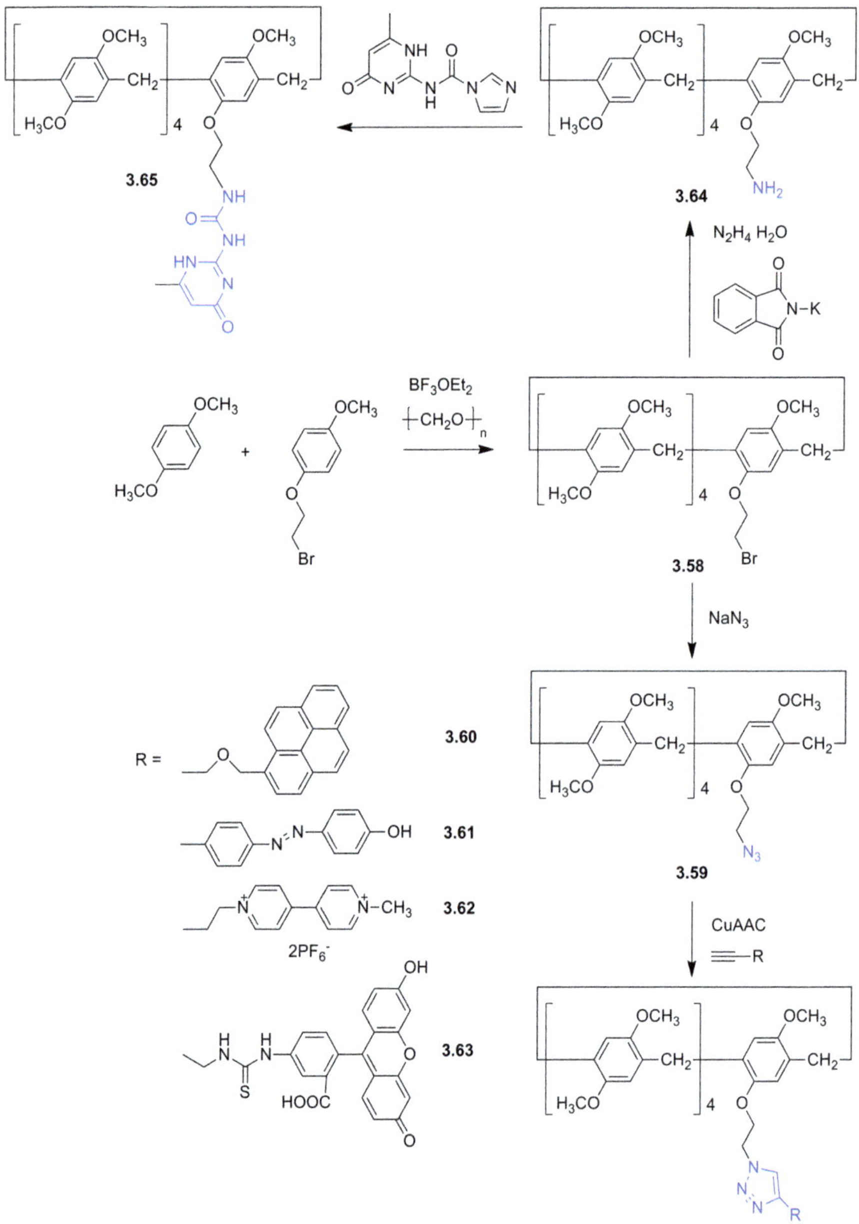

Scheme 3.7 Synthesis of mono-functionalized pillar[5]arenes from pillar[5]arene with one bromide.

a fluorescence quenching sensor for alkanediamines. A pillar[5]arene with one azobenzene moiety (**3.61**) was synthesized through a CuAAC reaction between **3.59** and 4-hydroxy-4′-ethynyl-azobenzene.[58] **3.61** showed a dynamic smart surface. A pillar[5]arene with one viologen moiety (**3.62**) was also synthesized by the CuAAC reaction between **3.59** and an alkyne-substituted viologen derivative.[59] **3.62** formed a self-inclusion complex in dilute concentrations, but supramolecular daisy chain oligomers and polymers formed with increasing concentrations. A fluorescein-pillar[5]arene conjugate (**3.63**), which was synthesized using **3.59** and a fluorescein derivative with an alkyne through a CuAAC reaction, allows the host to aggregate into vesicles that exhibit thermo-responsive emissions at different temperatures.[60]

Wang and co-workers used the mono-bromide **3.58** as a key compound to construct a pillar[5]arene-based supramolecular assembly. The mono-bromide **3.58** was substituted with an amino group by Gabriel synthesis. A mono-ureidopyrimidinone-functionalized pillar[5]arene (**3.65**) was synthesized from the mono-amine **3.64.**[61] **3.65** formed a supramolecular dimer by a quadruple hydrogen bond between ureidopyrimidinone moieties, and addition of a bisparaquat derivative afforded supramolecular polymers at high concentration.

The second method is the mono-deprotection method. We first reported the mono-deprotection method for synthesis of mono-functionalized pillar[5]arenes (Scheme 3.8).[62] Controlling the de-*O*-methylation of pre-formed per-methylated pillar[5]arene **3.1** gave mono-deprotected pillar[5]arene **3.66** in a moderate yield (22%). Etherification between mono-deprotected pillar[5]-arene **3.66** and alkyl-halide gives various mono-functionalized pillar[5]arenes. The introduction of mono-trimethylammonium moieties affords a host–guest conjugate **3.68.** This compound mainly forms a self-inclusion complex in $CDCl_3$. A pillar[5]arene dimer (**3.69**) can be prepared by linking mono-hydroxylated pillar[5]arenes with 1,4-bis(bromomethyl)benzene.[63] Li *et al.* synthesized other pillar[5]arene dimers, bridged by a flexible 1,4-dibromo-butane or relatively rigid 1,2-bis(bromomethyl)benzene.[64] We synthesized a pillar[5]arene with one alkyne moiety (**3.70**) by etherification of mono-hydroxylated pillar[5]arene and propargylbromide. The mono-alkyne **3.70** is a useful key compound to synthesize various mono-functionalized pillar[5]arenes through a CuAAC reaction.[65] Cao *et al.* reported a pillar[*n*]arene derivative with one phosphoryl group (**3.71**), which can be prepared using the reactivity of the phenolic moiety of **3.66.**[66] We obtained mono-hydroxylated pillar[6]arene **3.67** by mono-deprotection of per-ethylated pillar[6]arene **3.2** (yield 46%).[67] Mono-functionalized pillar[6]arene **3.67** could not be obtained by the co-cyclization method because cyclic hexamers are minor products in many cases. Therefore, the mono-deprotection approach is useful for the synthesis of various mono-functionalized pillar[6]arenes.

Zhou *et al.* synthesized the mono-functionalized pillar[6]arene **3.72** by etherification between excess pillar[6]arene with 12 bromides (**3.43**) and a naphthalene derivative with one phenolic moiety.[68] The other 11 bromides remained unreacted, therefore other functional groups could be installed. A cationic water-soluble pillar[5]arene (**3.73**) was synthesized with one naphthalene and nine trimethylammonium bromide moieties by a cationation reaction between the 11 bromides and trimethylamine (Scheme 3.9).

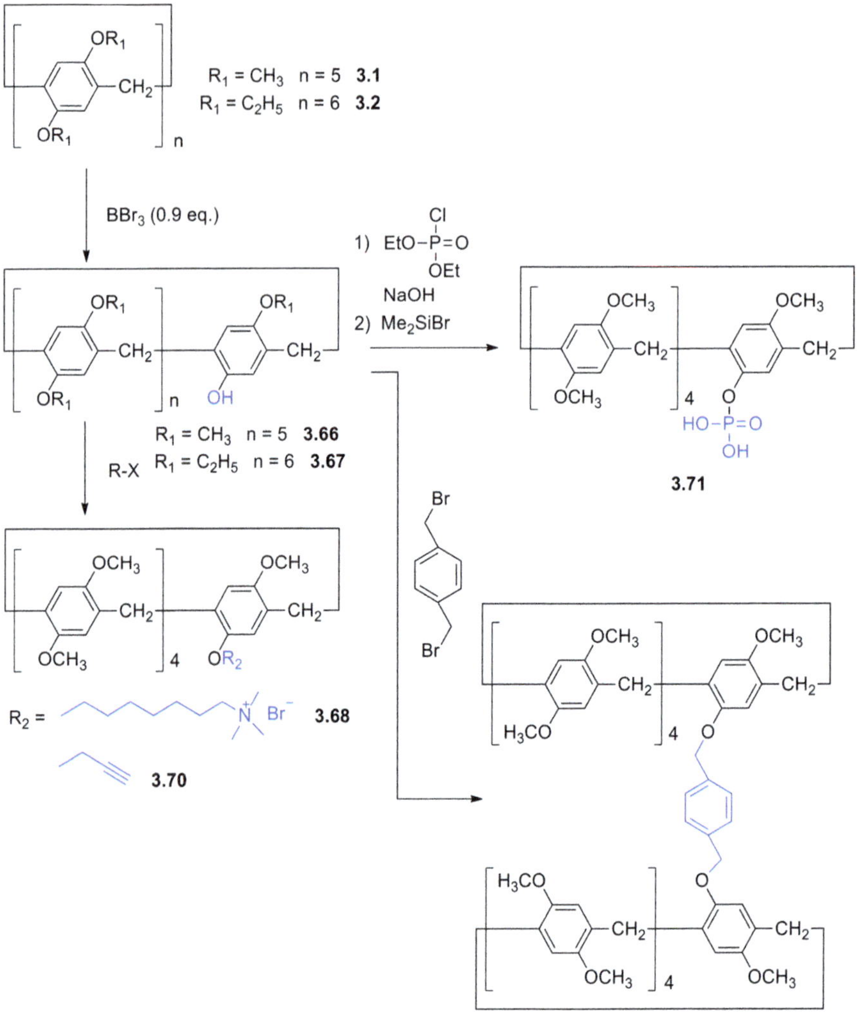

Scheme 3.8 Synthesis of mono-functionalized pillar[*n*]arenes and pillar[5]arene dimer from pillar[5]arene with one phenolic moiety.

3.5 Di-functionalized Pillar[5]- and Pillar[6]arenes

There is only one conformational isomer in the per- and mono-functionalized pillar[5]- and pillar[6]arenes. However, pillar[5]- and pillar[6]arenes with two functional groups afford various constitutional isomers; di-functionalized pillar[5]- and pillar[6]arenes have five (A1/B2, A1/C1, A1/A2, A1/B1 and A1/C2) and seven isomers (A1/B1, A1/C1, A1/D1, A1/A2, A1/B2, A1/C2 and A1/D2), respectively. Therefore, new synthetic strategies are required for their synthesis. First, di-functionalized pillar[5]arene was obtained by chance.[21]

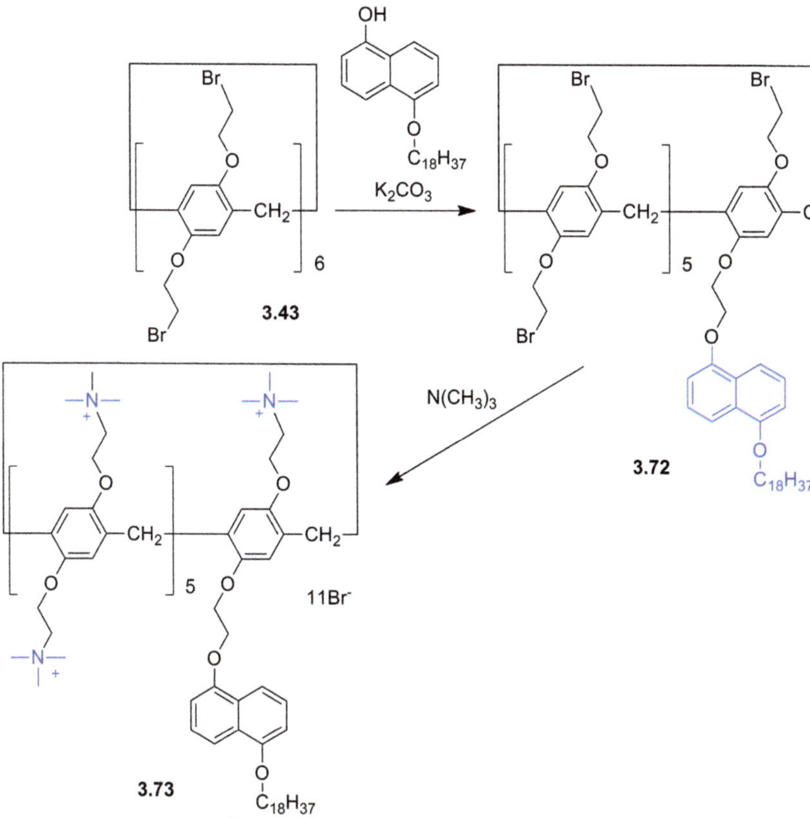

Scheme 3.9 Synthesis of pillar[6]arenes with two different functional groups by etherification and cationation.

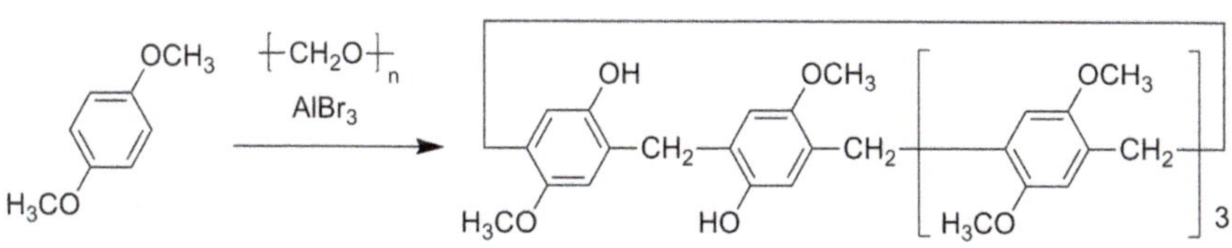

Scheme 3.10 Synthesis of A1/B2 dihydroxylated pillar[5]arene by cyclization of 1.4-dimethoxybenzene with $AlBr_3$.

The reaction of 1,4-dimethoxybenzene and paraformaldehyde with $AlBr_3$ afforded multiple-deprotected pillar[5]arenes. The deprotection reaction took place concurrently with cyclization (Scheme 3.10). A regioselective di-protected pillar[5]arene at the A1/B2 positions (**3.74**) can be isolated by silica gel chromatography and washing with methanol. The other method for the synthesis of di-functionalized pillar[5]arenes is co-cyclization (Scheme 3.11).[24] First, co-pillar[5]arenes were synthesized by Huang and

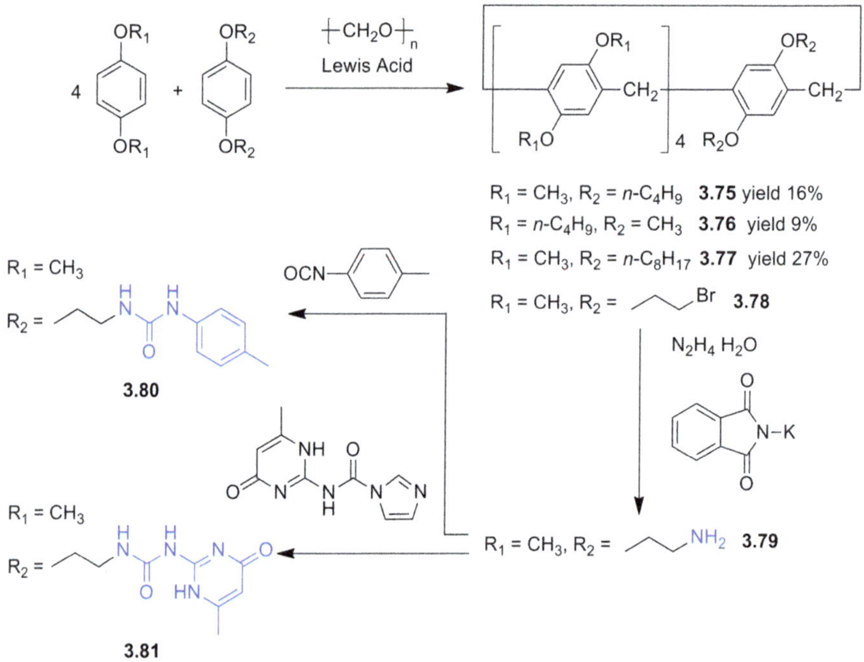

Scheme 3.11 Synthesis of A1/A2 di-functionalized pillar[5]arenes by the co-cyclization method.

co-workers[24] by co-cyclization of 1,4-dimethoxybenzene (DMB) and 1,4-dibutoxybenzene (DBB) or 1,4-dioxtyloxybenzene (DOB). In the feed ratio DMB : DBB = 4 : 1, a co-pillar[5]arene consisting of four DMB and one DBB units (**3.75**) was obtained in 16% yield by silica gel chromatography. When the ratio of the two monomers was reversed (DMB : DBB = 1 : 4), a co-pillar[5]arene containing one DMB and four DBB units (**3.76**) was obtained in 9% yield. When DOB was used as the monomer instead of DBB, a co-pillar[5]arene consisting of four DMB and one DOB units (**3.77**) was obtained (yield 27%).

Wang and co-workers synthesized a reactive di-bromo-functionalized pillar[5]arene by co-cyclization of 1,4-bis(2-bromoethoxy)benzene with DMB.[69] The two bromide moieties on pillar[5]arene **3.78** were converted to amino moieties **3.79** by Gabriel synthesis. Using the diamine **3.79** as a key compound, di-ureidopyrimidinone (**3.80**) and di-urea (**3.81**) functionalized pillar[5]arene derivatives were prepared.

Stoddart and co-workers synthesized the di-bromo-functionalized pillar[5]arene **3.82** by co-cyclization of 1,4-bis(3-bromopropoxy)benzene with DMB. Elimination of bromide moieties afforded diallyl ether-substituted pillar[5]arene **3.83**. De-protection of the allyl moiety afforded A1/A2-di-hydroxylated pillar[5]arene **3.84**. Reaction of **3.84** with triflic anhydride gave a ditriflated pillar[5]arene (**3.85**). A ter-phenyl substituted pillar[5]arene (**3.86**) was prepared by Pd-catalyzed Suzuki coupling between ditriflated

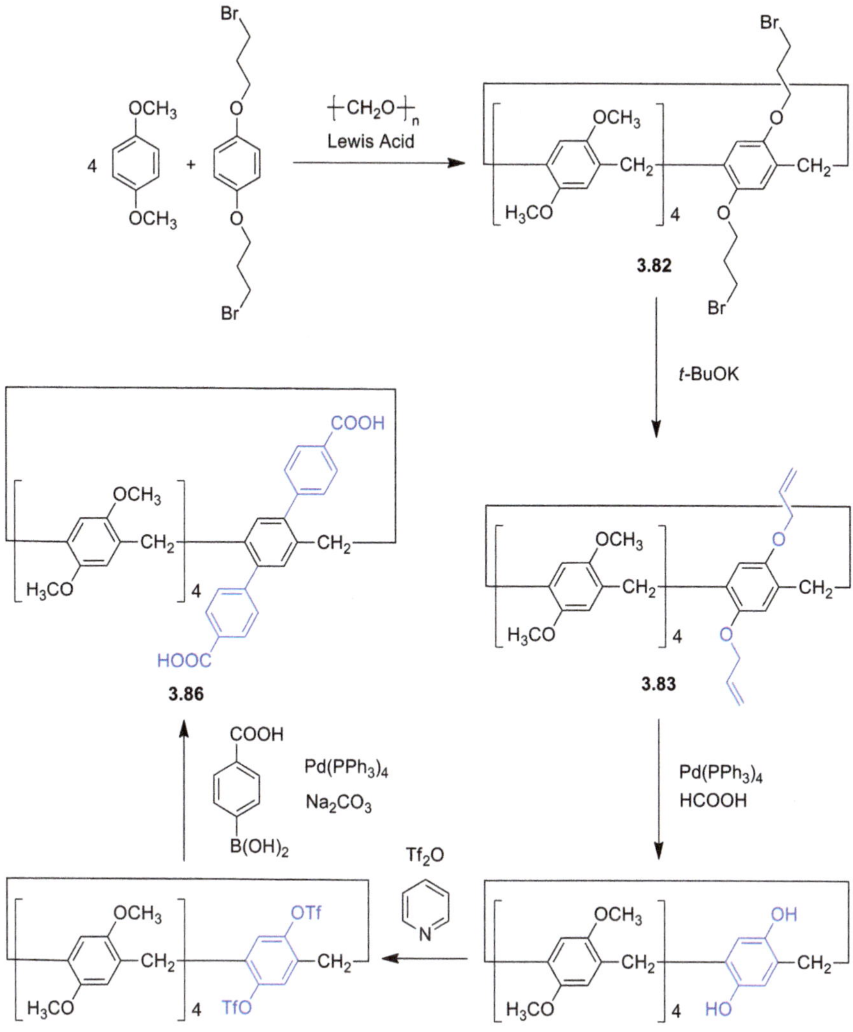

Scheme 3.12 Synthesis of A1/A2 di-functionalized pillar[5]arenes by combination of the co-cyclization method and Pd catalyzed reaction.

pillar[5]arene **3.85** and 4-(methoxycarbonyl)phenylboronic acid (Scheme 3.12). The di-carboxylic acid pillar[5]arene **3.86** can be used as building blocks for metal organic frameworks containing pillar[5]arenes.[70]

Co-cyclization of 1,4-dimethoxybenzene and dialkoxybenzene with different reactive moieties affords pillar[5]arenes with two different moieties at the A1/A2 positions. Yang *et al.* synthesized a pillar[5]arene with one bromide and one alkyne moiety at the A1 and A2 positions (**3.87**) by the co-cyclization method.[71] A pillar[5]arene with one anthracene and one cationic trimethyl ammonium moiety, at the A1 and A2 positions, can be synthesized by a CuAAC reaction between the alkyne and anthracence with an azide moiety **3.88**, and then a cationic reaction between the bromide and trimethylamine **3.89**. A combination of the co-cyclization method and methods for the synthesis of mono-functionalized pillar[5]arene achieved the synthesis of di-functionalized pillar[5]arenes with different functional groups at the A1 and A2 positions (Scheme 3.13).

The co-cyclization of different monomers represents a useful approach for the synthesis of functionalized pillar[5]arenes by a one-step reaction. However, this method cannot be applied for the synthesis of functionalized pillar[6]arenes because pillar[5]arenes are obtained as major products when dichloroethane is used as solvent. We found that chlorocyclohexane is a useful guest solvent for the synthesis of pillar[6]arenes (details in Chapter 2), therefore we synthesized a co-pillar[6]arene by co-cyclization of 1,4-bis(methylcyclohexyl)benzene and 1,4-bis(4-bromobutoxy)benzene.[72] A co-pillar[6]arene (**3.90**) was obtained in a yield of 9% along with a homopillar[5]arene (**3.20**; yield 1%) and a homopillar[6]arene (**3.91**; yield 8%).

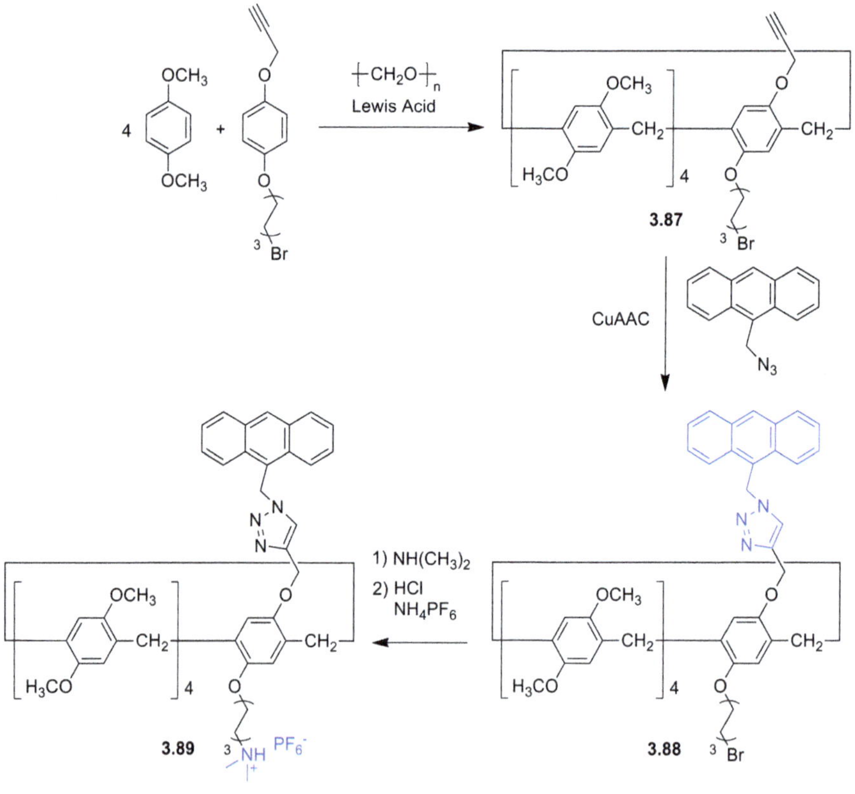

Scheme 3.13 Synthesis of A1/A2 di-functionalized pillar[5]arene with two different functional groups by co-cyclization, CuAAC and cationic reactions.

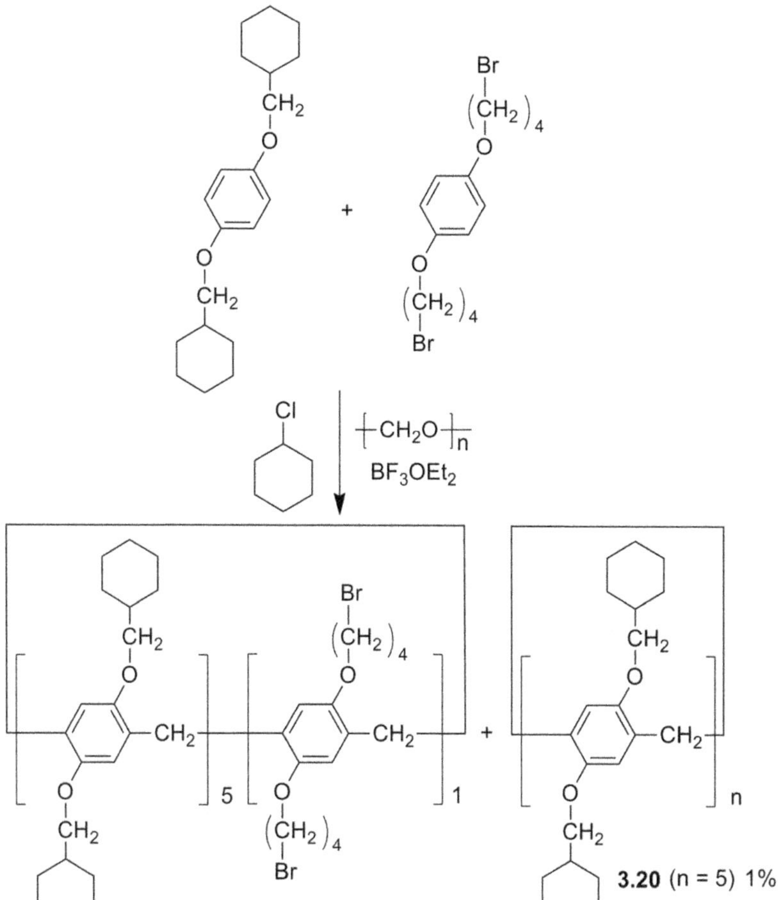

Scheme 3.14 Synthesis of A1/A2 di-functionalized pillar[6]arene with two bromide groups by the co-cyclization method.

The bromide moieties on **3.90** could be changed to clickable azide moieties, thus **3.90** was used as a key compound to synthesize A1/A2 di-functionalized pillar[6]arenes (Scheme 3.14).

Another new route for the selective synthesis of di-functionalized pillar[5]arenes is oxidation and reduction of pillar[*n*]arene units (Scheme 3.15). Our group[22] and Huang's group[73] reported that the oxidation of one pillar[5]arene unit afforded pillar[5]arenes containing one benzoquinone unit (**3.92** and **3.93**). Reduction of the benzoquinone unit yielded dihydroxylated pillar[5]arenes at the A1/A2 positions (**3.84** and **3.95**). This method avoids the formation of many constitutional isomers and can be used for pillar[5]arenes with one benzoquinone and two hydroxyl groups at the A1/A2 positions. A clickable di-functionalized pillar[5]arene was prepared by modification of the alkyne moieties.[22] The A1/A2 di-alkyne

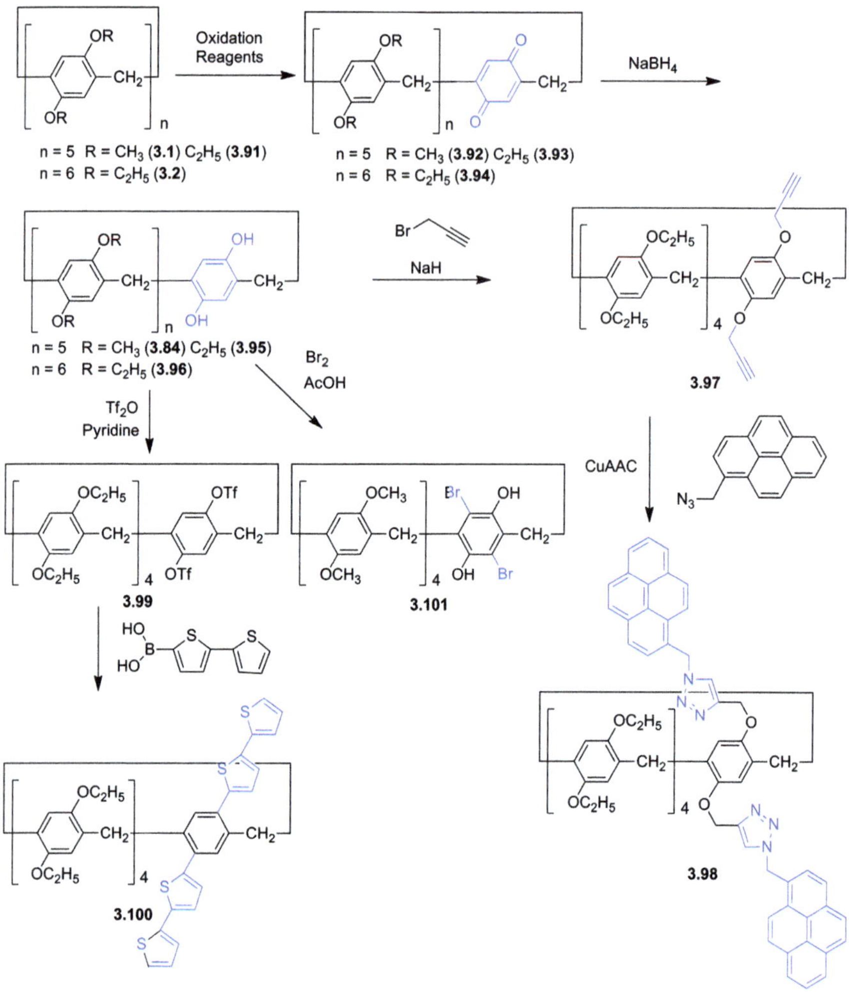

Scheme 3.15 Synthesis of A1/A2 di-functionalized pillar[5]arenes by oxidation–reduction of units.

pillar[5]arene **3.97** is also a good key compound for synthesis of bi-functional pillar[5]arenes. The fluorescence sensor pillar[5]arene with two pyrene moieties at the A1/A2 positions **3.98** can be prepared by a CuAAC reaction between A1/A2 dialkyne **3.97** and a pyrene with azide moiety. The A1/A2 di-triflated pillar[5]arene **3.99** was obtained by reaction between di-hydroxylated pillar[5]arene **3.95** and trifluoromethanesulfonic anhydride. A pillar[5]arene derivative containing one π-conjugated unit (**3.100**) was prepared by Pd-catalyzed cross-coupling of ditriflate **3.99** and (5-bithienyl)boronic acid.[74]

Functionalization can be increased at the *ortho* positions (A1′ and A2′) to both the methylene bridges and oxygen substituents. Stoddert *et al.* reported the introduction of bromide moieties at the *ortho* positions by bromination of the benzene moieties in pillar[5]arene.[75] Selective bromination of pillar[5]arene at the A1′ and A2′ positions (**3.101**) can be achieved using a pillar[5]arene containing one hydroquinone unit because the reactivity at the *ortho* positions of the hydroquinone unit with bromine is higher than that of the dimethoxybenzene units. Aryl halides can be functionalized by effective palladium-catalyzed reactions, such as Sonogashira, Suzuki and Heck couplings. Therefore, the dibromide is a useful key compound to create functionalized pillar[5]arenes.

3.6 Functionalized Pillar[5]arenes with More Than Three Different Substituents

Pillar[*n*]arenes containing more than three different substituents have many structural isomers. Our group and Huang's group separately reported the first tetra-functionalized pillar[5]arene obtained by the oxidation–reduction method (Scheme 3.16).[22] Oxidation of 1,4-dialkoxybenzene units in pillar[5]arene gave a pillar[5]arene with one to five hydroquinone units along with unreacted per-alkylated pillar[5]arene. Pillar[5]arenes containing two benzoquinone units at the A, B (**3.102** and **3.103**) or the A, C units (**3.104** and **3.105**) can be isolated by silica gel chromatography. Reduction of these benzoquinone units afforded A1/A2/B1/B2 (**3.106** and **3.107**) and A1/A2/C1/C2 tetra-hydroxylated pillar[5]arenes (**3.108** and **3.109**).

Huang and co-workers recently prepared two 2 : 3 co-pillar[5]arene constitutional isomers (Scheme 3.17).[76] To synthesize co-pillar[5]arenes containing one different unit, a 1 : 4 feed ratio is suitable. The 1 : 3 feed ratio afforded two 2 : 3 co-pillar[5]arene constitutional isomers, A1/A2/B1/B2 (**3.110**) and A1/A2/C1/C2 co-pillar[5]arenes (**3.111**), in 8% yields.

Pan and Xue reported pillar[5]arene derivatives with three different kinds of repeating unit (Scheme 3.18).[77] Per-alkylated pillar[5]arenes are oxidized upon addition of oxidants, but per-methoxycarbonyl-methoxy-substituted pillar[5]arene was not oxidized under the same oxidation conditions as the per-alkylated pillar[5]arenes. This indicates that 1,4-methoxycarbonyl-methoxybenzene units exhibited oxidative resistance when compared with 1,4-dialkoxybenzenes. Based on this finding, Pan and Xue first synthesized four methoxycarbonyl-methoxy-substituted pillar[5]arene (**3.112**) from A1/A2/C1/C2 tetra-hydroxylated pillar[5]arene (**3.108**). Oxidation of **3.112** afforded pillar[5]arenes with one (**3.113**), two (**3.114**) and three benzoquinone moieties (**3.115**). Because the methoxycarbonyl-methoxy-substituted units show oxidative resistance, the A and C units, which modified methoxy-carbonyl-methoxy groups, were not oxidized.[77]

Penta-functionalized pillar[5]arenes can be prepared from non-symmetrical monomers with reactive moieties (Scheme 3.19). Huang *et al.* synthesized a pillar[5]arene with five alkynes on the same rim (**3.116**). A CuAAC reaction between **3.116** and a naphthalene containing an azide

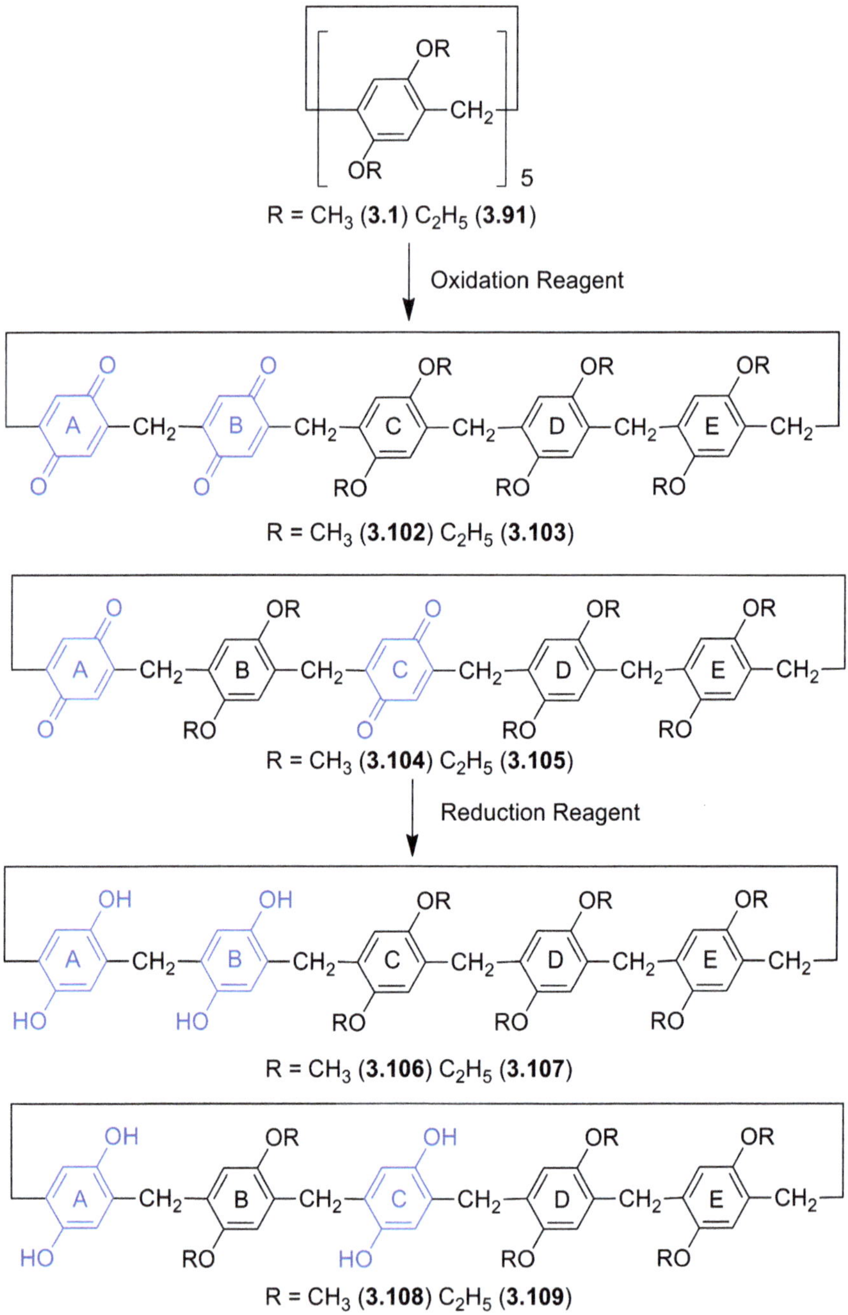

Scheme 3.16 Synthesis of A1/A2/B1/B2 and A1/A2/C1/C2 tetra-functionalized pillar[5]arenes by oxidation–reduction of units.

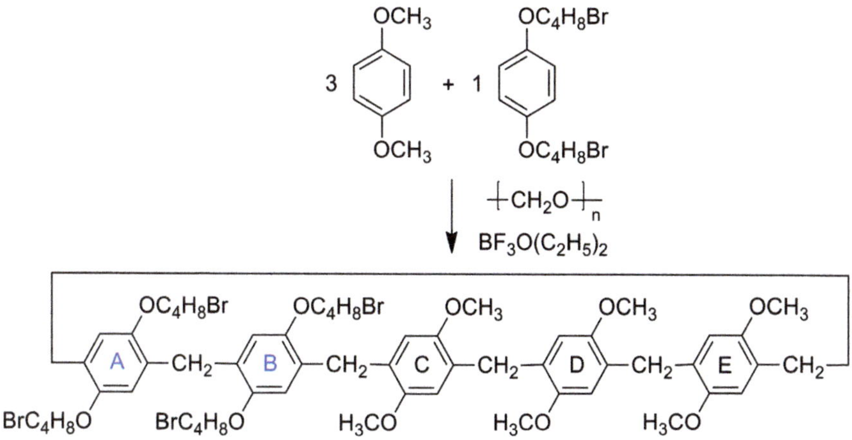

3.110

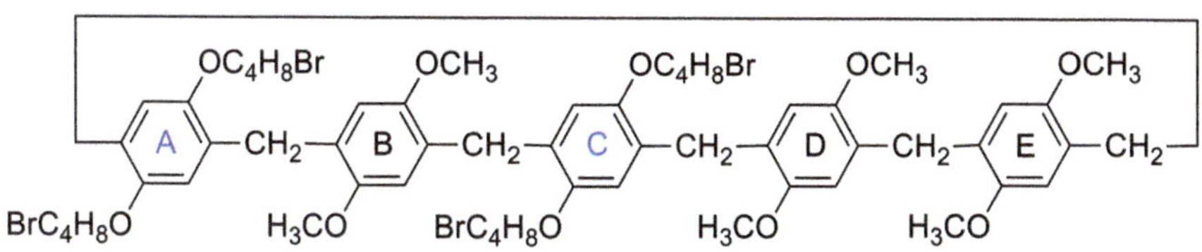

3.111

Scheme 3.17 Synthesis of A1/A2/B1/B2 and A1/A2/C1/C2 tetra-functionalized pillar[5]arenes by the co-cyclization method.

moiety afforded a pillar[5]arene with five naphthalene moieties.[78] There was a mixture of four conformational isomers, therefore a pillar[5]arene with five naphthalene moieties at the same rim (**3.117**) was isolated by silica gel chromatography.

3.7 Conclusion and Outlook

Based on the various organic synthetic methods, mono-, di-, tetra-, penta- and per-functionalized pillar[*n*]arenes have been obtained. The high solubility of pillar[*n*]arenes in different organic media enables their functionalization by various organic reactions. Although several methods of functionalization have been uncovered, new functionalization methods based on organic reactions should appear in the near future. A combination of different functionalization methods should lead to the synthesis of multi-functionalized pillar[*n*]arenes. In addition, their highly symmetrical structure makes characterization of multi-functionalized pillar[*n*]arenes easy, whereas characterization of multi-functionalized macrocycles based on the typical host molecules is quite difficult. Therefore, pillar[*n*]arenes are truly versatile macrocyclic compounds. Their versatile functionality is one of the

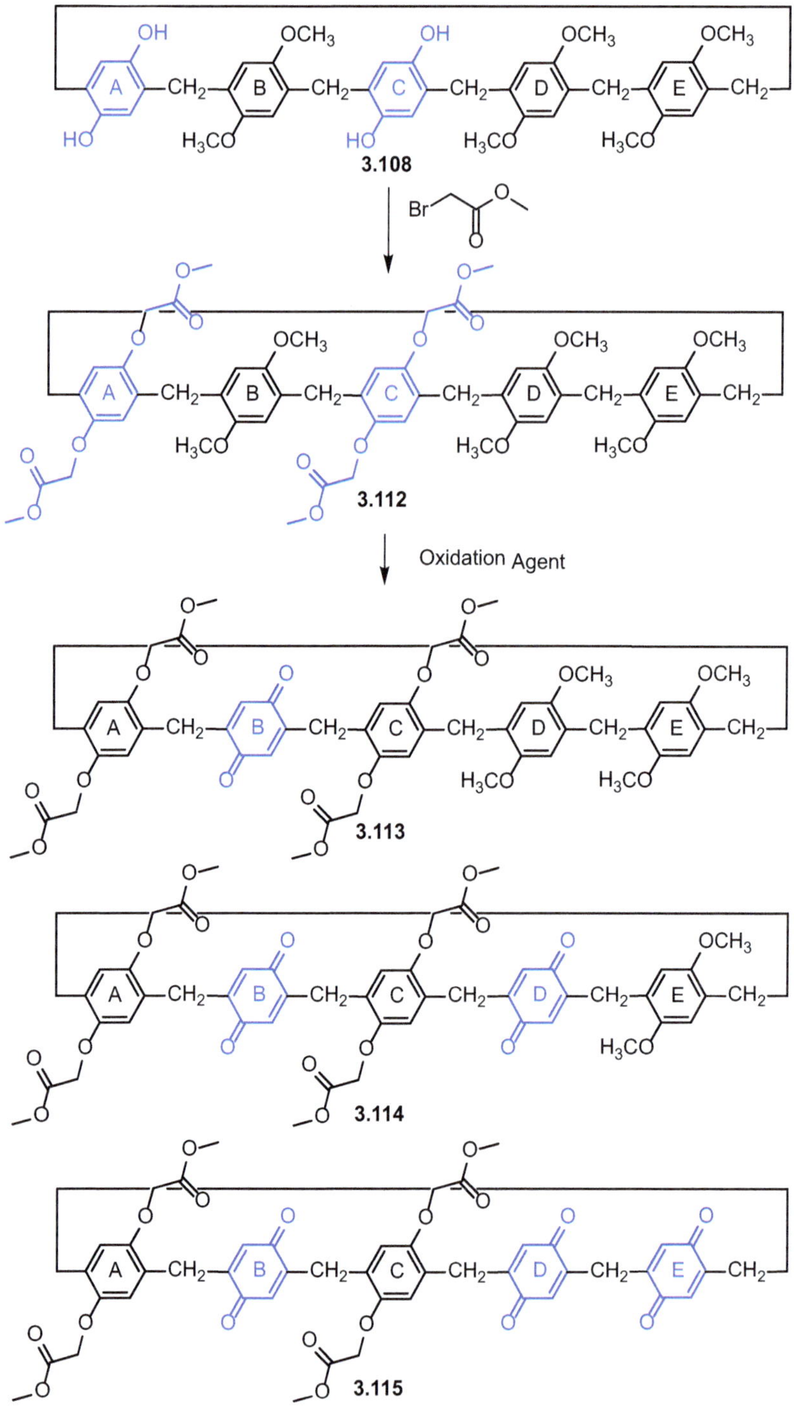

Scheme 3.18 Synthesis of pillar[5]arenes with one, two and three benzoquinone units by selective oxidation.

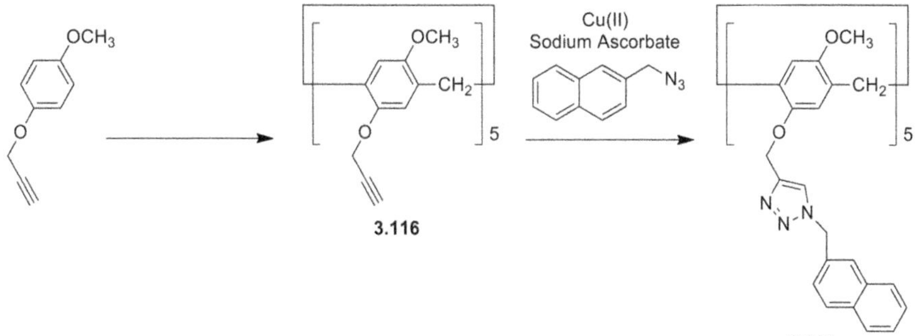

Scheme 3.19 Synthesis of penta-functionalized pillar[5]arene from a pillar[5]arene with five ethynyl groups, at A1, B1, C1, D1 and E1.

important aspects of pillar[*n*]arene chemistry, and should lead to a dramatic expansion of pillar[*n*]arene chemistry in the future.

References

1. A. R. Khan, P. Forgo, K. J. Stine and V. T. D'Souza, *Chem. Rev.*, 1998, **98**, 1977–1996.
2. E. Engeldinger, D. Armspach and D. Matt, *Chem. Rev.*, 2003, **103**, 4147–4174.
3. F. Bellia, D. La Mendola, C. Pedone, E. Rizzarelli, M. Saviano and G. Vecchio, *Chem. Soc. Rev.*, 2009, **38**, 2756–2781.
4. K. Takahashi, K. Hattori and F. Toda, *Tetrahedron Lett.*, 1984, **25**, 3331–3334.
5. H. Ikeda, Y. Nagano, Y. Du, T. Ikeda and F. Toda, *Tetrahedron Lett.*, 1990, **31**, 5045–5048.
6. C. D. Gutsche, B. Dhawan, J. A. Levine, K. Hyun No and L. J. Bauer, *Tetrahedron*, 1983, **39**, 409–426.
7. P. D. Beer, *Chem. Commun.*, 1997, 2377–2378.
8. Z. Ho, M. Ku, C. Shu and L. Lin, *Tetrahedron*, 1996, **52**, 13189–13200.
9. S. Shirakawa, T. Kimura, S. Murata and S. Shimizu, *J. Org. Chem.*, 2009, **74**, 1288–1296.
10. B. Colasson and O. Reinaud, *J. Am. Chem. Soc.*, 2008, **130**, 15226–15227.
11. N. Zhao, G. O. Lloyd and O. A. Scherman, *Chem. Commun.*, 2012, **48**, 3070–3072.
12. D. Lucas, T. Minami, G. Iannuzzi, L. Cao, J. B. Wittenberg, P. Anzenbacher and L. Isaacs, *J. Am. Chem. Soc.*, 2011, **133**, 17966–17976.
13. T. Takata, *Polym. J.*, 2006, **38**, 1–20.
14. Y. Kohsaka, Y. Koyama and T. Takata, *Angew. Chem., Int. Ed.*, 2011, **50**, 10417–10420.
15. T. Oku, Y. Furusho and T. Takata, *Angew. Chem., Int. Ed.*, 2004, **43**, 966–969.
16. K. Koga, D. Q. Yuan and K. Fujita, *Tetrahedron Lett.*, 2000, **41**, 6855–6857.

17. M. Petrillo, L. Marinescu, C. Rousseau and M. Bols, *Org. Lett.*, 2009, **11**, 1983–1985.
18. D. Q. Yuan, T. Yamada and K. Fujita, *Chem. Commun.*, 2001, 2706–2707.
19. T. Ogoshi and T. Yamagishi, *Eur. J. Org. Chem.*, 2013, 2961–2975.
20. N. L. Strutt, H. C. Zhang, S. T. Schneebeli and J. F. Stoddart, *Acc. Chem. Res.*, 2014, **47**, 2631–2642.
21. T. Ogoshi, K. Kitajima, S. Fujinami and T. Yamagishi, *Chem. Commun.*, 2011, **47**, 10106–10108.
22. T. Ogoshi, D. Yamafuji, D. Kotera, T. Aoki, S. Fujinami and T. Yamagishi, *J. Org. Chem.*, 2012, 77, 11146–11152.
23. T. Ogoshi, M. Hashizume, T. Yamagishi and Y. Nakamoto, *Chem. Commun.*, 2010, **46**, 3708–3710.
24. Z. B. Zhang, B. Y. Xia, C. Y. Han, Y. H. Yu and F. H. Huang, *Org. Lett.*, 2010, **12**, 3285–3287.
25. T. Ogoshi, S. Kanai, S. Fujinami, T. Yamagishi and Y. Nakamoto, *J. Am. Chem. Soc.*, 2008, **130**, 5022–5023.
26. X. B. Hu, L. Chen, W. Si, Y. H. Yu and J. L. Hou, *Chem. Commun.*, 2011, **47**, 4694–4696.
27. Y. L. Sun, Y. W. Yang, D.-X. Chen, G. Wang, Y. Zhou, C.-Y. Wang and J. F. Stoddart, *Small*, 2013, **9**, 3224–3229.
28. T. Adiri, D. Marciano and Y. Cohen, *Chem. Commun.*, 2013, **49**, 7082–7084.
29. T. Ogoshi, R. Shiga and T. Yamagishi, *J. Am. Chem. Soc.*, 2012, **134**, 4577–4580.
30. T. Ogoshi, K. Masaki, R. Shiga, K. Kitajima and T. Yamagishi, *Org. Lett.*, 2011, **13**, 1264–1266.
31. T. Ogoshi, K. Umeda, T. Yamagishi and Y. Nakamoto, *Chem. Commun.*, 2009, 4874–4876.
32. Y. Ma, X. Chi, X. Yan, J. Liu, Y. Yao, W. Chen, F. Huang and J. L. Hou, *Org. Lett.*, 2012, **14**, 1532–1535.
33. G. Yu, M. Xue, Z. Zhang, J. Li, C. Han and F. Huang, *J. Am. Chem. Soc.*, 2012, **134**, 13248–13251.
34. T. Ogoshi, K. Kida and T. Yamagishi, *J. Am. Chem. Soc.*, 2012, **134**, 20146–20150.
35. X. D. Chi, M. Xue, Y. J. Ma, X. Z. Yan and F. H. Huang, *Chem. Commun.*, 2013, **49**, 8175–8177.
36. Z. T. Li, J. Yang, G. C. Yu, J. M. He, Z. Abliz and F. H. Huang, *Org. Lett.*, 2014, **16**, 2066–2069.
37. Z. T. Li, J. Yang, G. C. Yu, J. M. He, Z. Abliz and F. H. Huang, *Chem. Commun.*, 2014, **50**, 2841–2843.
38. X. D. Chi and M. Xue, *Chem. Commun.*, 2014, **50**, 13754–13756.
39. J. Yang, X. Chi, Z. Li, G. Yu, J. He, Z. Abliz, N. Li and F. Huang, *Org. Chem. Front.*, 2014, **1**, 630–633.
40. X. Chi, X. Ji, D. Xia and F. Huang, *J. Am. Chem. Soc.*, 2015, **137**, 1440–1443.
41. T. Ogoshi, R. Shiga, M. Hashizume and T. Yamagishi, *Chem. Commun.*, 2011, **47**, 6927–6929.

42. H. M. Deng, X. Y. Shu, X. S. Hu, J. Li, X. S. Jia and C. J. Li, *Tetrahedron Lett.*, 2012, **53**, 4609–4612.
43. H. Zhang, X. Ma, K. T. Nguyen and Y. Zhao, *ACS Nano*, 2013, 7, 7853–7863.
44. I. Nierengarten, M. Nothisen, D. Sigwalt, T. Biellmann, M. Holler, J. S. Remy and J. F. Nierengarten, *Chem. – Eur. J.*, 2013, **19**, 17552–17558.
45. I. Nierengarten, S. Guerra, M. Holler, L. Karmazin-Brelot, J. Barbera, R. Deschenaux and J. F. Nierengarten, *Eur. J. Org. Chem.*, 2013, 3675–3684.
46. I. Nierengarten, K. Buffet, M. Holler, S. P. Vincent and J. F. Nierengarten, *Tetrahedron Lett.*, 2013, **54**, 2398–2402.
47. Y. Chang, K. Yang, P. Wei, S. Huang, Y. Pei, W. Zhao and Z. Pei, *Angew. Chem., Int. Ed.*, 2014, **53**, 13126–13130.
48. T. Ogoshi, N. Ueshima, T. Yamagishi, Y. Toyota and N. Matsumi, *Chem. Commun.*, 2012, **48**, 3536–3538.
49. Y. Yao, M. Xue, X. D. Chi, Y. J. Ma, J. M. He, Z. Abliz and F. H. Huang, *Chem. Commun.*, 2012, **48**, 6505–6507.
50. W. Chen, Y. Zhang, J. Li, X. Lou, Y. Yu, X. Jia and C. Li, *Chem. Commun.*, 2013, **49**, 7956–7958.
51. Y. J. Ma, X. F. Ji, F. Xiang, X. D. Chi, C. Y. Han, J. M. He, Z. Abliz, W. X. Chen and F. H. Huang, *Chem. Commun.*, 2011, **47**, 12340–12342.
52. Y. J. Ma, M. Xue, Z. B. Zhang, X. D. Chi and F. H. Huang, *Tetrahedron*, 2013, **69**, 4532–4535.
53. B. Gomez, V. Francisco, F. Fernandez-Nieto, L. Garcia-Rio, M. Martin-Pastor, M. R. Paleo and F. J. Sardina, *Chem. – Eur. J.*, 2014, **20**, 12123–12132.
54. K. C. Jie, Y. Yao, X. D. Chi and F. H. Huang, *Chem. Commun.*, 2014, **50**, 5503–5505.
55. T. Ogoshi, N. Ueshima and T.-a. Yamagishi, *Org. Lett.*, 2013, **15**, 3742–3745.
56. F. Zhang, X. L. Cao, D. M. Tian and H. B. Li, *Chin. J. Chem.*, 2015, **33**, 368–372.
57. N. L. Strutt, R. S. Forgan, J. M. Spruell, Y. Y. Botros and J. F. Stoddart, *J. Am. Chem. Soc.*, 2011, **133**, 5668–5671.
58. H. Zhang, N. L. Strutt, R. S. Stoll, H. Li, Z. Zhu and J. F. Stoddart, *Chem. Commun.*, 2011, **47**, 11420–11422.
59. N. L. Strutt, H. C. Zhang, M. A. Giesener, J. Y. Lei and J. F. Stoddart, *Chem. Commun.*, 2012, **48**, 1647–1649.
60. H. Zhang, X. Ma, J. Guo, K. T. Nguyen, Q. Zhang, X.-J. Wang, H. Yan, L. Zhu and Y. Zhao, *RSC Adv.*, 2013, **3**, 368–371.
61. Y. F. Guan, M. F. Ni, X. Y. Hu, T. X. Xiao, S. H. Xiong, C. Lin and L. Y. Wang, *Chem. Commun.*, 2012, **48**, 8529–8531.
62. T. Ogoshi, K. Demachi, K. Kitajima and T. Yamagishi, *Chem. Commun.*, 2011, **47**, 7164–7166.
63. T. Ogoshi, K. Demachi, K. Kitajima and T. Yamagishi, *Chem. Commun.*, 2011, **47**, 10290–10292.

64. C. Li, K. Han, J. Li, H. Zhang, J. Ma, X. Shu, Z. Chen, L. Weng and X. Jia, *Org. Lett.*, 2012, **14**, 42–45.
65. T. Ogoshi, K. Yoshikoshi, T. Aoki and T. Yamagishi, *Chem. Commun.*, 2013, **49**, 8785–8787.
66. Y. Chen, M. He, B. Li, L. Wang, H. Meier and D. Cao, *RSC Adv.*, 2013, **3**, 21405–21408.
67. T. Ogoshi, H. Kayama, D. Yamafuji, T. Aoki and T. Yamagishi, *Chem. Sci.*, 2012, **3**, 3221–3226.
68. J. Zhou, M. Chen and G. W. Diao, *Chem. Commun.*, 2014, **50**, 11954–11956.
69. Q. P. Duan, W. Xia, X. Y. Hu, M. F. Ni, J. L. Jiang, C. Lin, Y. Pan and L. Y. Wang, *Chem. Commun.*, 2012, **48**, 8532–8534.
70. N. L. Strutt, D. Fairen-Jimenez, J. Iehl, M. B. Lalonde, R. Q. Snurr, O. K. Farha, J. T. Hupp and J. F. Stoddart, *J. Am. Chem. Soc.*, 2012, **134**, 17436–17439.
71. K. Wang, C. Y. Wang, Y. M. Zhang, S. X. A. Zhang, B. Yang and Y. W. Yang, *Chem. Commun.*, 2014, **50**, 9458–9461.
72. T. Ogoshi, N. Ueshima, T. Akutsu, D. Yamafuji, T. Furuta, F. Sakakibara and T. A. Yamagishi, *Chem. Commun.*, 2014, **50**, 5774–5777.
73. C. Y. Han, Z. B. Zhang, G. C. Yu and F. H. Huang, *Chem. Commun.*, 2012, **48**, 9876–9878.
74. T. Ogoshi, D. Yamafuji, T. Akutsu, M. Naito and T. Yamagishi, *Chem. Commun.*, 2013, **49**, 8782–8784.
75. N. L. Strutt, H. C. Zhang, S. T. Schneebeli and J. F. Stoddart, *Chem. – Eur. J*, 2014, **20**, 10996–11004.
76. P. Wei, X. Yan, J. Li, Y. Ma and F. Huang, *Chem. Commun.*, 2013, **49**, 1070–1072.
77. M. G. Pan and M. Xue, *Rsc Adv.*, 2014, **4**, 260–263.
78. G. Yu, Z. Zhang, C. Han, M. Xue, Q. Zhou and F. Huang, *Chem. Commun.*, 2012, **48**, 2958–2960.

CHAPTER 4

Conformation and Planar Chirality of Pillar[n]arenes

TOMOKI OGOSHI* AND TADA-AKI YAMAGISHI

Graduate School of Natural Science and Technology, Kanazawa University, Kakuma-machi, Kanazawa 920-1192, Japan
*Email: ogoshi@se.kanazawa-u.ac.jp

4.1 Introduction

The conformational mobility of macrocyclic hosts is very important because the solubility and self-assembly properties of the macrocyclic have a large influence on the ability and selectivity of the host–guest complex.[1–3] Macrocyclic molecules can be divided into two categories according to their conformational flexibility: conformationally flexible or fixed macrocyclic compounds. Curcurbit[*n*]urils and deep cavitands are designated as conformationally fixed macrocyclic compounds owing to the double-bridging between constituent units (Figure 4.1a and b).[4–6]

Their low conformational change during host–guest complexation is favorable for a decrease in entropy loss and contributes to a very stable host–guest complexation and guest-shape selective recognition. Therefore, increasing the conformational rigidity is a useful method to enhance the ability of host–guest complexation. However, rigid and highly symmetrical structures, in many cases, lead to a low solubility in solvents. This is because self-assembly between macrocyclic molecules with rigid and highly symmetrical structures is easy, and leads to hard solvation. Introducing bulky and/or highly compatible substituents is favorable for overcoming the low

Monographs in Supramolecular Chemistry No. 18
Pillararenes
Edited by Tomoki Ogoshi
© The Royal Society of Chemistry 2016
Published by the Royal Society of Chemistry, www.rsc.org

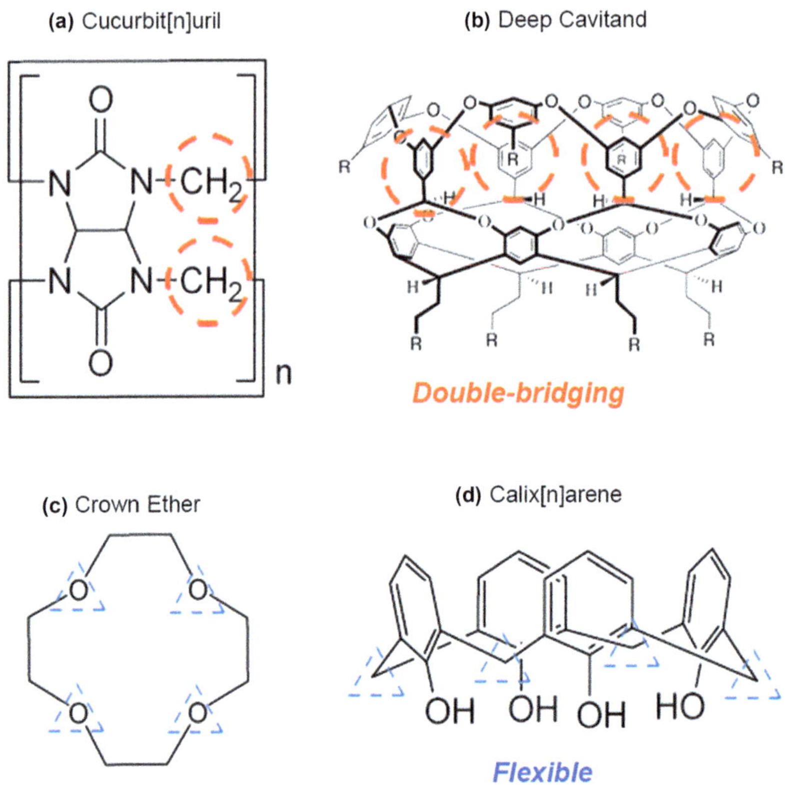

Figure 4.1 Chemical structures of conformationally fixed or flexible macrocyclic hosts.

solubility. For example, Kim *et al.* reported the production of soluble cucurbit[*n*]urils by the introduction of cyclohexyl moieties into their structures.[7] Cyclohexyl moiety-substituted cucurbit[*n*]urils are soluble in common solvents, such as water, methanol, DMF and DMSO, and are more soluble in water ($\sim 2\times10^{-1}$ M) than non-substituted cucurbit[*n*]urils (the solubility of non-substituted cucurbit[*n*]urils is $<3\times10^{-2}$ M). Gibb and Gibb synthesized a water-soluble deep-cavity cavitand by introducing carboxylic acid moieties.[8] The cavitand is well soluble in aqueous media, and can easily capture guest molecules.

Crown ethers, calix[*n*]arenes and calix[*n*]pyrroles are designated as conformationally flexible macrocyclic hosts (Figure 4.1c and d).[9,10] In terms of the host–guest complexation, their high flexibility is unfavorable because conformational change (entropy loss) occurs during the complexation process. However, this dynamic structural change of the macrocyclic shape is beneficial for their use as sensors, switching of the molecular recognition ability and their self-assembly. For example, crown ethers are conformationally flexible macrocycles, therefore incorporation of stimuli-responsive parts, such as photo-responsive azobenzene moieties and bipyridyl ligands, enables change in the conformational flexibility in response to stimulation.[11,12]

Depending on the variety of the pillar[*n*]arene, in many cases the phenolic units of pillar[*n*]arenes can rotate and form various conformers in solution, therefore pillar[*n*]arenes are classified as flexible macrocyclic compounds.[13] This chapter deals with the conformational mobility of pillar[*n*]arenes in solution and in the crystal state. The substituents on the rims, cavity size and number of rings of pillar[*n*]arenes dramatically influence the conformational mobility and the most stable structures of pillar[*n*]arenes in solution and in the solid state.

4.2 Conformation of Pillar[5]arenes

4.2.1 Conformation of Pillar[5]arenes with Simple Alkyl Substituents

Calix[*n*]arenes[14] and calix[*n*]pyrroles[10] can dramatically change their conformation. Vase-shaped calix[*n*]arenes have two rotation modes of the phenolic units: the oxygen-through-the-annulus rotation and the *para*-substituent-through-the-annulus rotation (Figure 4.2).[15]

The rotation of the units enables many conformers of calix[4]arene: cone, partial cone, 1,2-alternate and 1,3-alternate. In contrast, pillar[*n*]arenes have highly symmetrical structures when compared with calix[*n*]arenes, therefore the rotation mode for the phenolic units of pillar[*n*]arenes is only the oxygen-through-the-annulus rotation around the methylene bridges as the axis.[13] The rotation of the phenolic units around 180° results in a change in the positions of the substituents. The positional changes of the substituents give rise to the planar chirality of the pillar[*n*]arene.

Figure 4.3a shows a mono-substituted-[*n*]paracyclophane as a typical planar-chiral compound.[16,17]

Hence, when the bridge linker in mono-substituted-[*n*]paracyclophanes is sufficiently long, the benzene moiety can rotate around 180°. In contrast,

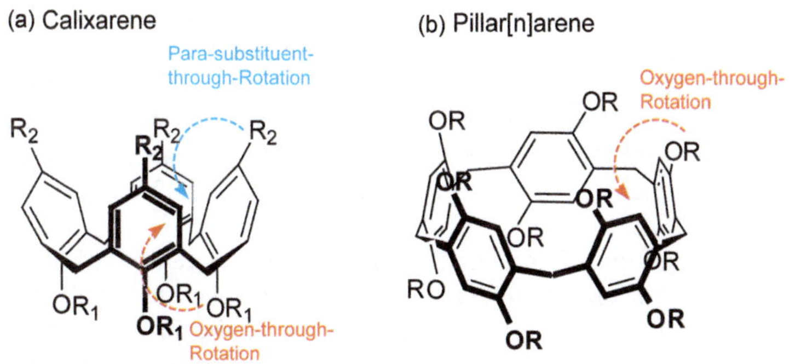

Figure 4.2 Rotational modes of (a) calixarene and (b) pillar[*n*]arene.

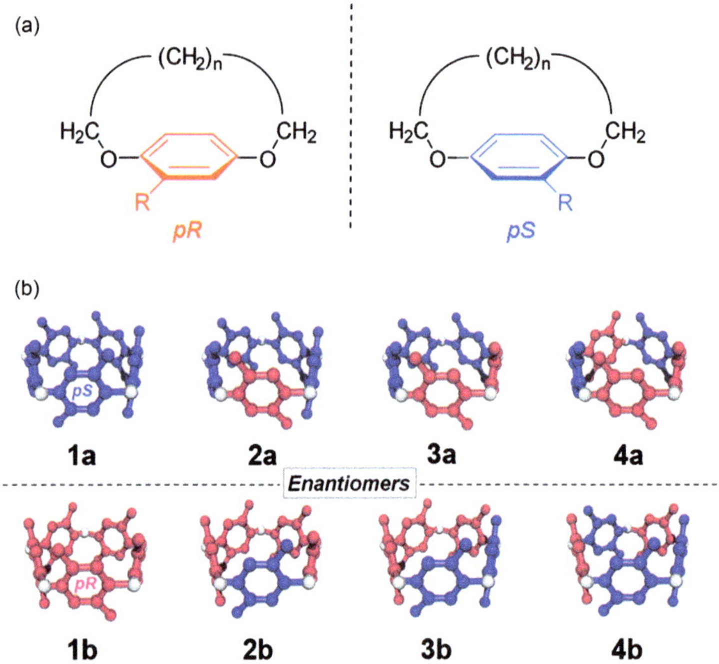

Figure 4.3 Planar chirality of (a) typical substituted paracyclophane and (b) pillar[5]arene.

short linkers prevent rotation of the benzene moiety. The inhibition of the rotation of the benzene unit results in planar chirality.

When this concept is applied to the stereochemistry of pillar[*n*]arene, rotation of the units triggers the planar-chiral inversion from *pS* to *pR* or from *pR* to *pS*. Pillar[5]arenes are cyclic pentamers, therefore they have eight possible conformers: (*pS*, *pS*, *pS*, *pS*, *pS*), (*pR*, *pR*, *pR*, *pR*, *pR*), (*pR*, *pS*, *pS*, *pS*, *pS*), (*pS*, *pR*, *pR*, *pR*, *pR*), (*pR*, *pR*, *pS*, *pS*, *pS*), (*pS*, *pS*, *pR*, *pR*, *pR*), (*pR*, *pR*, *pS*, *pR*, *pS*) and (*pS*, *pS*, *pR*, *pS*, *pR*), namely **1a**/**b**, **2a**/**b**, **3a**/**b** and **4a**/**b**, respectively. In many cases, these conformers interconvert at 25 °C. Thus, inhibition of the rotation of the units contributes to the separation of these planar-chiral pillar[5]arene enantiomers. In calix[*n*]arenes, modification of alkyl substituents can slow the rotational motion of the phenolic units, and can immobilize the conformation of calix[*n*]arenes.[18] Thus, we first introduced various lengths of alkyl chains, such as ethyl, propyl, butyl, pentyl, hexyl and dodecyl chains to both rims of the pillar[5]arene.[19]

The conformational characteristics of the per-alkylated pillar[5]arenes are investigated by variable temperature ^{1}H NMR measurements (Figure 4.4). In the case of per-hexylated pillar[5]arene (**4.1**), the proton signals from the methylene moieties adjacent to O atoms were observed as two peaks with an integration ratio of 1 : 1 at −30 °C. The split methylene peaks

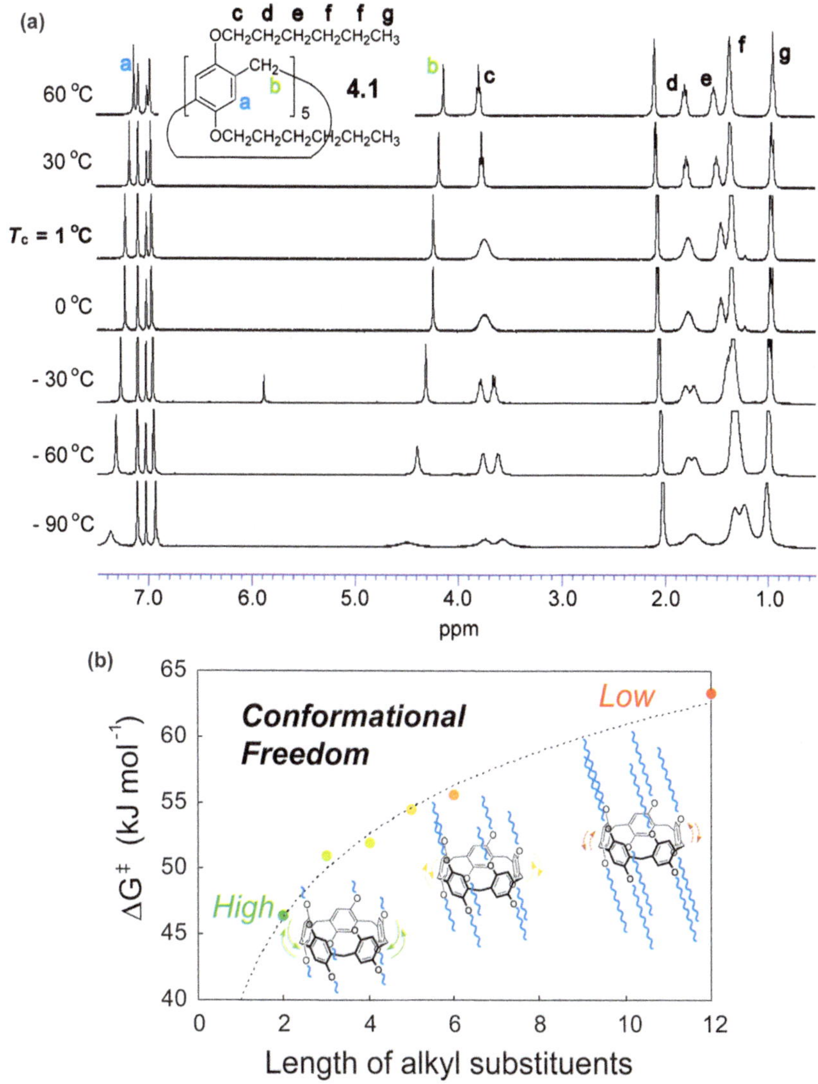

Figure 4.4 (a) Variable-temperature ^{1}H NMR spectra of **4.1** (10 mM) in toluene-d_8. (b) Plot of the $\Delta G^{\ddagger}$ values for pillar[5]arenes with different lengths of alkyl chains in toluene-d_8.
Reproduced with permission from ref. 19. Copyright 2011 American Chemical Society.

coalesced at 1 °C. At 30 °C, the peak was observed as one peak. The change in peak shape corresponds to the speed of rotation of the unit. The rotation of the units at low temperatures was slow on the NMR timescale, therefore the methylene protons are diastereotopic at low temperatures. At 1 °C, the rotational speeds of the unit are the same on the NMR timescale, and over 1 °C, the rotation was fast on the NMR timescale. Thus, splitting the proton signals can be used as a probe to determine the rotational barrier of the units. The rotational barrier (ΔG) values are calculated using the following equation:

$$\Delta G = 8.314T_c[22.96 + \log(T_c/\delta\nu)] \quad (4.1)$$

where ΔG is the free energy of activation for the rotation of the units and $\delta\nu$ is the chemical shift difference in the frozen structures (at low temperatures). The ΔG values increase as the length of the alkyl chain increases. However, even by introducing long dodecyl chains at both rims, the rotation of the units cannot be suppressed. Therefore, pillar[5]arenes carrying more bulky substituents, such as cyclohexylmethyl (**4.2**) and cyclohexylethyl pillar[5]arenes (**4.3**), have been synthesized (Figure 4.5).[20]

In **4.3**, the proton signals from the methylene moieties coalesced upon heating (Figure 4.5a), indicating heat-induced rotation of the units. However, the methylene protons in **4.2** did not coalesce upon heating within the temperature range investigated (Figure 4.5b), indicating no rotation or only slow rotation of the units on the NMR timescale. To clarify whether rotation of the units in **4.2** takes place, **4.2** was investigated by chiral column chromatography (Figure 4.5c), upon which **4.2** showed two peaks with equal areas. The fractions were collected separately, and then the first and second fractions were re-injected. In both cases, the original peak was detected but the paired peak from an enantiomer was not, indicating an absence of racemization. Figure 4.5d shows the UV-vis and CD spectra of these fractions. These fractions had an absorption band at 270–320 nm, which is assigned to the phenyl moieties of the pillar[5]arene units. The CD signals of these fractions observed in this range are mirror images, indicating isolation of enantiopure *pS* and *pR* from enantiomers. Pillar[5]arenes have four diastereomers, therefore the question is which pair of enantiomers is isolated. From ^{1}H NMR measurements, the proton signal of the phenyl and methylene bridge moieties of **4.2** was observed as a singlet. The equivalent signal indicates that the isolated enantiomer structure of **4.2** can be assigned as C_5 symmetrical **1a** and **1b**.

The other diastereomers did not form because the **1a** and **1b** forms are the most stable structures among all the diastereomers. Energies [B3LYP/6-31G(d,p) in kJ mol^{-1}] of these different diastereomers of the simple per-methylated pillar[5]arene **4.4** (Figure 4.6) relative to its low-energy diastereomers **1a/b** were calculated as: **2a/b** 73.1 < **3a/b** 97.2 < **4a/b** 167.4. To interconvert from **1a** to **1b** forms or from **1b** to **1a** forms, **4.4** needs to go through the other diastereomers. The energies of the other diastereomers are relatively high compared with that of the stable conformers **1a/b**, therefore the interconversion should take place by communication between units (domino effect). The single crystal structures of pillar[5]arenes determine which diastereomers are the most stable structures. The effect of substituents on the X-ray crystal structures of pillar[5]arenes has been investigated. Figure 4.6 shows the X-ray crystal structures of other per-alkylated pillar[5]arenes.

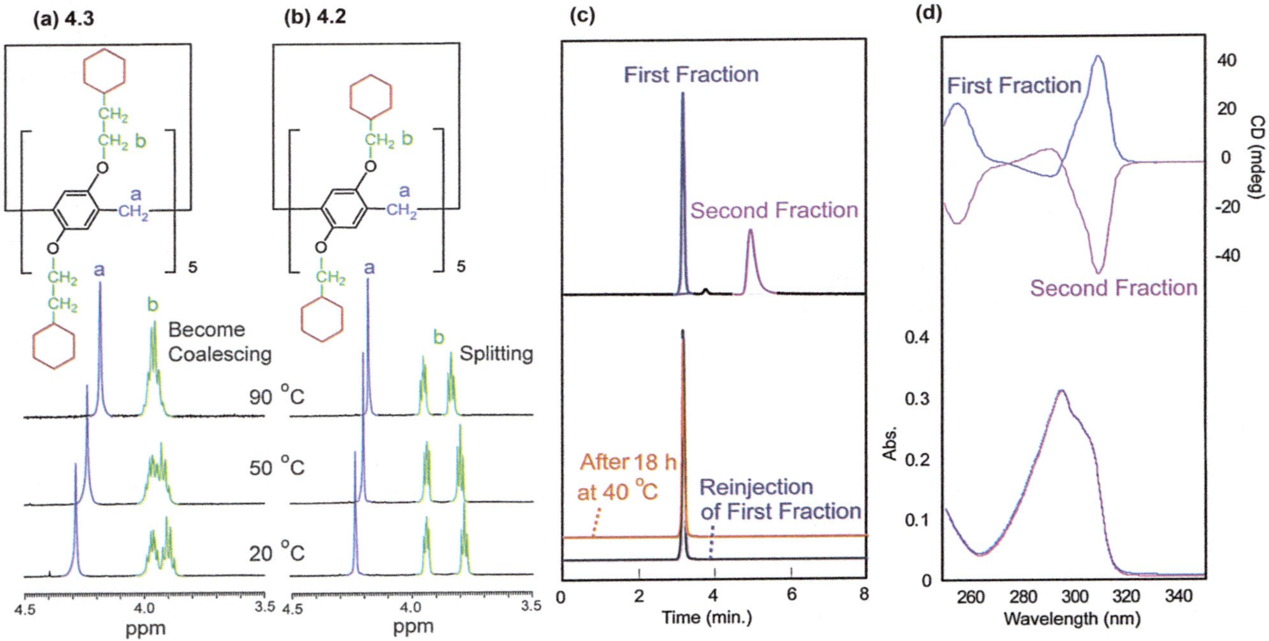

Figure 4.5 Variable-temperature partial ^{1}H NMR spectra of (a) **4.3** and (b) **4.2** in toluene-d_8. (c) Chiral HPLC traces of **4.2** and the first fraction of **4.2** by holding at 40 °C for 18 h. Hexane : EtOH = 97 : 3 (vol%) was used as the eluent. (d) Ultraviolet-visible (UV-vis) and circular dichroism (CD) spectra of the first and second fractions of **4.2** (14 μL mol^{-1} cm^{-1}) in hexane at 25 °C. Reproduced with permission from ref. 20. Copyright 2011 American Chemical Society.

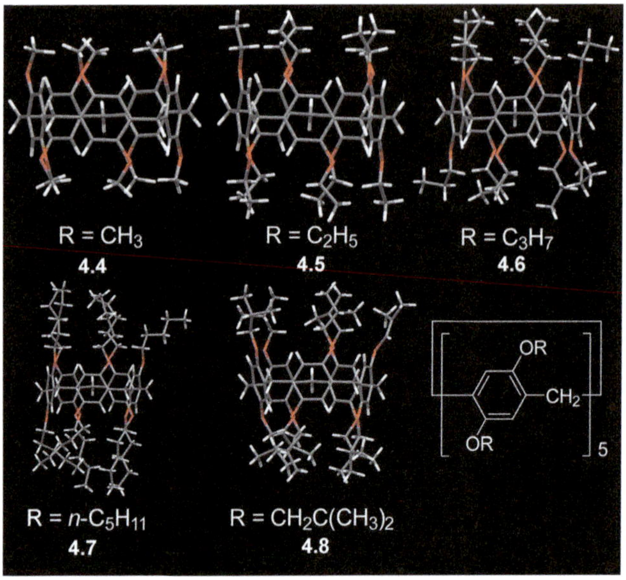

Figure 4.6 X-ray crystal structures of simple per-alkylated pillar[5]arenes.

The structures of per-methylated (**4.4**),[21] per-ethylated (**4.5**),[22] per-propylated (**4.6**),[23] per-pentylated (**4.7**)[19] and per-diisobutylated pillar[5]arenes (**4.8**)[24] are cyclic pentamers with the constituent units connected by methylene bridges at the *para* positions. When the substituents are simple alkyl chains, there are no interactions between alkyl substituents on adjacent units. Therefore, only the most stable enantiopure conformers, in which all of the substituents at both rims have the same orientation, are observed. In the case of pillar[5]arenes with short alkyl chains (**4.4**, **4.5** and **4.6**), 1,4-dialkoxybenzene units are uniformly arranged in a vertical direction. However, the introduction of bulky substituents distorts the pentagon and pillar architecture (**4.7** and **4.8**). This arises from the reduction of the steric hindrance between adjacent units.

4.2.2 Conformation of Pillar[5]arenes with One π-Conjugated Unit

The other method used to inhibit the rotation of the units is direct incorporation of the π-conjugated unit. Stoddart *et al.* reported that the introduction of only two bulky mono-substituted phenyl rings at the A1/A2 positions of pillar[5]arene (**4.9**) can inhibit the rotation of the constituent units, and contributes to the separation of planar-chiral pillar[5]arenes.[25]

Figure 4.7 shows the X-ray crystal structure of **4.9**. All of the methoxy substituents at both rims show the same orientation of the π-conjugated unit. Therefore, as with simple per-alkylated pillar[5]arenes, C_5 symmetrical enantiomers **1a/b** are isolated in pillar[5]arenes with one π-conjugated unit.

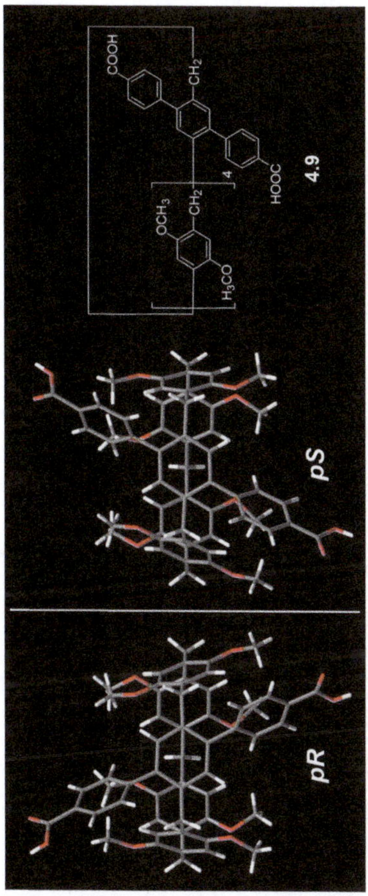

Figure 4.7 X-ray crystal structure and chemical structure of planar chiral pillar[5]arene with one π-conjugated unit **4.9**.

We also reported a planar-chiral pillar[5]arene derivative containing one π-conjugated unit.[26] Introduction of bithienyl moieties to the pillar[5]arene at the A1/A2 unit (**4.10**) also prevented the rotation of the units, and resulted in the separation of enantiomers using a chiral HPLC column. Figure 4.8 shows UV-vis and CD spectra of *pS*-**4.10** and *pR*-**4.10**; **4.10** has two absorption bands, at 270–320 nm and 320–400 nm, which are assigned as the absorption from the phenyl moieties of the pillar[5]arene units and the π-conjugated unit.

As with **4.2**, the CD signals of *pS*-**4.10** and *pR*-**4.10** are mirror images. Negative and positive CD signals were also observed in the region of the π-conjugated unit (320–400 nm). This means that CD signals covered the absorption region of the π-conjugated unit because this unit is directly incorporated in the pillar[5]arene units. The phenyl proton signals of **4.10** were observed as five singlet peaks from ^{1}H NMR measurement, hence the conformation of **4.10** can be assigned as a pair of C_5 symmetrical enantiomers **1a/b**. **4.10** showed fluorescence at 450 nm from the π-conjugated unit. Therefore, we measured the circularly polarized luminescence of **4.10**. However, no Circularly Polarized Luminescence (CPL) signal was observed. The CD anisotropy factor, g_{abs}, for *pR*-**4.10** was $+1.3\times10^{-6}$, which is unfortunately not sufficient to provide an observable CPL signal.

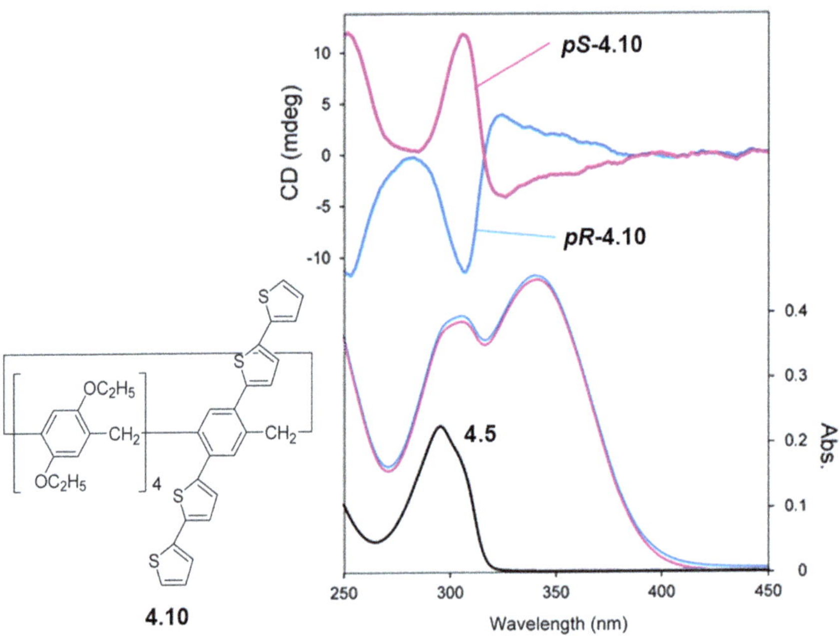

Figure 4.8 Chemical structure, UV-vis and CD spectra of *pS* and *pR* forms of pillar[5]arene with one π-conjugated unit (**4.10**) and per-ethylated pillar[5]arene (**4.5**).
Reproduced from ref. 26 with permission from The Royal Society of Chemistry.

4.2.3 Conformation of Hydroxylated Pillar[5]arenes

Native cyclodextrins are conformationally rigid host molecules because their structures are stabilized by an intra-molecular hydrogen bond network.[27] In contrast, structures of per-methylated cyclodextrins are flexible, owing to the lack of an intra-molecular hydrogen bond network.[28–31] Thus, the intra-molecular hydrogen bond network dramatically contributes to the conformational flexibility of cyclodextrins. As with native cyclodextrins, per-hydroxylated pillar[5]arene **4.11**, which has 10 phenol groups, would form an intra-molecular hydrogen bond network.[13] Therefore, its conformational flexibility was investigated by variable temperature ^{1}H NMR (Figure 4.9).

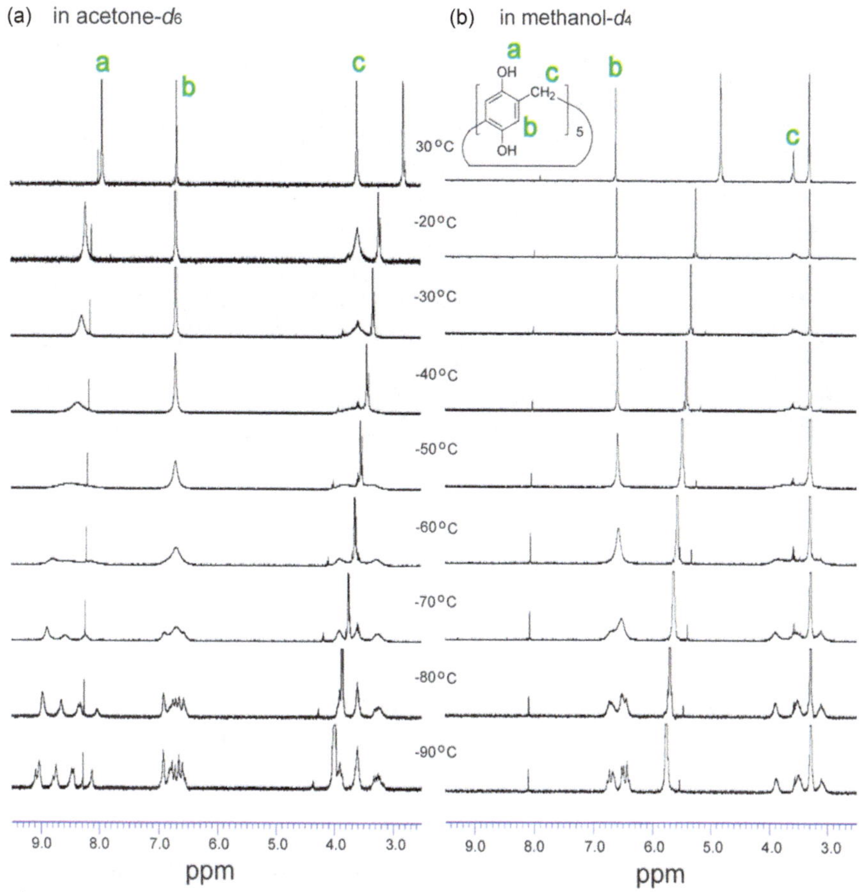

Figure 4.9 Variable-temperature ^{1}H NMR spectra of per-hydroxylated pillar[5]arene **4.11** (10 mM) in (a) acetone-d_6 and (b) methanol-d_4.
Reproduced with permission from ref. 13. Copyright 2010 American Chemical Society.

At 25 °C, the signals from phenol, phenyl and methylene are observed as singlets. As the measurement temperatures decreased, these signals became broad then split, indicating that the conformational interconversion of **4.11** was slow on the NMR timescale. In the case of per-methylated pillar[5]arene (**4.4**), the most stable conformations are the C_5 symmetrical diastereomers **1a/b**. However, the split signals of **4.11** at low temperature suggest that the non-symmetrical diastereomers are the most stable structure of **4.11** in the frozen state. Figure 4.10 shows the X-ray single crystal structure of **4.11**.[32]

Intra-molecular hydrogen bonds arranged in an alternating up-and-down manner induce flipping of two phenolic units at the A and C units. However, pillar[5]arene is composed of an odd number of units, therefore the two

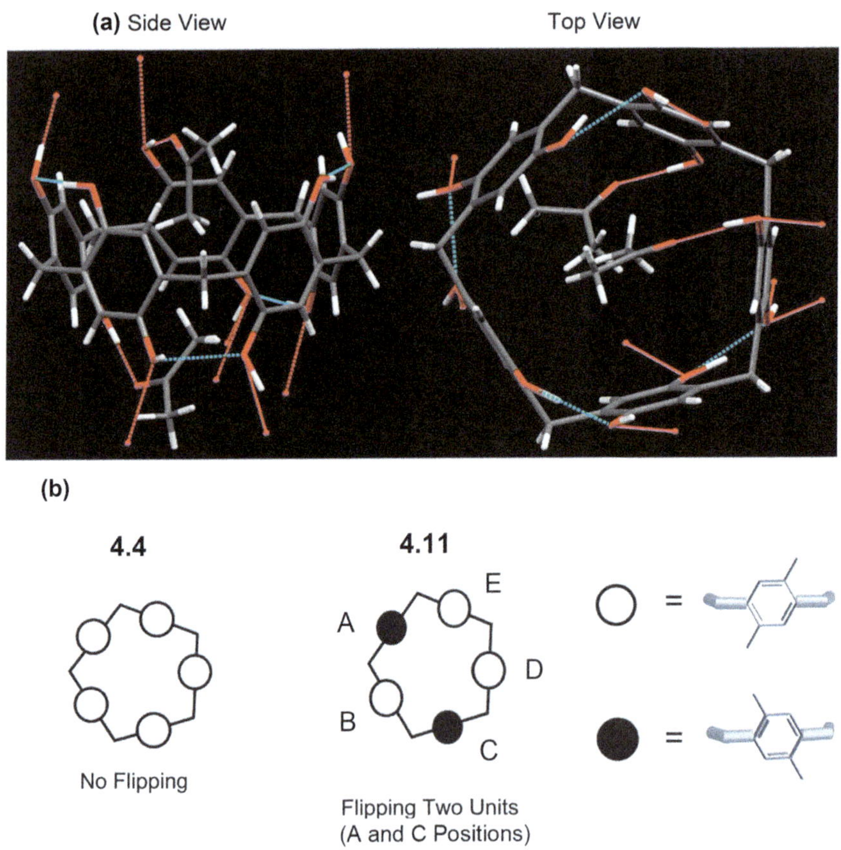

Figure 4.10 (a) X-ray crystal structure of per-hydroxylated pillar[5]arene **4.11**. The blue and red lines indicate intra- and inter-molecular hydrogen bonds, respectively. (b) Conformations of per-methylated (**4.4**) and per-hydroxylated (**4.11**) pillar[5]arenes in the crystal state.
Reproduced with permission from ref. 32. Copyright 2011 American Chemical Society.

phenol moieties at D and E units cannot form intra-molecular hydrogen bonds, and they form inter-molecular hydrogen bonds with two acetone molecules (solvent). Therefore, the non-symmetrical diastereomers **4a/b** represent the most stable structure of **4.11** in the solid state. Peerannawar and Gejji reported the relative energies of these diastereomers.[33] Energies [B3LYP/6-31G(d,p) in kJ mol^{-1}] of these different diastereomers relative to their low-energy diastereomers **4a/b** are calculated to be: **2a/b**, 24.1< **3a/b**, 36.1< **1a/b** 49.2. The most stable and unstable diastereomers of **4.11** are the exact opposites of those of the simple per-methylated pillar[5]arene **4.4. 4.11** forms up-and-down **4a/b** conformations to form an intra-molecular hydrogen bond network between units, while **4.4** forms highly regular **1a/b** conformations to reduce the steric hindrance between substituents.

Di-hydroxylated pillar[5]arene has five possible constitutional conformers: A1/B1, A1/C1, A1/A2, A1/B2 and A1/C2.[34] Among them, A1/A2 and A1/B2 di-hydroxylated pillar[5]arenes have been synthesized. The crystal structure of A1/A2 di-hydroxylated pillar[5]arene (**4.12**) was reported by Stoddart *et al.* (Figure 4.11).[25]

Flipping of one phenolic unit at position A is observed to form two intra-molecular hydrogen bonds, indicating that the **2a/b** forms are the most stable structures in **4.12**. We reported the crystal structure of A1/B2 di-hydroxylated pillar[5]arene **4.13** (Figure 4.12).[35] Two phenolic units, at positions A and B, were flipped in the crystal structure of A1/B2 di-hydroxylated pillar[5]arene **4.13** to form intra-molecular hydrogen bonds. Therefore, the **3a/b** forms are the most stable structures in **4.13**, and di- and per-hydroxylated pillar[5]arene derivatives are the most stable conformations to form intra-molecular hydrogen bonds.

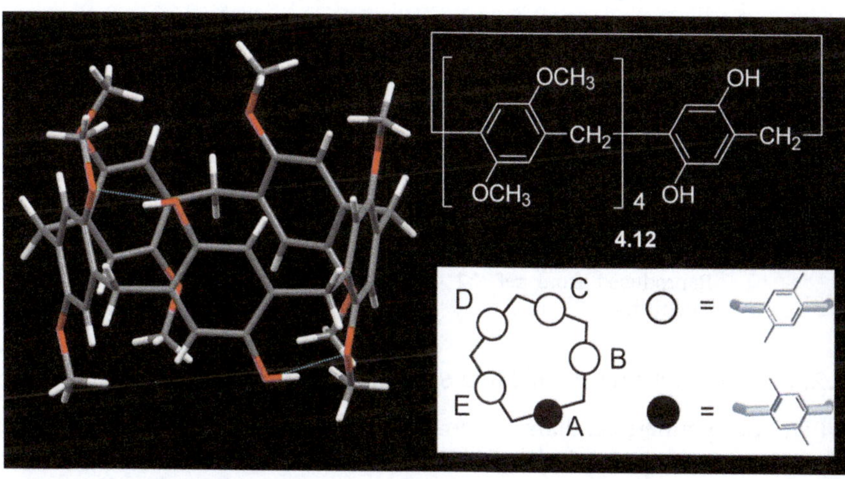

Figure 4.11 X-ray crystal structure and chemical structure of A1/A2 di-hydroxylated pillar[5]arene **4.12**.
Reproduced with permission from ref. 25. Copyright 2012 American Chemical Society.

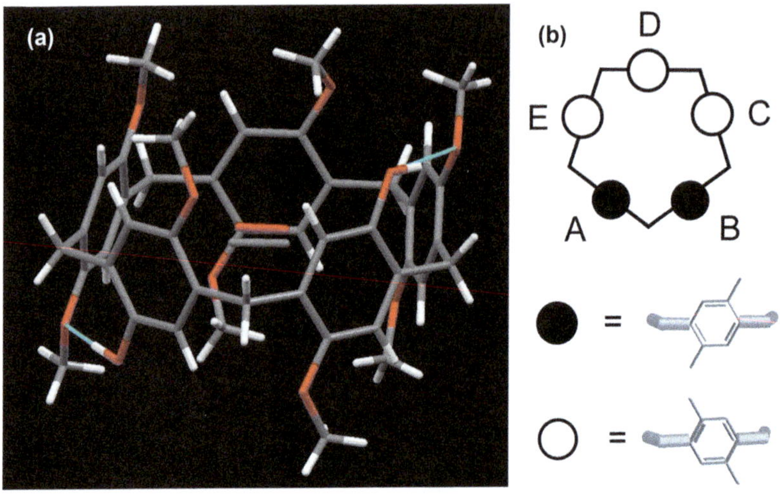

Figure 4.12 X-ray crystal structure and chemical structure of A1/B2 di-hydroxylated pillar[5]arene **4.13**.
Reproduced from ref. 35 with permission from The Royal Society of Chemistry.

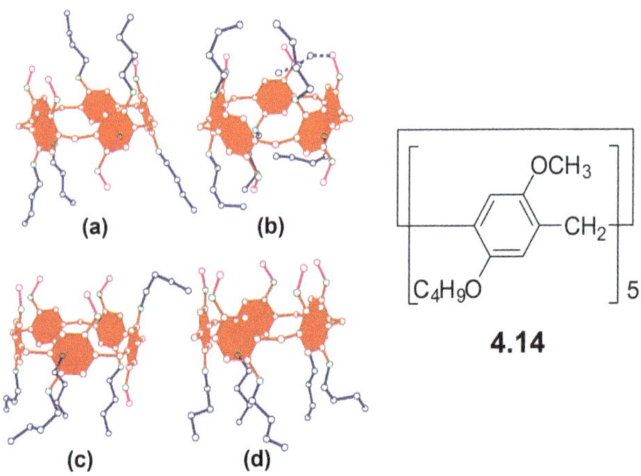

Figure 4.13 Chemical and X-ray crystal structure of non-symmetrical pillar[5]arene **4.14**.
Reproduced from ref. 37 with permission from The Royal Society of Chemistry.

4.2.4 Conformation of Non-symmetrical Pillar[5]arenes

When non-symmetrical dialkoxybenzenes are used as a monomer, a mixture of four constitutional isomers is obtained.[36,37] Figure 4.13 shows the chemical structures and X-ray crystalline structures of the four isomers of methoxybutoxypillar[5]arene **4.14**.

As with per-alkylated pillar[5]arenes, the most stable conformations are the highly regular diastereomers **1a/b**, owing to the lack of interaction between alkoxy substituents on the rims. The non-symmetrical structure is a useful platform for the synthesis of amphiphilic molecules (see Chapter 9 for details).

4.3 Conformations of Pillar[6]arenes

4.3.1 Conformation of Symmetrical Per-alkylated Pillar[6]arenes

Cyclic hexamers, pillar[6]arenes, have 13 conformers, as shown in Figure 4.14.[38]

There are five diastereomers×two enantiomers. In addition, pillar[6]arenes have an even number of rings, therefore there are three *meso*-forms of pillar[6]arenes. Because rotation of the units can be prevented by modification of the bulky cyclohexylmethyl groups in the cyclic pentamer, pillar[5]arene, we synthesized per-cyclohexylmethylated pillar[6]arene **4.15** to inhibit the rotation of the pillar[6]arene units and separate the conformers (Figure 4.15).[38]

A conformational study of **4.15** was also performed using variable temperature ^{1}H NMR. At 25 °C, the proton signal from the methylene moiety adjacent to an O atom was observed as a singlet. The signal was split into two peaks at −60 °C. At −60 °C, the rotation of the units is slow on the NMR timescale, thus the protons become diastereotopic. Therefore, even by modification of the bulky cyclohexylmethyl groups, we cannot inhibit the rotation of the units. The size of the cyclohexylmethyl groups is sufficient to inhibit the rotation of pillar[5]arene units, but insufficient to inhibit the rotation of the pillar[6]arene units. This is because the cavity size of

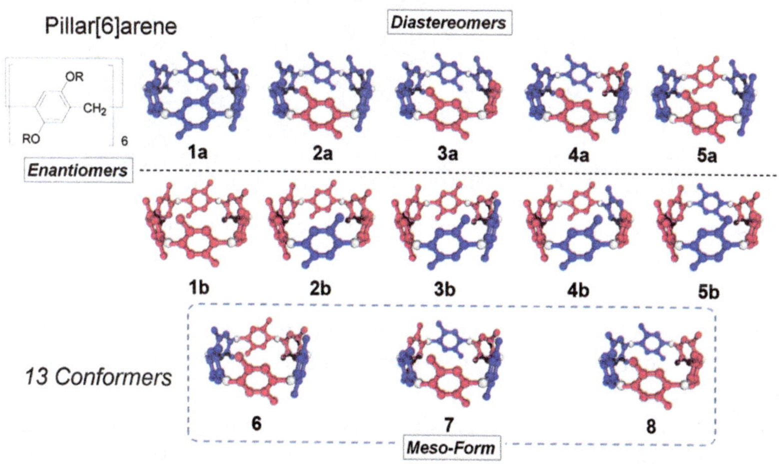

Figure 4.14 Conformers of pillar[6]arenes.
Reproduced from ref. 38 with permission from The Royal Society of Chemistry.

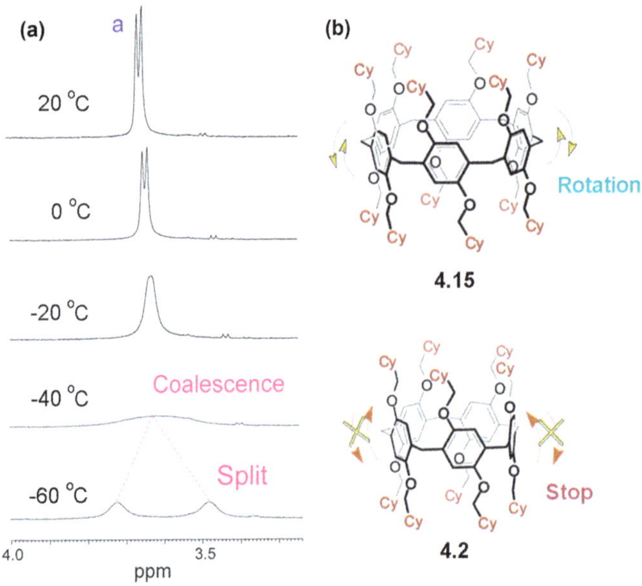

Figure 4.15 (a) Partial variable-temperature ^{1}H NMR spectra of **4.15** in toluene-d_8. (b) Rotation movements of per-cyclohexyl methylated pillar[*n*]arenes **4.2** ($n = 5$) and **4.15** ($n = 6$).
Reproduced from ref. 38 with permission from The Royal Society of Chemistry.

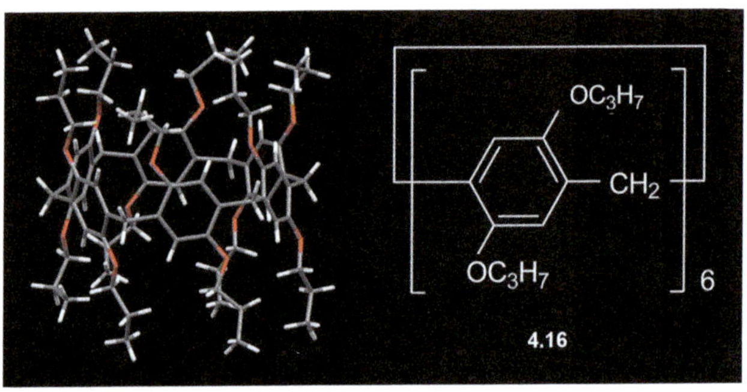

Figure 4.16 Chemical and X-ray crystal structure of per-propylated pillar[6]arene **4.16**.

pillar[6]arenes is larger than that of pillar[5]arenes. Modification of the more bulky substituents or new concepts are needed to stop the rotation of the pillar[6]arene units.

Figure 4.16 shows the X-ray single crystal structure of a simple per-alkylated pillar[6]arene, per-propylated pillar[6]arene (**4.16**).[23] As with the simple per-alkylated pillar[5]arenes, the stable parts of the conformation of **4.16** are the C_6 symmetrical enantiomers **1a/b**.

Figure 4.17 Chemical and X-ray crystal structure of per-hydroxylated pillar[6]arene **4.17**.

4.3.2 Conformation of Per-hydroxylated Pillar[6]arene

As with per-hydroxylated pillar[5]arene (**4.11**), per-hydroxylated pillar[6]arene (**4.17**), which contains 12 hydroxyl groups, also forms an intra-molecular hydrogen bond network. The intra-molecular hydrogen bond network induced flipping of the units.[39] In the case of per-hydroxylated pillar[5]arene (**4.11**), the intra-molecular hydrogen bond network is not continuous because pillar[5]arene has an odd number of macrocyclic rings.[32] In contrast, pillar[6]arene has an even number of macrocyclic rings. Therefore, the intra-molecular hydrogen bond network is completely continuous in an alternating up-and-down manner and induces flipping of the three phenolic units at the A, C and E units (Figure 4.17).

Peerannawar and Gejji reported the relative stabilization energies of different conformers.[40] The *meso*-form of conformer 7 is the most stable, and the energies (in kJ mol^{-1}) of the different conformers relative to conformer 7 are calculated to be: 21.4 (**4a/b**)< 41.3 (**3a/b**)< 45.5 (**2a/b**)< 61.9 (**8**)< 72.4 (**1a/b**). The stable *meso*-form of conformer 7 has a highly symmetrical structure, therefore their structure can be completely fixed by bridging between the hydroquinone units (see Chapter 12 for details). The highly regular pentagonal structure of **4.17** assembles into a one-dimensional channel structure in the crystal state, and these can be used as building blocks to form highly packed two-dimensional hexagonal sheets (see Chapter 6 for details).[41]

4.4 Conclusion and Outlook

In this chapter, we have described the conformations and planar chirality of pillar[*n*]arenes. In many cases, 1,4-dialkoxybenzene units in pillar[*n*]arenes can rotate freely, therefore pillar[*n*]arenes can form various conformers. Simple pillar[5]- and pillar[6]arenes with alkyl substituents form C_5 and C_6 symmetrical conformations (**1a**/**b**), respectively. However, pillar[*n*]arenes with phenolic moieties form various conformers to yield intra-molecular hydrogen bonds. Substituents have a large effect on the conformation of pillar[*n*]arenes. Introducing bulky substituents or one rigid π-conjugated unit inhibited the rotation of the units in pillar[5]arene, and contributes to the isolation of the C_5 enantiomers in the **1a**/**b** forms. Separation of the other pairs of enantiomers is one of the next challenges to be overcome. The conversion of conformationally flexible to fixed pillar[5]arenes should also be a future target for investigation; conformationally fixed macrocyclic hosts can form very stable host–guest complexes because rigid structures are favorable for the decrease in entropy that occurs during host–guest complexation.

References

1. J. Labuta, J. P. Hill, S. Ishihara, L. Hanyková and K. Ariga, *Acc. Chem. Res.*, 2015, **48**, 521–529.
2. R. Misra and T. K. Chandrashekar, *Acc. Chem. Res.*, 2008, **41**, 265–279.
3. M. X. Wang, *Acc. Chem. Res.*, 2012, **45**, 182–195.
4. J. W. Lee, S. Samal, N. Selvapalam, H. J. Kim and K. Kim, *Acc. Chem. Res.*, 2003, **36**, 621–630.
5. J. Lagona, P. Mukhopadhyay, S. Chakrabarti and L. Isaacs, *Angew. Chem., Int. Ed.*, 2005, **44**, 4844–4870.
6. Z. Laughrey and B. C. Gibb, *Chem. Soc. Rev.*, 2011, **40**, 363–386.
7. J. Zhao, H. J. Kim, J. Oh, S. Y. Kim, J. W. Lee, S. Sakamoto, K. Yamaguchi and K. Kim, *Angew. Chem., Int. Ed.*, 2001, **40**, 4233–4235.
8. C. L. D. Gibb and B. C. Gibb, *J. Am. Chem. Soc.*, 2004, **126**, 11408–11409.
9. C. J. Pedersen, *J. Am. Chem. Soc.*, 1967, **89**, 7017–7036.
10. S. K. Kim and J. L. Sessler, *Acc. Chem. Res.*, 2014, **47**, 2525–2536.
11. S. Shinkai, T. Minami, Y. Kusano and O. Manabe, *J. Am. Chem. Soc.*, 1983, **105**, 1851–1856.
12. J. Rebek, J. E. Trend, R. V. Wattley and S. Chakravorti, *J. Am. Chem. Soc.*, 1979, **101**, 4333–4337.
13. T. Ogoshi, K. Kitajima, T. Aoki, T. Yamagishi and Y. Nakamoto, *J. Phys. Chem. Lett.*, 2010, **1**, 817–821.
14. M. Takeuchi, M. Ikeda, A. Sugasaki and S. Shinkai, *Acc. Chem. Res.*, 2001, **34**, 865–873.
15. A. Ikeda and S. Shinkai, *Chem. Rev.*, 1997, **97**, 1713–1734.
16. *Cyclophane Chemistry*, ed. F. Vögtle, Wiley, Chichester, 1993.
17. F. Diederich, *Cyclophane*, Royal Chemical Society, Cambridge, 1991.
18. K. Iwamoto, K. Araki and S. Shinkai, *J. Org. Chem.*, 1991, **56**, 4955–4962.
19. T. Ogoshi, K. Kitajima, T. Aoki, S. Fujinami, T. Yamagishi and Y. Nakamoto, *J. Org. Chem.*, 2010, **75**, 3268–3273.

20. T. Ogoshi, K. Masaki, R. Shiga, K. Kitajima and T. Yamagishi, *Org. Lett.*, 2011, **13**, 1264–1266.
21. T. Ogoshi, S. Kanai, S. Fujinami, T. Yamagishi and Y. Nakamoto, *J. Am. Chem. Soc.*, 2008, **130**, 5022–5023.
22. T. Ogoshi, R. Sueto, K. Yoshikoshi, Y. Sakata, S. Akine and T. Yamagishi, *Angew. Chem., Int. Ed.*, 2015, **54**, 9849–9852.
23. G. Yu, C. Han, Z. Zhang, J. Chen, X. Yan, B. Zheng, S. Liu and F. Huang, *J. Am. Chem. Soc.*, 2012, **134**, 8711–8717.
24. C. Y. Han, F. Y. Ma, Z. B. Zhang, B. Y. Xia, Y. H. Yu and F. H. Huang, *Org. Lett.*, 2010, **12**, 4360–4363.
25. N. L. Strutt, D. Fairen-Jimenez, J. Iehl, M. B. Lalonde, R. Q. Snurr, O. K. Farha, J. T. Hupp and J. F. Stoddart, *J. Am. Chem. Soc.*, 2012, **134**, 17436–17439.
26. T. Ogoshi, D. Yamafuji, T. Akutsu, M. Naito and T. Yamagishi, *Chem. Commun.*, 2013, **49**, 8782–8784.
27. A. Miyawaki, P. Kuad, Y. Takashima, H. Yamaguchi and A. Harada, *J. Am. Chem. Soc.*, 2008, **130**, 17062–17069.
28. K. Kano, R. Nishiyabu and R. Doi, *J. Org. Chem.*, 2005, **70**, 3667–3673.
29. K. Kano, T. Ishimura and S. Negi, *J. Inclusion Phenom. Mol. Recognit. Chem.*, 1995, **22**, 285–298.
30. T. Steiner and W. Saenger, *Angew. Chem., Int. Ed.*, 1998, **37**, 3404–3407.
31. M. R. Caira, S. A. Bourne, W. T. Mhlongo and P. M. Dean, *Chem. Commun.*, 2004, 2216–2217.
32. T. Ogoshi, T. Aoki, K. Kitajima, S. Fujinami, T. Yamagishi and Y. Nakamoto, *J. Org. Chem.*, 2011, **76**, 328–331.
33. S. R. Peerannawar and S. P. Gejji, *Comput. Theor. Chem.*, 2012, **999**, 169–178.
34. T. Ogoshi, D. Yamafuji, D. Kotera, T. Aoki, S. Fujinami and T. Yamagishi, *J. Org. Chem.*, 2012, **77**, 11146–11152.
35. T. Ogoshi, K. Kitajima, S. Fujinami and T. Yamagishi, *Chem. Commun.*, 2011, **47**, 10106–10108.
36. Y. H. Kou, H. Q. Tao, D. R. Cao, Z. Y. Fu, D. Schollmeyer and H. Meier, *Eur. J. Org. Chem.*, 2010, **2010**, 6464–6470.
37. Z. Zhang, Y. Luo, B. Xia, C. Han, Y. Yu, X. Chen and F. Huang, *Chem. Commun.*, 2011, **47**, 2417–2419.
38. T. Ogoshi, K. Demachi, K. Masaki and T. A. Yamagishi, *Chem. Commun.*, 2013, **49**, 3952–3954.
39. Y. Ma, X. Chi, X. Yan, J. Liu, Y. Yao, W. Chen, F. Huang and J. L. Hou, *Org. Lett.*, 2012, **14**, 1532–1535.
40. S. R. Peerannawar and S. P. Gejji, *Phys. Chem. Chem. Phys.*, 2012, **14**, 8711–8722.
41. T. Ogoshi, K. Yoshikoshi, R. Sueto, H. Nishihara and T. Yamagishi, *Angew. Chem., Int. Ed.*, 2015, **55**, 6466–6469.

CHAPTER 5

Host–Guest Properties of Pillar[n]arenes

TOMOKI OGOSHI* AND TADA-AKI YAMAGISHI

Graduate School of Natural Science and Technology,
Kanazawa University, Kakuma-machi, Kanazawa 920-1192, Japan
*Email: ogoshi@se.kanazawa-u.ac.jp

5.1 Introduction

Macrocyclic compounds, such as cyclodextrins, calix[*n*]arenes (CDs), cucurbit-[*n*]urils (CB[*n*]s), blue boxes and crown ethers, play a major role in supramolecular chemistry because they have a well-defined cavity with a size on the angstrom scale. They can capture guest molecules by various physical interactions depending on their cavity sizes, the chemical structure of constituent units and functional groups. For example, CDs, which have been investigated since the early 1900s and are widely used, consist of water-soluble glucopyranose units. Therefore, they are soluble in aqueous media and can accommodate hydrophobic guests in aqueous media.[1] Crown ethers, which consist of ethylene oxide moieties, form host–guest complexes not only with metal ions by multi-coordination bonds but also with organic ammonium salts by multiple NH–O hydrogen bonds.[2] Blue boxes, which were introduced by Stoddart and co-workers,[3] comprise viologen moieties. The components of the viologen moieties in the blue boxes are electron-poor cationic molecules, therefore their cavities are also an electron-poor space. Electron-rich molecules, such as dialkoxybenzenes, naphthalenes and tetrathiafulvalene derivatives, are therefore good guest molecules for blue boxes.[4–6] In this chapter, we discuss the host–guest properties of

Monographs in Supramolecular Chemistry No. 18
Pillararenes
Edited by Tomoki Ogoshi
© The Royal Society of Chemistry 2016
Published by the Royal Society of Chemistry, www.rsc.org

pillar[*n*]arenes. The structures and electrostatic effects are very important factors for their molecular recognition ability, therefore we first discuss their structures and electron potential profiles. Second, we discuss their host–guest complexation ability in organic and aqueous media. Finally, the host–guest properties of larger pillar[*n*]arene homologs ($n \geq 7$) are discussed.

5.2 X-ray Crystal Structure and Electron Potential Profiles of Pillar[*n*]arenes

Figure 5.1 shows the chemical and X-ray crystal structures, and calculated electrostatic potential profiles, of pillar[5]-, pillar[6]arenes and a related macrocycle, calix[5]arene.

The inner surfaces of the pillar[5]- and pillar[6]arene cavities are negative because alkoxy groups on the rims are electron-donating moieties. Compared with an open-ended calix[5]arene structure, their pillar-shaped structures contribute to enhance the π-electron density of their cavity. Therefore, pillar[5]- and pillar[6]arenes prefer to accommodate electron-poor molecules, such as cationic molecules and molecules containing electron-with-drawing groups, in their cavities.[7]

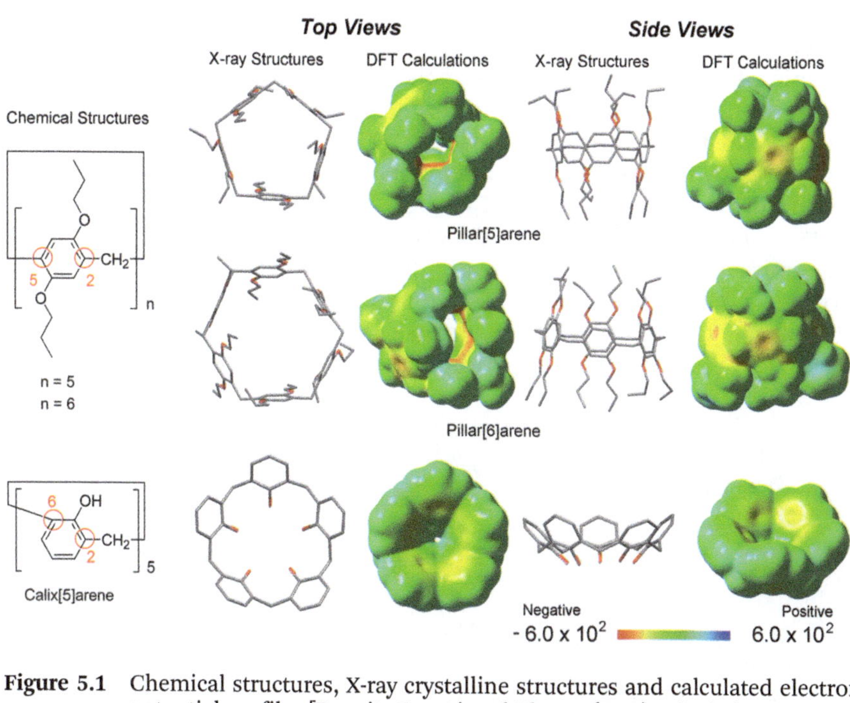

Figure 5.1 Chemical structures, X-ray crystalline structures and calculated electron potential profiles [Density Functional Theory (DFT) calculations, B3LYP/6-31G(d,p)] of pillar[5]arene, pillar[6]arene, and calix[5]arene. Reproduced from ref. 7 with permission from The Royal Society of Chemistry.

5.3 Host–Guest Properties of Pillar[5]arenes

5.3.1 per-Hydroxylated Pillar[5]arene

The size of the pillar[5]arene cavity is approximately 5 Å, which is the same cavity size as α-CD and CB[6]. α-CD and CB[6] can accommodate linear-shaped hydrocarbons and simple aromatic molecules because their cavity is sufficiently large to incorporate the size of these guest molecules.[1,8] Pillar[5]arene derivatives also form host–guest complexes with the linear hydrocarbons and mono-aromatic molecules. Pillar[5]arenes have a total of 10 substituents on both their rims. The type of substituents contributes to the solubility of pillar[5]arenes. A pillar[5]arene with 10 phenol moieties (**H5.1**; Figure 5.2) is soluble in polar solvents, such as THF, acetone, acetonitrile, methanol, DMF and DMSO, apart from water, and is not soluble in nonpolar solvents such as chloroform, hexane and toluene because the phenolic moiety is a polar functional group. Therefore, acetone and methanol are mainly used as the complexation media for **H5.1**. Hydroquinone units composed of **H5.1** are electron-donating moieties; therefore, **H5.1** forms host–guest complexes with electron-accepting molecules such as viologen, pyridinium and imidazolium cations. The complexation was first investigated by ^{1}H NMR (Figure 5.3).[9]

By mixing **H5.1** with viologen salt (**G5.1**) or pyridinium salt (**G5.2**), upfield shifts were observed in the proton signals from the viologen or pyridinium moieties. In contrast, the proton signals from **H5.1** showed downfield shifts. These results indicate that the electron-poor molecules **G5.1** and **G5.2** were included in the electron-rich cavity of **H5.1**. The complexation was also monitored by fluorescence measurements. Figure 5.4 shows the florescence spectra of **H5.1** upon addition of viologen salt **G5.1**.

H5.1 showed an emission at 330 nm following excitation at 280 nm. The emission derived from the hydroquinone units of **H5.1**. Addition of the viologen **G5.1** induced fluorescence quenching. The association constant of viologen **G5.1** with **H5.1**, determined by fluorescence quenching, was 10 times higher than that of pyridinium **G5.2** with **H5.1** (Table 5.1, runs 1

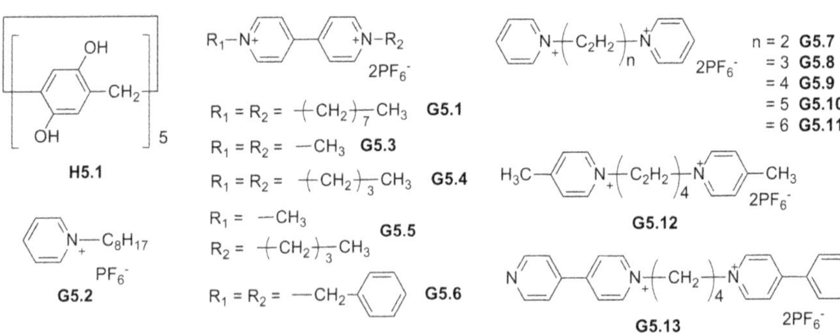

Figure 5.2 Chemical structures of per-hydroxylated pillar[5]arene **H5.1** and guests for **H5.1**.

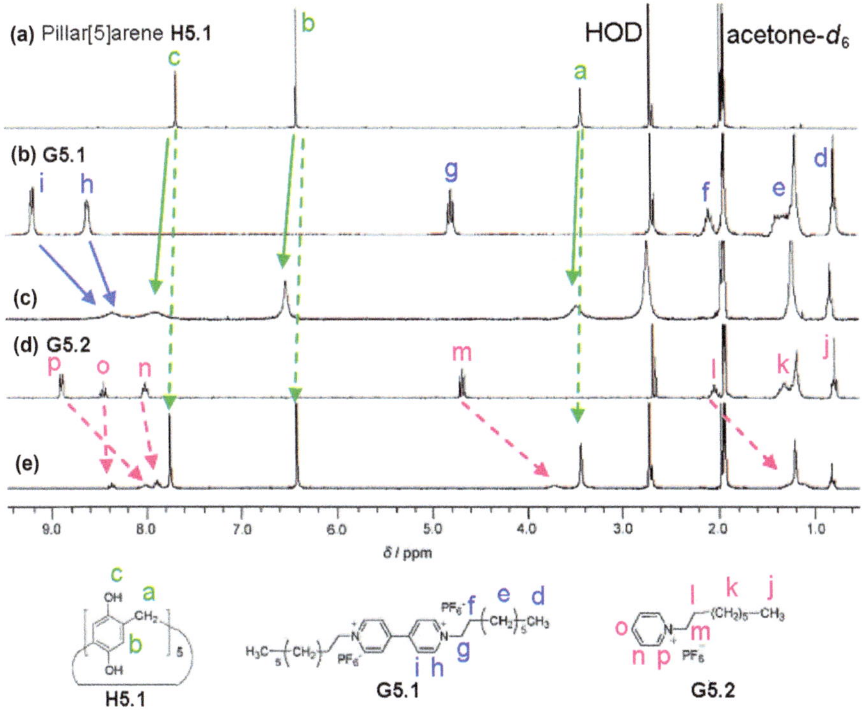

Figure 5.3 ^{1}H NMR spectra (5 mM, 25 °C) of pillar[5]arene (**H5.1**), (b) viologen guest **G5.1**, (c) a 1 : 1 mixture of **H5.1** and **G5.1**, (d) pyridinium guest **G5.2** and (e) a 1 : 1 mixture of **H5.1** and **G5.2** in acetone-d_6.
Reproduced with permission from ref. 9. Copyright 2008 American Chemical Society.

and 2), because viologen salts show stronger electron-accepting abilities than pyridinium salts. Li *et al.* investigated the host–guest complexes of viologen and pyridinium salts with **H5.1** by UV-vis measurements (Figure 5.5).[10]

Both solutions of **H5.1** and viologen salts (**G5.1**, **G5.3–G5.6**) in DMSO were colorless, therefore they exhibited no absorption over 400 nm. The solution changed from colorless to red by the addition of these viologen salts to **H5.1**. UV-vis spectra of the mixtures showed a new band at about 450 nm. The band was assigned as a charge-transfer band, indicating complexation of these viologen salts with **H5.1**. However, a mixture of a viologen salt with benzyl moieties (**G5.6**) and **H5.1** did not afford a significant charge-transfer (CT) band. A possible reason for this is that the benzyl moieties are too bulky to slip over the pillar[5]arene cavity. Mixtures of dipyridinium guests (**G5.7–G5.11**) and **H5.1** did not afford significant CT bands. This is because pyridinium salts are weaker electron acceptors than viologen salts. The host–guest complexation ability between bis(pyridinium) derivatives **G5.7–G5.13** and **H5.1** was investigated by UV-vis and ^{1}H NMR titrations (Table 5.1, runs 3–9). **G5.9**, containing a C4 linker, formed the most stable

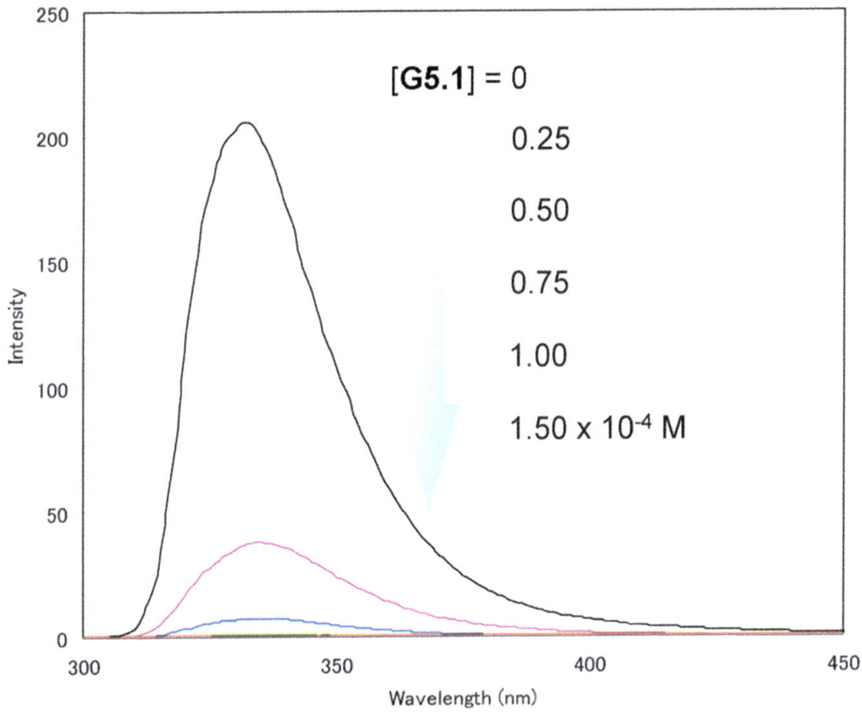

Figure 5.4 Emission spectra (excited at 279 nm) of **H5.1** (1.00×10^{-4} M) upon addition of **G5.1** at 25 °C.

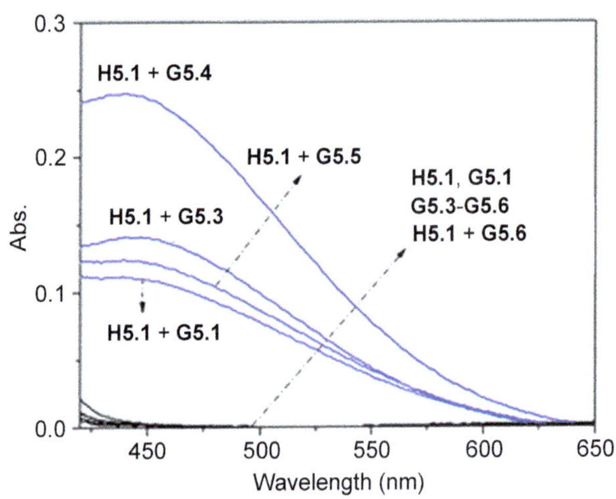

Figure 5.5 UV-vis spectra of **G5.1, G5.3–G5.6** (1.9 mM–2.0 mM) in the presence of **H5.1** (4.0 mM) in DMSO at 25 °C.
Reproduced from ref. 10 with permission from The Royal Society of Chemistry.

Table 5.1 Host–guest complexes of per-hydroxylated pillar[5]arene.

Run	Host	Guest	Solvent	Association constant (K) (M^{-1})	Method	Ref.
1	**H5.1**	**G5.1**	Methanol	$(1.2\pm0.2)\times10^4$	Fl[b]	9
2	**H5.1**	**G5.2**	Methanol	$(1.2\pm0.3)\times10^3$	Fl[b]	9
3	**H5.1**	**G5.7**	DMSO	—[a]	UV	10
4	**H5.1**	**G5.8**	DMSO	$8.8\pm0.7\times10$	UV	10
5	**H5.1**	**G5.9**	DMSO	$4.5\pm0.4\times10^2$	UV	10
6	**H5.1**	**G5.10**	DMSO	$3.7\pm0.3\times10^2$	UV	10
7	**H5.1**	**G5.11**	DMSO	$1.2\pm0.1\times10^2$	UV	10
8	**H5.1**	**G5.12**	DMSO	$4.0\pm0.3\times10^2$	UV	10
9	**H5.1**	**G5.13**	DMSO-d_6	$7.4\pm0.3\times10^2$	NMR	10
10	**H5.1**	**G5.14**	Acetone-d_6	$K_1=1.1\times10^2$, $K_2=2.0\times10$	NMR	11
11	**H5.1**	**G5.15**	Acetone-d_6	$K_1=7.0\times10^2$, $K_2=1.7$	NMR	11
12	**H5.1**	**G5.16**	Acetone-d_6	$4.6\pm0.6\times10^2$	NMR	12
13	**H5.1**	**G5.17 (*trans*)**	Acetone-d_6	$2.34\pm0.33\times10^2$	NMR	13
14	**H5.1**	**G5.17 (*cis*)**	Acetone-d_6	$2.05\pm0.08\times10^2$	NMR	13

[a]The K value was too small to be calculated.
[b]Fl; Fluorescence measurement.

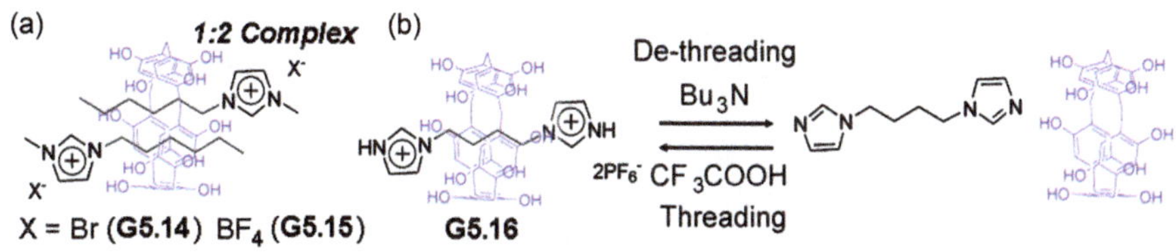

Figure 5.6 (a) Formation of 1:2 host–guest complexes between **H5.1** and ionic liquids. (b) Base–acid controlled threading/de-threading process.

complexes with **H5.1** among all the derivatives **G5.7–G5.11** having different chain lengths (C2–C6), because the length of the C4 linker fits well to the **H5.1** cavity. Introduction of methyl (**G5.12**, run 8) does not affect the association constant when compared with that obtained with the non-substituted compound (**G5.9**, run 5), but the substitution of pyridyl for hydrogen (**G5.13**, run 9) increased the association constant by 1.6 times, owing to the introduction of π-stacking between the pyridyl of **G5.13** and the benzene moiety of **H5.1**.

Ionic liquids are also included in the cavity of **H5.1** (Figure 5.6a).[11] Typical ionic liquids of 1-hexyl-3methylimidazoliums with Br^- (**G5.14**) and BF_4^- anions (**G5.15**) formed 1:2 host–guest complexes with **H5.1** (Table 5.1, runs 10 and 11).

Bis(imidazolium)dication linked by the C4 linker **G5.16** also formed 1:1 complexes with **H5.1** in polar solvents (run 12; Figure 5.6b).[12] The threading/de-threading process can be reversibly controlled upon the addition of acid and base to the **H5.1⊃G5.16** complex. Upon addition of n-Bu_3N, deprotonation of the guest occurred. This was the trigger for the de-threading of the guest molecule. Addition of CF_3COOH then caused re-protonation of the guest, which then led to the formation of the original host–guest complex.

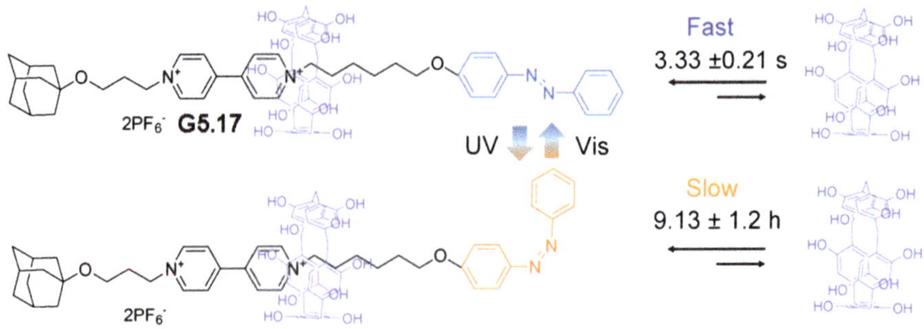

Figure 5.7 Photo-switching transformation between second and hour timescales: threading of **H5.1** onto the azobenzene end of the viologen derivative (**G5.17**).

We reported the photo-responsive shuttling behavior of the complex between a viologen axle with an azobenzene end (**G5.17** and **H5.1**) (Table 5.1, runs 13 and 14).[13] When the azobenzene end was in the *trans* form, the threading process was very quick. The half-life ($t_{1/2}$), estimated by two-dimensional (2D) Exchange Spectroscopy (EXSY) NMR, was 3.33 ± 0.21 s (Figure 5.7). In contrast, the threading of the azobenzene end in the *cis* form, whose photo-isomerization was induced by UV light irradiation, was slow. The half-life, determined by time-dependent ^{1}H NMR change, was 9.13 ± 1.2 h. Photo-isomerization from *trans* to *cis* forms increased the free energy of activation (22.8 ± 0.24 kJ mol^{-1}), which induced the transformation of the timescale for the threading of the azobenzene end from seconds to hours.

From the single X-ray crystal analysis, **H5.1** formed intra- and intermolecular hydrogen bonds, which resulted in the formation of 1D extrinsic channels with a diameter of 7.10 Å, slit channels of 5.14 Å by 9.14 Å, and an intrinsic cavity of 6.76 Å. The multiple hydrogen bonds stabilized the supramolecular structure even after removal of the trapped solvents by heating under reduced pressure. The supramolecular assembly of the activated **H5.1** took up CO_2, but did not accommodate CH_4 and N_2, indicating that the activated **H5.1** can selectively capture CO_2. The activated porous crystals of **H5.1** showed remarkable CO_2 selectivity over CH_4 (375 : 1) and N_2 (339 : 1).[14]

5.3.2 per-Alkylated Pillar[5]arenes

5.3.2.1 Cationic Guests

Host–guest complexation between per-alkylated pillar[5]arene and cationic guests was first investigated by Huang and co-workers.[15] The association constant of the host–guest complex between per-butylated pillar[5]arene (**H5.5**) and chloroform-soluble octatrimethyl ammonium hexafluorophosphate (**G5.18**) determined by ^{1}H NMR titration was $(4.1 \pm 0.4) \times 10^3$ M^{-1} in 1 : 1 stoichiometry (Figure 5.8 and Table 5.2, run 15).

We investigated the host–guest complexation of **G5.18** with pillar[5]arenes with different substituents. The *K* values for the complex of **G5.18** with 10

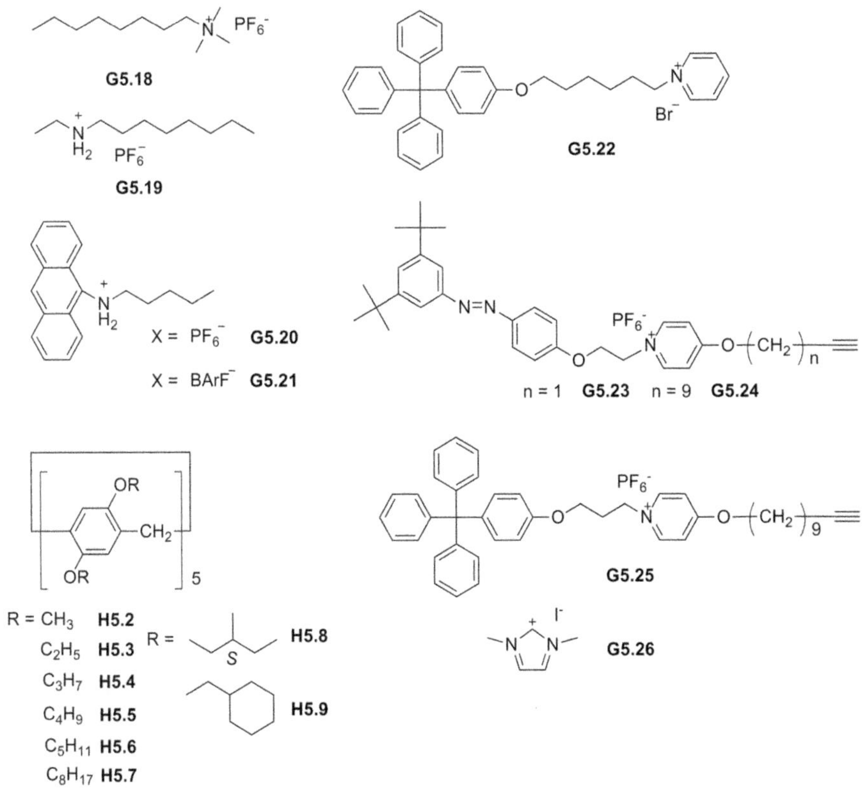

Figure 5.8 Chemical structures of simple per-alkyalted pillar[5]arenes and cationic guests for these per-alkylated pillar[5]arenes.

Table 5.2 Host–guest complexes of per-alkylated pillar[5]arenes with cationic guests.

Run	Host	Guest	Solvent	Association constant (*K*) (M^{-1})	Method	Ref.
15	**H5.5**	**G5.18**	$CDCl_3$	$(4.1 \pm 0.4)\times 10^3$	NMR	15
16	**H5.8**	**G5.18**	$CDCl_3$	1.3×10^3	NMR	16
17	**H5.9**	**G5.18**	$CDCl_3$	8.3×10^2	NMR	17
18	**H5.2**	**G5.19**	$CDCl_3$	$1.09 \pm 0.41\times 10^3$	NMR	18
19	**H5.6**	**G5.19**	$CDCl_3$	$2.40 \pm 0.20\times 10^3$	NMR	18
20	**H5.3**	**G5.20**	$CDCl_3$	$6.1 \pm 0.8\times 10$	NMR	19
21	**H5.3**	**G5.21**	$CDCl_3$	$3.4 \pm 0.4\times 10^4$	NMR	19
22	**H5.2**	**G5.22**	$CDCl_3$	$1.35 \pm 0.075\times 10^3$	NMR	20
23	**H5.3**	**G5.23**	$CDCl_3$	$2.7 \pm 0.4\times 10^4$	NMR	21
24	**H5.3**	**G5.24**	$CDCl_3$	$7.6 \pm 1.1\times 10^4$	NMR	21
25	**H5.3**	**G5.25**	$CDCl_3$	$1.01 \pm 0.21\times 10^4$	NMR	22
26	**H5.4**	**G5.26**	$CDCl_3$	$1.04 \pm 0.15\times 10^3$	NMR	23

(*S*)-methylbutyl (**H5.8**)[16] and 10 cyclohexylmethyl substituents (**H5.9**)[17] determined by ^1H NMR measurements were $1.3\times 10^3\ M^{-1}$ and 830 M^{-1}, respectively (runs 16 and 17), which is lower than for **H5.5** [$K = (4.1 \pm 0.4)\times 10^3\ M^{-1}$].

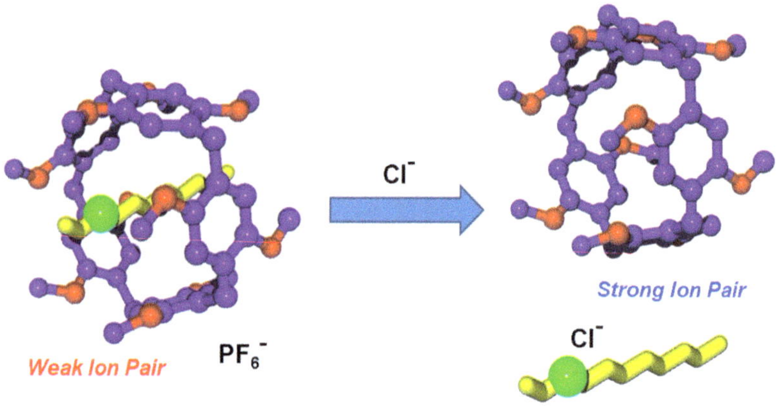

Figure 5.9 Complexation switch off by adding Cl^-.

Increasing the steric hindrance around the rims is an unfavorable condition for the formation of the host–guest complexes. The main driving force for the complexation was cation/π interactions between the ammonium cation moiety of **G5.18** and benzene moieties of the per-alkylated pillar[5]arenes.

The host–guest properties of a secondary ammonium salt guest, *n*-octylmethyl ammonium hexafluorophosphate (**G5.19**), with per-alkylated pillar[5]arenes were also investigated by Huang and co-workers.[18] The *K* values of the complexes of **G5.19** with decamethane (**H5.2**) and decapentane (**H5.6**), determined by ^{1}H NMR titrations, were all in the vicinity of 10^3 M^{-1} (runs 18 and 19). The CH/N and CH/π interactions were the main driving forces for formation of the host–guest complexes. The complexation can be switched off by the addition of a chloride anion (Figure 5.9). The ion pairing between the secondary ammonium cation and the PF_6^- anion is not strong enough to prevent complexation, but that between the secondary ammonium cation and the chloride anion is too strong to form the complex. When cationic molecules were chosen as a guest, the type of counter anions had a large effect on the complexation.

Li and co-workers investigated the counter anion effect on the ability of a secondary ammonium cation to form host–guest complexes with per-ethylated pillar[5]arene (**H5.3**).[19] The *K* value of the complex between the secondary ammonium cation with a hexafluorophosphate anion (**G5.20**) and **H5.3** was 61 ± 8 M^{-1} (run 20). In contrast, for the secondary ammonium cations with a tetrakis[3,5-bis(trifluoromethyl)phenyl]borate (BArF) anion (**G5.21**), the *K* value was $(3.4 \pm 0.4) \times 10^4$ M^{-1} (run 21), which is 560 times larger than that obtained for the hexafluorophosphate anion. The weakly coordinating BArF anion gave the free "naked" secondary ammonium cation; therefore, the free secondary ammonium cation was very strongly included in the pillar[5]arene cavity.

Pyridinium cations are also good guests for pillar[5]arenes. The *K* values of the host–guest complexes between a pyridinium cation derivative with a bromide anion (**G5.22**) and per-methylated pillar[5]arene (**H5.2**) were

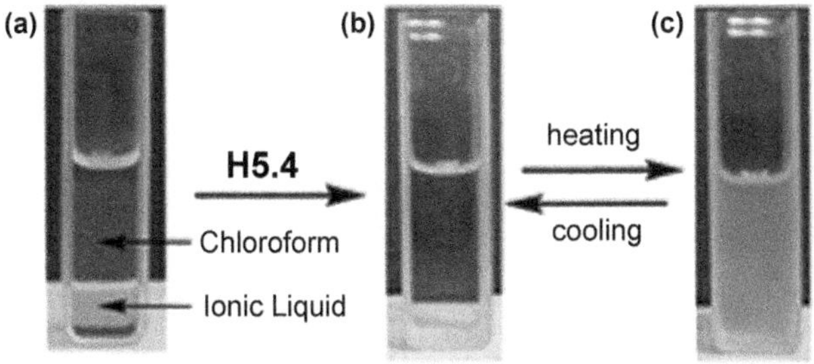

Figure 5.10 LCST-type phase changes induced by the host–guest complexation between pillar[5]arene **H5.4** and ionic liquid in chloroform. (a) Phase separation between ionic liquid and chloroform; (b) the transparent solution of **H5.4** and ionic liquid; (c) LCST-type phase separation after heating.
Reproduced with permission from ref. 23. Copyright 2014, Wiley-VCH Verlag Gmbh & Co. KGaA, Weinheim.

approximately 10^3 M^{-1} (run 22).[20] In contrast, pyridinium cations with PF_6^- (**G5.23–G.25**) formed more stable host–guest complexes with per-ethylated pillar[5]arene **H5.3** (*ca.* 10^4 M^{-1}),[21,22] which is approximately 10 times larger than that with a bromide anion (runs 23–25). This is because the ion pairing between a pyridinium cation and a bromide anion is stronger than that between a pyridinium cation and a PF_6^- anion. The stable host–guest complexation can be used for the synthesis of mechanically interlocked molecules (details in Chapter 8) and construction of supramolecular polymers (details in Chapter 7).

Pillar[5]arene can form a host–guest complex with an ionic liquid composed of an imidazolium cation and iodide anion (**G5.26**).[23] The *K* value of the complex between the ionic liquid and per-propylated pillar[5]arene (**H5.4**) in chloroform was $(1.04 \pm 0.15) \times 10^3$ M^{-1} (run 26). The solubility of **G5.26** in chloroform was low, therefore a mixture of **G5.26** and chloroform showed phase separation (Figure 5.10a). However, when **H5.4** was added to the mixture, the immiscible phase changed to a homogeneous solution due to the host–guest complexation (Figure 5.10b). The solution changed to a turbid solution with heating (Figure 5.10c), and became a clear and transparent solution again with cooling. Thus, the host–guest complex showed a lower critical solution temperature (LCST) in chloroform. The host–guest complexation could be weakened at high temperature. Disassembly of the host–guest complexes at high temperature is the reason why the complex showed LCST in chloroform.

5.3.2.2 *Neutral Guests*

Inclusion of cationic molecules by hosts in organic solvents is also observed with other host molecules, such as crown ethers and calixarenes.[24–26] Cation/π and coordination interactions are the main driving forces in these complexes. However, the formation of host–guest complexes with neutral

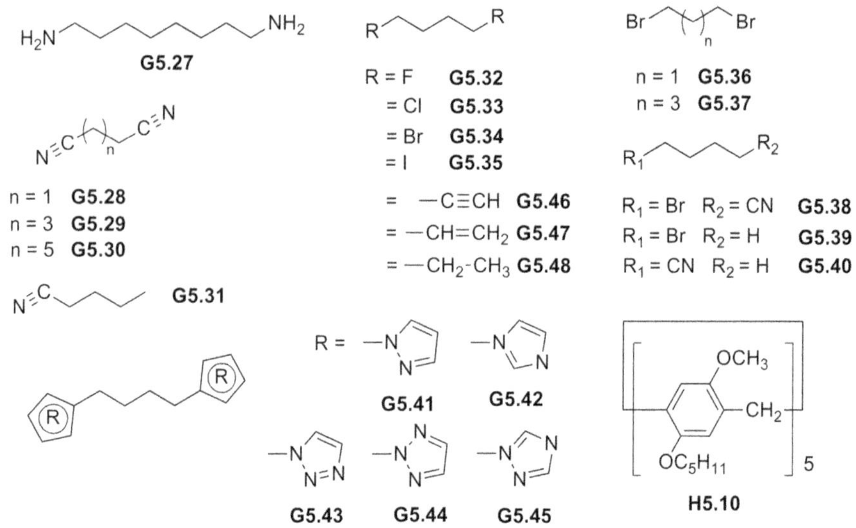

Figure 5.11 Chemical structures of neutral guests for per-alkylated pillar[5]arenes and non-symmetric pillar[5]arene **H5.10**.

guest molecules in organic media is very rare because it is difficult to introduce physical interactions between the host and the neutral guest molecules.[27–29] Neutral guest recognition by artificial synthetic receptors is therefore a challenging topic.

Stoddart and co-workers first investigated the neutral guest recognition ability of pillar[5]arene using neutral alkanediamines as a guest.[30] The association constant of the pillar[5]arene **H5.2**$\supset$1,8-diaminooctane (Figure 5.11, **G5.27**) complex in $CDCl_3$ determined by ^1H NMR titration was $70 \pm 10\ M^{-1}$ (Table 5.3, run 27). Multiple CH/π interactions should be the main driving force for the complexation.

Formation of stable host–guest complexes between neutral guests and alkyl-substituted pillar[5]arenes was investigated by Li and co-workers. The ability of neutral dinitriles to form stable host–guest complexes with per-alkylated pillar[5]arenes depended on the length of the alkyl chain between two cyano groups.[31] 1,4-Dicyanobutane (**G5.29**), possessing four methylene units in its linker, is the most suitable dicyanoalkane for the formation of stable host–guest complexes with pillar[5]arenes (run 29). In contrast, in the case of dinitriles with long chains and mononitriles, the binding constants clearly decreased. The length of the alkyl chain on the rims of pillar[5]arene also affected the host–guest ability. Short substituted per-methyl- (**H5.2**) and per-ethyl- (**H5.3**) pillar[5]arenes formed more stable host–guest complexes with dicyanobutane (runs 29 and 32) than long substituted per-butyl- (**H5.5**, run 33) and per-octyl- (**H5.7**, run 34) pillar[5]arenes. This can be explained by the difference of dipole–dipole interactions between alkoxy moieties in pillar[5]arenes and cyano moieties in dicyanobutane. The C–O bond of the

Table 5.3 Host–guest complexes of per-alkylated pillar[5]arenes with neutral guests.

Run	Host	Guest	Solvent	Association constant (K) (M^{-1})	Method	Ref.
27	**H5.2**	**G5.27**	$CDCl_3$	$7.0 \pm 1.0 \times 10$	NMR	30
28	**H5.3**	**G5.28**	DMSO-d_6–$CDCl_3$ (1:9)	^{a}	NMR	31
29	**H5.3**	**G5.29**	DMSO-d_6–$CDCl_3$ (1:9)	$1.5 \pm 0.3 \times 10^4$	NMR	31
30	**H5.3**	**G5.30**	DMSO-d_6–$CDCl_3$ (1:9)	$6.5 \pm 0.5 \times 10^4$	NMR	31
31	**H5.3**	**G5.31**	$CDCl_3$	$4.0 \pm 0.4 \times 10^2$	NMR	31
32	**H5.2**	**G5.29**	DMSO-d_6–$CDCl_3$ (1:9)	$2.4 \pm 0.4 \times 10^4$	NMR	31
33	**H5.5**	**G5.29**	DMSO-d_6–$CDCl_3$ (1:9)	$5.4 \pm 0.3 \times 10^3$	NMR	31
34	**H5.7**	**G5.29**	DMSO-d_6–$CDCl_3$ (1:9)	$3.3 \pm 0.4 \times 10^3$	NMR	31
35	**H5.3**	**G5.32**	$CDCl_3$	$8.6 \pm 0.5 \times 10$	NMR	32
36	**H5.3**	**G5.33**	$CDCl_3$	$1.9 \pm 0.2 \times 10^3$	NMR	32
37	**H5.3**	**G5.34**	$CDCl_3$	$4.9 \pm 0.3 \times 10^3$	NMR	32
38	**H5.3**	**G5.35**	$CDCl_3$	$1.0 \pm 0.1 \times 10^4$	NMR	32
39	**H5.3**	**G5.36**	$CDCl_3$	$3.0 \pm 0.2 \times 10^2$	NMR	32
40	**H5.3**	**G5.37**	$CDCl_3$	$1.4 \pm 0.1 \times 10^3$	NMR	32
41	**H5.2**	**G5.34**	$CDCl_3$	$1.6 \pm 0.1 \times 10^3$	NMR	32
42	**H5.5**	**G5.34**	$CDCl_3$	$4.3 \pm 0.4 \times 10^3$	NMR	32
43	**H5.7**	**G5.34**	$CDCl_3$	$4.1 \pm 0.3 \times 10^3$	NMR	32
44	**H5.2**	**G5.38**	$CDCl_3$	$1.7 \pm 0.2 \times 10^4$	NMR	33
45	**H5.10**	**G5.38**	$CDCl_3$	$8.4 \pm 0.6 \times 10^4$	NMR	33
46	**H5.6**	**G5.38**	$CDCl_3$	$1.6 \pm 0.3 \times 10^4$	NMR	33
47	**H5.2**	**G5.39**	$CDCl_3$	$4.1 \pm 0.3 \times 10$	NMR	33
48	**H5.10**	**G5.39**	$CDCl_3$	$7.5 \pm 0.2 \times 10$	NMR	33
49	**H5.6**	**G5.39**	$CDCl_3$	$8.2 \pm 0.4 \times 10$	NMR	33
50	**H5.2**	**G5.40**	$CDCl_3$	$5.7 \pm 0.4 \times 10^2$	NMR	33
51	**H5.10**	**G5.40**	$CDCl_3$	$5.5 \pm 0.3 \times 10^2$	NMR	33
52	**H5.6**	**G5.40**	$CDCl_3$	$2.2 \pm 0.1 \times 10^2$	NMR	33
53	**H5.3**	**G5.41**	$CDCl_3$	$6.6 \pm 0.2 \times 10$	NMR	34
54	**H5.3**	**G5.42**	$CDCl_3$	$2.0 \pm 0.3 \times 10^4$	NMR	34
55	**H5.3**	**G5.43**	$CDCl_3$	$1.6 \pm 0.3 \times 10^4$	NMR	34
56	**H5.3**	**G5.44**	$CDCl_3$	$1.1 \pm 0.1 \times 10$	NMR	34
57	**H5.3**	**G5.45**	$CDCl_3$	$6.6 \pm 0.4 \times 10^3$	NMR	34
58	**H5.2**	**G5.45**	$CDCl_3$	$1.2 \pm 0.3 \times 10^3$	NMR	34
59	**H5.5**	**G5.45**	$CDCl_3$	$7.3 \pm 0.5 \times 10^3$	NMR	34
60	**H5.3**	**G5.46**	$CDCl_3$	$8.2 \pm 0.5 \times 10$	NMR	35
61	**H5.3**	**G5.47**	$CDCl_3$	$1.2 \pm 0.2 \times 10$	NMR	35
62	**H5.3**	**G5.48**	$CDCl_3$	<2	NMR	35

[a]The K value was too high to be calculated.

short substituents, methoxy and ethoxy, has a larger dipole moment than that of the long substituents, butyloxy and octyloxy, and therefore affords stronger dipole–dipole forces to the cyano groups.

Complexation of neutral 1,4-dihaloalkanes with per-alkylated pillar[5]arenes was also investigated by Li and co-workers.[32] As with dinitriles, the suitable chain length for the complexation was a C4 linker. The association constants increased in the order of F < Cl < Br < I (runs 35–38), indicating that the dominant driving force for the complexation can be explained by the dispersion force depending on the movement of electrons to produce temporary dipoles. This explanation is consistent with the host–guest complexation of 1,4-dibromobutane (**G5.34**) with pillar[5]arene with different chain lengths (runs 37 and 41–43). **H5.2** gave the smallest *K* value among per-ethyl- (**H5.3**), per-butyl- (**H5.5**) and per-octyl- (**H5.7**) pillar[5]arenes. This is because there are more dispersion interactions of long alkyl-substituted pillar[5]arenes with 1,4-dibromobutane than for short methyl-substituted pillar[5]arene.

This is therefore the opposite result to that observed for the complexation between dinitriles and per-alkylated pillar[5]arenes. Based on these results, Li and co-workers reported the regio-selective complexation of a non-symmetrical guest 5-bromovaleronitrile (**G5.38**) by a non-symmetrical pillar[5]arene having different methyl and pentyl rims (Figure 5.12, **H5.10**).[33]

The *K* value for **G5.38** with the non-symmetrical pillar[5]arene **H5.10** was $(8.4 \pm 0.6) \times 10^4$ M^{-1} (run 45), which is 4.9 and 5.3 times higher than those obtained with symmetrical per-methylated (**H5.2**, run 44) and per-pentylated (**H5.6**, run 46) pillar[5]arenes. The selective binding system is the main reason for the regio-selective complexation. The methyl rim of pillar[5]arene has a preference for the CN side of the guest over the Br side, while the pentyl rim of pillar[5]arene has a preference for the Br side of the guest over the CN side. The X-ray crystal structure of the complex also supports the regio-selective complexation.

The stable host–guest complexation between per-alkylated pillar[5]arenes and diheterocyclic-substituted alkane guests was also investigated by Li and co-workers.[34] As was observed for dinitriles and dihalo-alkanes, the C4 linker is the best linker length for diheterocyclic-substituted alkane guests. 1,4-Diheterocyclic-substituted butylene guests having different numbers and positions of nitrogen atoms exhibited different association constants. The *K* values of guests with outer nitrogen atoms for per-alkylated pillar[5]arenes (runs 54, 55 and 57) are much higher than for those without outer nitrogen atoms (runs 53 and 56). Therefore, one of the dominant driving forces for the complexation is CH–N hydrogen bonding interactions between the alkyl

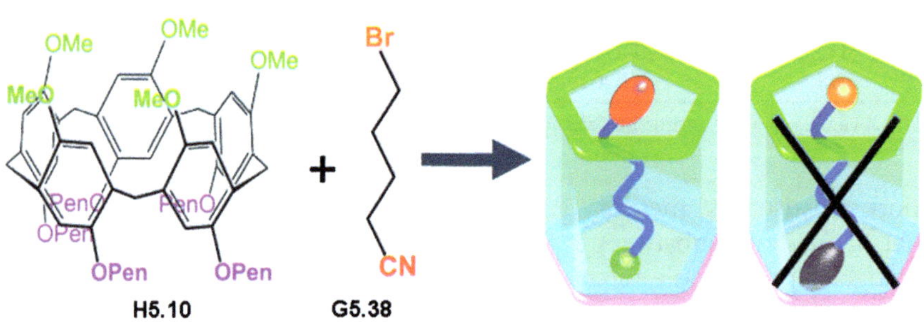

Figure 5.12 Regio-selective formation of the host–guest complex between non-symmetrical pillar[5]arene **H5.10** and non-symmetrical neutral guest **G5.38**.
Reproduced from ref. 33 with permission from The Royal Society of Chemistry.

moieties of the wheel on the rims and the outer nitrogen atoms of the heterocycles. The CH–N hydrogen bonding interaction can be explained by the complexation behavior between a diheterocyclic butylene guest and pillar[5]arenes with different alkyl substituents. The *K* value for guest **G5.45** with per-butylated pillar[5]arene **H5.5** (run 59) is 6.1 times higher than that with per-methylated pillar[5]arene **H5.2** (run 58). This is because there are more CH–N interactions between butyl groups of pillar[5]arene **H5.5** and **G5.45** than between methyl groups of **H5.2** and **G5.45**. Among these heterocyclic groups, triazole moieties can be prepared by a simple CuAAC reaction. Therefore, they are useful stations for the construction of various pillar[5]arene-based supramolecular assemblies.

The ability of hydrocarbon guests to form host–guest complexes with peralkylated pillar[5]arenes was investigated by Li's and our groups.[35] They investigated the complexation behavior of saturated and unsaturated hydrocarbons, that is, *n*-octane (**G5.48**), 1,7-octadiene (**G5.47**) and 1,7-octadiyne (**G5.46**), by per-ethylated pillar[5]arene **H5.3**. The host–guest binding ability of these guests by per-ethylated pillar[5]arene increases in the following order: 1,7-octadiene > 1,7-octadiyne > *n*-octane (runs 60–62). The electron-negativity of the terminal carbon atom in these guests caused an increase in the acidity of the methylene protons, making them better hydrogen bonding donors. Thus, the dominant driving forces in these complexes are CH/O and CH/π interactions.

Our groups reported a clear shape selectivity of pillar[5]arene for alkanes (Figure 5.13). Linear alkanes, including *n*-hexane, formed host–guest complexation with per-methylated pillar[5]arene **H5.2**, although we could not determine the *K* value for the complex because of the low stability of the host–guest complex.

In contrast, bulky cyclohexane and branched 2- and 3-methylhexane hardly interacted with per-methylated pillar[5]arene **H5.2**. This is because the

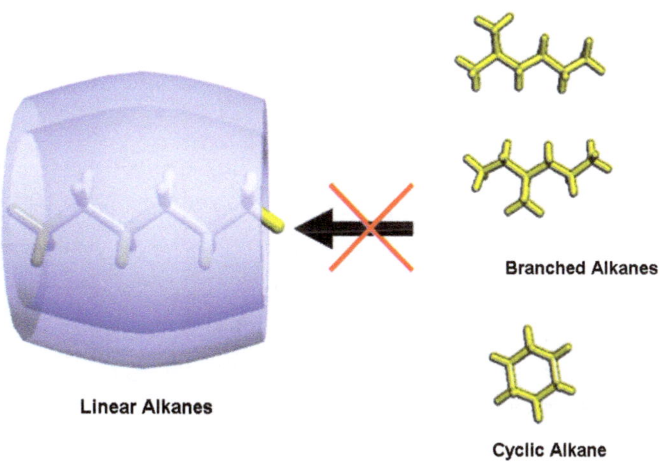

Figure 5.13 Alkane-shape selective host–guest complexation.

pillar[5]arene cavity has a narrow and well-defined cylindrical structure. The high selectivity for the shape of alkanes is not observed in other host molecules, thus it is one of the interesting host–guest features of pillar[5]arenes.

5.3.3 Water-soluble Pillar[5]arenes

Introduction of water-soluble substituents, such as anionic, cationic and non-ionic moieties, is a useful way to obtain water-soluble pillar[5]arenes. The first water-soluble pillar[5]arene synthesized by our group was a pillar[5]arene with 10 carboxylate moieties **H5.11** (Figure 5.14).[36]

H5.11 accommodated with paraquat **G5.49** with a high association constant [$K = (8.2 \pm 1.7) \times 10^4$ M^{-1}, Table 5.4, run 63], which is *ca.* 70 times higher than the corresponding association constant for the complex between per-hydroxylated pillar[5]arene **H5.1** and paraquat **G5.1** in methanol (Table 5.1, run 1). In aqueous media, a hydrophobic–hydrophilic interaction synergistically stabilizes host–guest complexation along with electrostatic and charge-transfer interactions. Many groups then investigated host–guest complexation in aqueous media using **H5.11**.

Li and co-workers screened host–guest interactions of **H5.11** with a series of cationic 1,4-bis(pyridinium)butane guests (runs 64–70).[37] 1,4-Bis(pyridinium)butane derivatives, except for the bulky 2,6-dimethyl-substituted guest (**G5.54**, run 68), formed very stable host–guest complexes ($K > 10^5$ M^{-1}). **G5.54** formed an external complex with **H5.11**, and the association constant of the complex was reduced more than 100 times compared with the other 1,4-bis(pyridinium)butane guests. These results indicate that encapsulation of the guest part is necessary to stabilize the host–guest complex formation.

Li and co-workers also investigated host–guest interactions of **H5.11** with 20 basic amino acids.[38] **H5.11** showed stronger binding abilities for Lys (**G5.57**), Arg (**G5.58**) and His (**G5.59**) (runs 71–73, $K = ca.$ 10^3 M^{-1}) compared

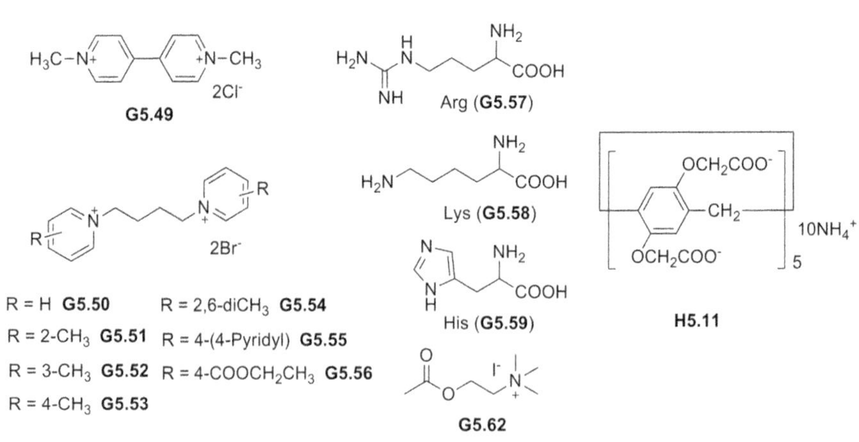

Figure 5.14 Chemical structures of water-soluble anionic pillar[5]arene **H5.11** and guests for **H5.11**.

Table 5.4 Host–guest complexes of water-soluble pillar[5]arenes.

Run	Host	Guest	Solvent	Association constant (K) (M^{-1})	Method	Ref.
63	**H5.11**	**G5.49**	H_2O	$8.2 \pm 1.4 \times 10^4$	Fl	36
64	**H5.11**	**G5.50**	H_2O	$1.1 \pm 0.2 \times 10^6$	Fl	37
65	**H5.11**	**G5.51**	H_2O	$1.9 \pm 0.4 \times 10^4$	Fl	37
66	**H5.11**	**G5.52**	H_2O	$3.5 \pm 0.1 \times 10^5$	Fl	37
67	**H5.11**	**G5.53**	H_2O	$3.6 \pm 0.5 \times 10^5$	Fl	37
68	**H5.11**	**G5.54**	H_2O	$2.4 \pm 0.3 \times 10^3$	Fl	37
69	**H5.11**	**G5.55**	H_2O	$7.0 \pm 0.7 \times 10^5$	Fl	37
70	**H5.11**	**G5.56**	H_2O	$2.4 \pm 0.2 \times 10^5$	Fl	37
71	**H5.11**	**G5.57**	D_2O	$5.9 \pm 0.4 \times 10^3$	NMR	38
72	**H5.11**	**G5.58**	D_2O	$1.5 \pm 0.2 \times 10^3$	NMR	38
73	**H5.11**	**G5.59**	D_2O	$1.8 \pm 0.2 \times 10^3$	NMR	38
74	**H5.11**	**G5.60**	D_2O	$1.28 \pm 0.42 \times 10^2$	NMR	40
75	**H5.11**	**G5.61**	D_2O	$1.32 \pm 0.25 \times 10^5$	NMR	40
76	**H5.11**	**G5.62**	H_2O	$5.05 \pm 0.13 \times 10^4$	ITC[a]	41
77	**H5.12**	**G5.63**	D_2O	$1.33 \pm 0.94 \times 10^4$	NMR	42
78	**H5.12**	**G5.64**	D_2O	$1.03 \pm 0.37 \times 10^2$	NMR	43
79	**H5.12**	**G5.65**	D_2O	$4.61 \pm 0.43 \times 10$	NMR	43
80	**H5.12**	**G5.66**	D_2O	$2.07 \pm 0.59 \times 10$	NMR	43
81	**H5.12**	**G5.67**	D_2O	$1.90 \pm 0.33 \times 10$	NMR	43
82	**H5.12**	**G5.68**	D_2O	$9.68 \pm 0.60 \times 10$	NMR	43
83	**H5.12**	**G5.69**	D_2O	$4.26 \pm 0.52 \times 10^2$	NMR	43
84	**H5.14**	**G5.70**	H_2O	$4.3 \pm 0.5 \times 10^3$	UV	45

[a]ITC; Isothermal Titration Calorimetry.

with other α-amino acids ($K < 20\ M^{-1}$). The electrostatic interaction between cationic lateral chains of these amino acids and carboxylate anions on the rims of **H5.11** results in the molecular selective binding of these amino acids.

Cohen *et al.* reported encapsulation of xenon into the cavity of **H5.11** and detection of a hydrophobic guest in the **H5.11** cavity using ^{129}Xe NMR (Figure 5.15).[39]

Xenon is hydrophobic, therefore its solubility in water is poor. However, xenon gas was bubbled in an aqueous solution of **H5.11**. The peak from xenon was observed by ^{129}Xe NMR, indicating that xenon was dissolved in aqueous media by encapsulation into the cavity of **H5.11**. The NMR chemical shift of xenon is highly sensitive to the local environment. The chemical shift of the xenon atom showed a downfield shift by addition of hydrophobic guest *n*-hexane, indicating inclusion of *n*-hexane into the pillar[5]arene cavity.

Xue and co-workers reported the inclusion of a fluorescent compound, 10-methylacridinium iodide (**G5.60**), in the **H5.11** cavity in aqueous media (Figure 5.16).[40]

Upon addition of **H5.11** to the aqueous solution of **G5.60**, the fluorescence was quenched, suggesting the formation of the host–guest complex. The fluorescence quenching is caused by photoinduced electron transfer from anionic carboxylate moieties of **H5.11** to the excited state of **G5.60**. The association constant, determined by ^{1}H NMR titration, was $K = (1.28 \pm 0.42) \times 10^2\ M^{-1}$ (run 74). When paraquat (**G5.61**) was added to the non-fluorescent **H5.11**$\supset$**G5.60** complex in water, the florescence intensity

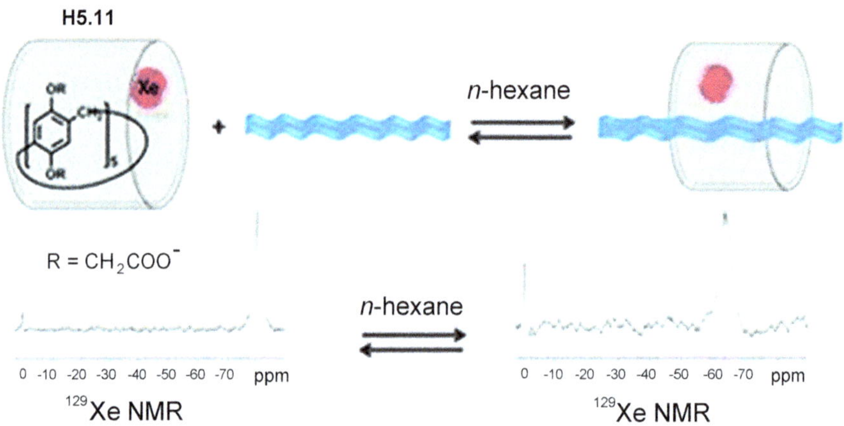

Figure 5.15 Detection of inclusion of *n*-hexane by changing signal from ^{129}Xe signal. Reproduced from ref. 39 with permission from The Royal Society of Chemistry.

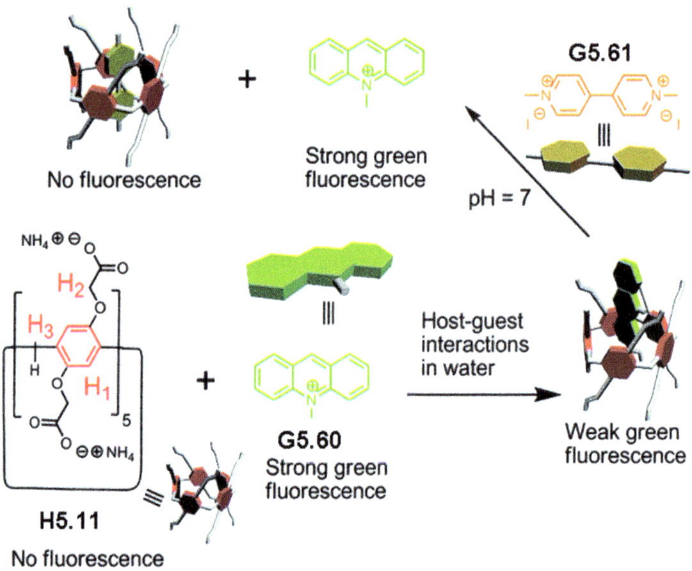

Figure 5.16 Illustration of fluorescence change detection of cyanide (**G5.60**) and paraquat (**G5.61**).
Reproduced from ref. 40 with permission from The Royal Society of Chemistry.

increased. The *K* value of the complex between **H5.11** and **G5.61** measured by ^{1}H NMR titration was $(1.32 \pm 0.25) \times 10^5$ M^{-1} (run 75), which is about 1000 times higher than that of the complex between **H5.11** and **G5.60**. Thus, the "turn-on" fluorescence change was caused by the loss of **G5.60** from the cavity of **H5.11** upon addition of **G5.61**.

Yu and co-workers reported formation of the host–guest complex of **H5.11** with acetylcholine (**G5.62**), which is one of the neurotransmitters in the autonomic nervous system.[41] The *K* value of the complex, determined by ITC measurements, was $(5.05 \pm 0.13) \times 10^4$ M^{-1} (run 76). The binding affinity of the host–guest complex is derived from multiple co-operative electrostatic interactions between the carboxylate anionic groups on the rims of **H5.11** and the cationic trimethylammonium moiety of **G5.62**, and hydrophobic interactions between the alkyl chain of **G5.62** and the pillar[5]arene cavity.

Introduction of cationic moieties into the pillar[5]arene rims also facilitates water-soluble pillar[5]arenes. Huang and co-workers synthesized a cationic water-soluble pillar[5]arene (**H5.12**) by introducing trimethyl ammonium groups at both rims (Figure 5.17).[42]

Owing to the presence of 10 cationic moieties on the rims, **H5.12** can form a stable host–guest complex with anionic molecules. 1-Octanesulfonate (**G5.63**) was included in the cavity of **H5.12**. The association constant of the complex, determined by ^{1}H NMR titration, was $(1.33 \pm 0.94) \times 10^4$ M^{-1} for 1:1 stoichiometry (run 77). Binding of neutral guests by the cationic pillar[5]arene **H5.12** was also investigated by Huang and co-workers (runs 78–83).[43] The higher guest hydrophobicity leads to the higher host affinity. **H5.12** can bind alcohols, such as propanol (**G5.64**), *n*-pentanol (**G5.65**) and *n*-butanol (**G5.66**), with association constants between 20 and 100 M^{-1}, but cannot bind alcohols with short alkyl moieties, such as methanol and ethanol. *tert*-Butanol and isopropanol did not form host–guest complexes with **H5.12** because of the steric hindrance. Mono-substituted benzene derivatives also form weak host–guest complexes with **H5.12** in aqueous media ($K = 20$–400 M^{-1}).

A water-soluble pillar[5]arene with 10 amine moieties **H5.13** was synthesized by Hou and co-workers.[44] **H5.13** formed host–guest complexes with linear diacids in aqueous media (Figure 5.18).

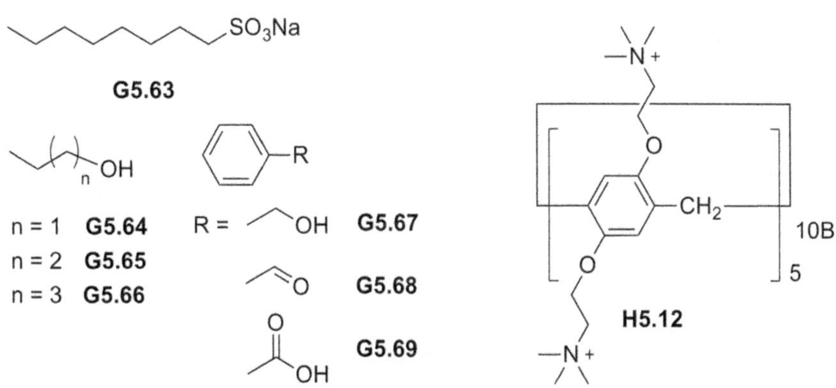

Figure 5.17 Chemical structures of water-soluble cationic pillar[5]arene **H5.12** and guests for **H5.12**.

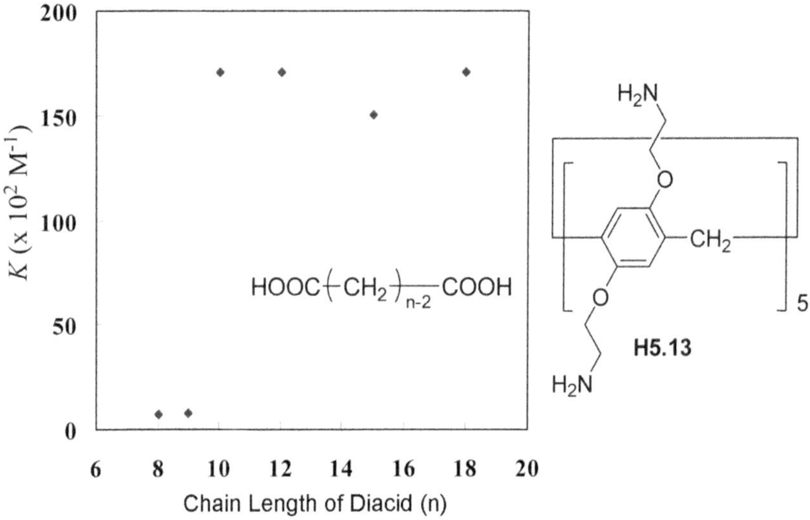

Figure 5.18 Plot of association constants (K) *vs.* the chain length of diacids (n), indicating that hydrophobic interaction contributed to the binding of **H5.13** with diacids.
Reproduced from ref. 44 with permission from The Royal Society of Chemistry.

The electrostatic interaction between the amino units of **H5.13** and the carboxylic acid units of the diacid is the main driving force for the complexation. The K values of longer diacids ($n \geq 10$) were larger ($K = ca.\ 10^4\ M^{-1}$) than those of shorter diacids ($K = ca.\ 10^2\ M^{-1}$). Therefore, for the longer diacids, a hydrophobic–hydrophilic interaction is also an important driving force to stabilize the complexation.

We synthesized a pillar[5]arene with non-ionic tri(ethylene oxide)-substituted pillar[5]arene **H5.14** (Figure 5.19).[45]

Addition of viologen guest **G5.70** to an aqueous solution of **H5.14** changed the color of the solution from colorless to yellow, indicating formation of the charge-transfer complex. The association constant of the complex, determined by UV titration, was $(4.3 \pm 0.5) \times 10^3\ M^{-1}$ for 1 : 1 stoichiometry (run 84). **H5.14** showed a clouding point at 42 °C, but addition of **G5.70** increased the clouding point to 60 °C. The electron repulsion force between the included viologen guests inhibits aggregation of **H5.70**. Viologen forms a very stable host–guest complex with CB[7] ($>10^5\ M^{-1}$); therefore, we added CB[7] to the host–guest complex. The **H5.14**$\supset$**G5.70** complex showed a clouding point at 60 °C, but addition of CB[7] to the complex resulted in a decrease of the clouding point to 46 °C. This phenomenon arises from the dissociation of the CT complex between **H5.14** and **G5.70**, and association of **G5.70** and CB[7]. Hence, we can tune the clouding point using host–guest systems.

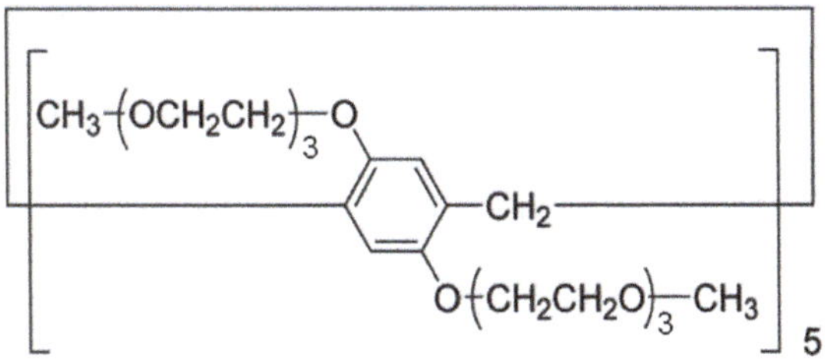

H5.14

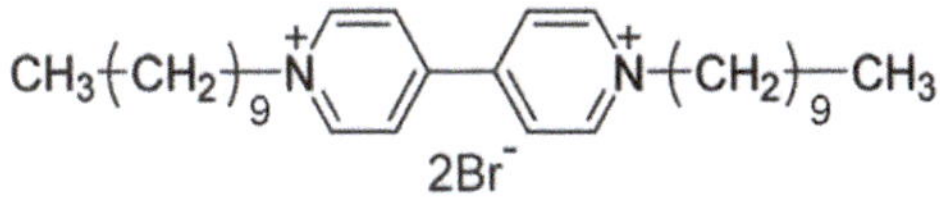

G5.70

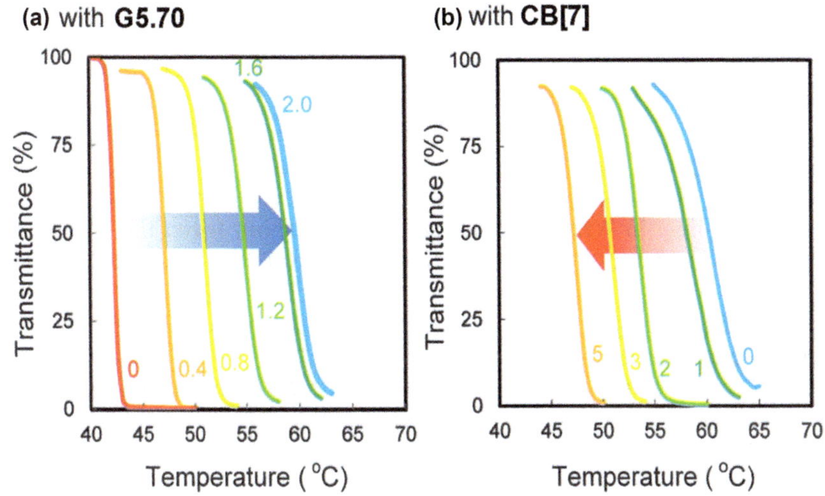

Figure 5.19 (a) Temperature dependence of light transmittance of an aqueous solution of **H5.14** (2 mM) upon addition of **G5.70** (0–2 mM) and heating. (b) Temperature dependence of light transmittance of a mixture of **H5.14** (2 mM) and **G5.70** (2 mM) in aqueous solution upon addition of CB[7] (0–5 mM) and heating.
Reproduced with permission from ref. 45. Copyright 2012 American Chemical Society.

5.3.4 Functionalized Pillar[5]arenes

One of the important characteristics of pillar[5]arene chemistry is their versatile functionality. By various synthetic approaches, position-selective mono-, di-, tetra-, penta- and per-functionalized pillar[5]arenes can be prepared (details shown in Chapter 3). Modification of the molecular recognition sites on pillar[5]arene scaffolds, based on their versatile functionality, is useful to enhance the host–guest ability of pillar[5]arenes compared with simple alkylated pillar[5]arenes. In addition, the functionalization of pillar[5]arenes enables application of the host–guest ability of pillar[5]arenes in other fields of chemistry. We discuss the host–guest properties of mono-, di-, penta- and per-functionalized pillar[5]arenes.

5.3.4.1 Host–Guest Properties of Mono-functionalized Pillar[5]arenes

Ion pairing between cations and anions is a barrier for the inclusion of cations into the pillar[5]arene cavity because strong ion-pairing prevents the complexation. Wang and co-workers synthesized a heteroditopic pillar[5]-arene containing one urea moiety (**H5.15**) for ion-pair recognition of alkylammonium cations and counter anions (Figure 5.20).[46]

The K values of the complexation of **H5.15** with different n-butylammonium cations (**G5.71–G5.73**) were approximately 10^3–10^4 M^{-1} (Table 5.5, runs 85–87). Meanwhile, the K values between these guests and per-methylated pillar[5]arene without the urea moiety (**H5.2**) were very low (K = 10–10^2 M^{-1}, runs 88–90). Therefore, the hydrogen bonding between the urea moiety in **H5.15** and the counter anion improved the recognition of the alkylammonium guest by the heteroditopic receptor **H5.15** when compared with the monotopic receptor **H5.2**.

Xia and Xue synthesized a pillar[5]arene with the monocarboxylic acid-functionalized pillar[5]arene **H5.16** (Figure 5.21).[47]

The K value of the complexation between **H5.16** and a monoamine derivative **G5.74** was $(1.52 \pm 0.075) \times 10^4$ M^{-1} (run 91). Proton transfer from the carboxylic acid moiety in **H5.16** to the amine moiety in **G5.74** stabilized the complexation.

We investigated alkane recognition by a monomeric per-methylated pillar[5]arene **H5.2** and pillar[5]arene dimer **H5.17**, which was prepared by connecting two mono-reactive pillar[5]arenes through a linker.[48] Determination of the K value of the complex between **H5.2** and n-hexane was difficult owing to the weak host–guest complex. In contrast, the K value between the dimer **H5.17** and n-hexane determined by ^{1}H NMR titration was 98 ± 12 M^{-1} (run 92). The dimerization is useful for enhancing CH/π interactions between pillar[5]arenes and alkanes.

Li and co-workers also synthesized pillar[5]arene dimers to improve the host–guest complexation ability toward neutral guests. A pillar[5]arene dimer bridged by a flexible aliphatic chain **H5.18** was synthesized.[49] The dimer linked by a rigid phenylene bridge (**H5.17**) showed a high binding ability to a bromo-substituted pyromellitic diimide derivative (**G5.76**). The K value of the complex of **G5.76** with **H5.17** was 17 times larger than that with **H5.18** linked by a flexible aliphatic chain (runs 93 and 94). This arises from the π/π interaction between the phenylene linker in **H5.17** and the

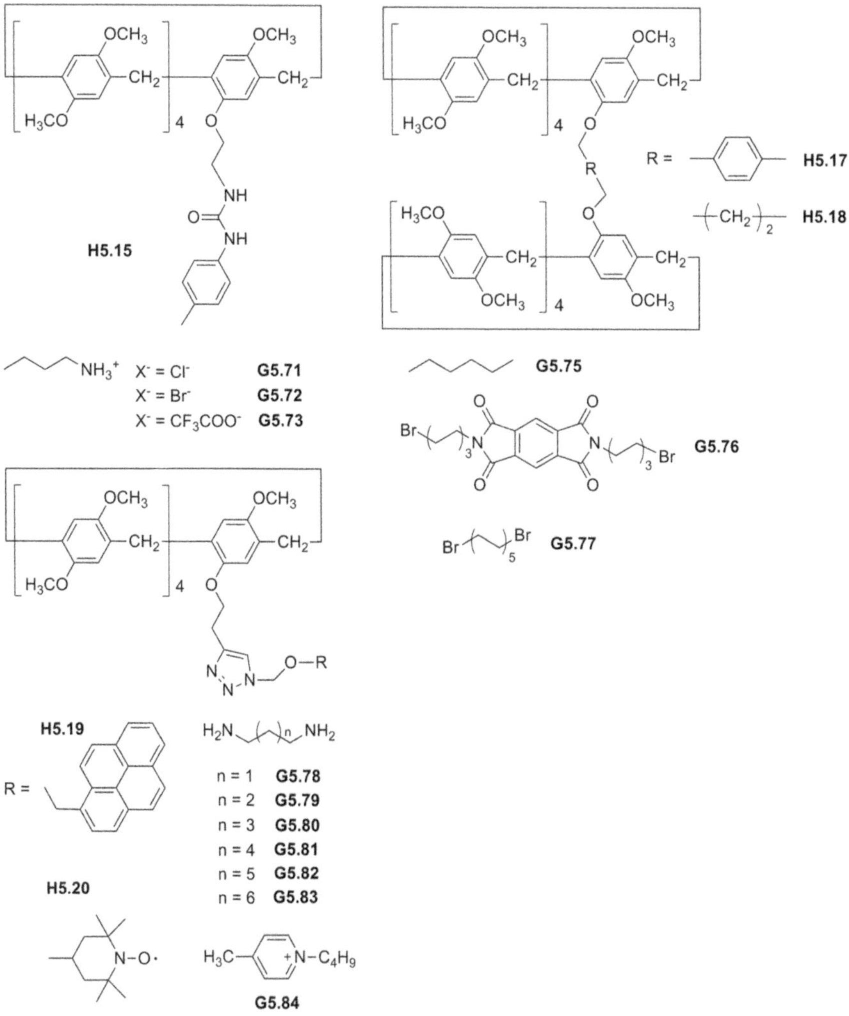

Figure 5.20 Chemical structures of mono-functionalizing pillar[5]arenes and guests for the mono-functionalizing pillar[5]arenes.

pyromellitic diimide moiety in **G5.76**, along with co-operative binding between the two pillar[5]arene moieties in **H5.17** and alkane moieties in **G5.76.** In contrast, **H5.18** showed strong binding ability with linear alkanes containing bromide group ends (**G5.77**), which was approximately 10 times higher than that for **H5.17** (runs 95 and 96). This is because **H5.18** is able to change its structure to accommodate the short chain length guest.

Stoddart and co-workers synthesized a monopyrene-functionalized pillar[5]arene (**H5.19**) to detect the complexation of alkyldiamines at low concentrations.[30] When alkyldiamines were added to the solution of **H5.19,** fluorescence quenching of the pyrene moiety in **H5.19** took place because primary amines quench the pyrene fluorescence through photoinduced

Table 5.5 Host–guest complexes of mono-functionalized pillar[5]arenes.

Run	Host	Guest	Solvent	Association constant (K) (M^{-1})	Method	Ref.
85	**H5.15**	**G5.71**	$CDCl_3$	$2.01 \pm 0.72 \times 10^3$	NMR	46
86	**H5.15**	**G5.72**	$CDCl_3$	$4.61 \pm 1.68 \times 10^3$	NMR	46
87	**H5.15**	**G5.73**	$CDCl_3$	$2.62 \pm 0.73 \times 10^4$	NMR	46
88	**H5.2**	**G5.71**	$CDCl_3$	$2.81 \pm 0.28 \times 10$	NMR	46
89	**H5.2**	**G5.72**	$CDCl_3$	$3.54 \pm 1.16 \times 10^2$	NMR	46
90	**H5.2**	**G5.73**	$CDCl_3$	$6.76 \pm 1.16 \times 10^2$	NMR	46
91	**H5.16**	**G5.74**	$CDCl_3$	$1.52 \pm 0.075 \times 10^4$	NMR	47
92	**H5.17**	**G5.75**	$CDCl_3$	$9.8 \pm 1.2 \times 10$	NMR	48
93	**H5.17**	**G5.76**	$CDCl_3$	$1.3 \pm 0.2 \times 10^3$	NMR	49
94	**H5.18**	**G5.76**	$CDCl_3$	$1.1 \pm 0.1 \times 10^2$	NMR	49
95	**H5.17**	**G5.77**	$CDCl_3$	$7.6 \pm 0.9 \times 10$	NMR	49
96	**H5.18**	**G5.77**	$CDCl_3$	$8.2 \pm 0.8 \times 10^2$	NMR	49
97	**H5.19**	**G5.78**	$MeCN/H_2O$ (1 : 1)	$1.00 \pm 0.20 \times 10^4$	Fl	30
98	**H5.19**	**G5.79**	$MeCN/H_2O$ (1 : 1)	$1.63 \pm 0.14 \times 10^4$	Fl	30
99	**H5.19**	**G5.80**	$MeCN/H_2O$ (1 : 1)	$2.12 \pm 0.10 \times 10^4$	Fl	30
100	**H5.19**	**G5.81**	$MeCN/H_2O$ (1 : 1)	$1.15 \pm 0.14 \times 10^4$	Fl	30
101	**H5.19**	**G5.82**	$MeCN/H_2O$ (1 : 1)	$1.91 \pm 0.10 \times 10^4$	Fl	30
102	**H5.19**	**G5.83**	$MeCN/H_2O$ (1 : 1)	$3.60 \pm 0.26 \times 10^4$	Fl	30
103	**H5.20**	**G5.84**	$CDCl_3$	2.57×10^3	NMR	50

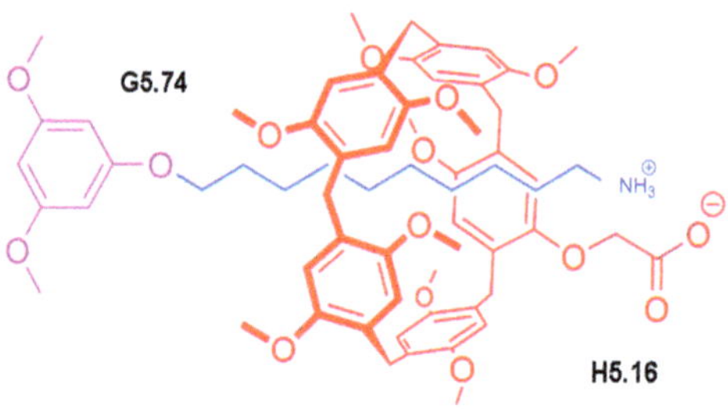

Figure 5.21 Host–guest complexation using ionic interactions between the mono-carboxylate **H5.16** and mono-amine guest **G5.74.**

electron transfer. The K values for the series of alkanediamines containing the C3 to C8 linker (**G5.78–G5.83**), determined by fluorescence quenching, were all in the vicinity of 10^4 M^{-1} (runs 97–102).

The method for detecting complexation using Electron Paramagnetic Resonance (EPR) spectroscopy was investigated by Mezzina and Lucarini and co-workers.[50] They synthesized a novel pillar[5]arene with a paramagnetic 2,2,6,6-tetramethylpiperidine *N*-oxide moiety (**H5.20**). **H5.20** formed a host–guest complex with 4-methyl-*N*-butylpyridinium hexafluorophosphate (**G5.84**). The K value of the complex, determined by ^1H NMR titrations, was 2572 M^{-1} (run 103). The EPR spectrum showed signals from 14N hyperfine splitting. Unfortunately, clear signal changes were not observed upon addition of **G5.84**. This indicates that the microenvironment of the radical moiety in **H5.20** did not change with **G5.84** because the radical moiety is located outside the recognition site of the pillar[5]arene cavity.

5.3.4.2 *Host–Guest Properties of Di- and Penta-functionalized Pillar[5]arenes*

A pillar[5]arene with two urea moieties in one unit (**H5.21**) was synthesized by Wang and co-workers (Figure 5.22).[51]

The **H5.21** formed host–guest complexes with diacidalkanes even in the highly polar solvent DMSO. The hydrogen bonding interactions between the bis-urea hydrogen in **H5.21** and dicarboxylate oxygens of the guests stabilized the host–guest complexation alongside the CH/π interactions between the pillar[5]arene cavity and linear alkyl chains of the guests. The chain length of diacidalkanes was the dominant factor for the complexation. Diacidalkanes containing C14 and C15 polymethylene chains formed more stable complexes (Table 5.6, runs 107 and 108, $K = ca.$ 150 M^{-1}) than those containing C11–C13 and C20 polymethylene chains (runs 104–106 and 109, $K = 28$–76 M^{-1}). This arises from the structural complementarity between diacidalkanes and **H5.21**.

Yu and co-workers synthesized a novel difunctionalized pillar[5]arene with two carboxylic acids in one unit (**H5.22**), and investigated host–guest complexation of **H5.22** with alkanediamines **G5.91–G5.96**.[52] The K values of the complexes were of the magnitude of 10^5 M^{-1} (runs 110–115), which is

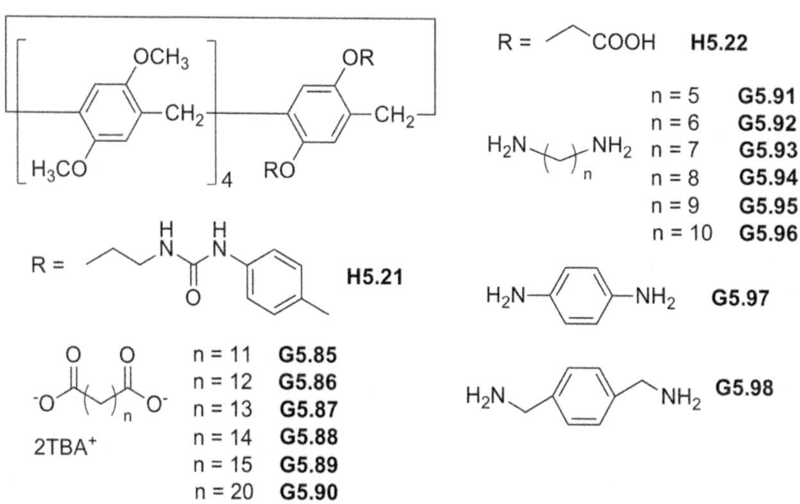

Figure 5.22 Chemical structures of di-functionalizing pillar[5]arenes and guests for the di-functionalizing pillar[5]arenes.

Table 5.6 Host–guest complexes of di-functionalized pillar[5]arenes.

Run	Host	Guest	Solvent	Association constant (K) (M^{-1})	Method	Ref.
104	**H5.21**	**G5.85**	DMSO-d_6	2.8×10	NMR	51
105	**H5.21**	**G5.86**	DMSO-d_6	6.9×10	NMR	51
106	**H5.21**	**G5.87**	DMSO-d_6	7.1×10	NMR	51
107	**H5.21**	**G5.88**	DMSO-d_6	1.4×10^2	NMR	51
108	**H5.21**	**G5.89**	DMSO-d_6	1.5×10^2	NMR	51
109	**H5.21**	**G5.90**	DMSO-d_6	7.6×10	NMR	51
110	**H5.22**	**G5.91**	$CHCl_3$: $CH_3OH=1:1$	$2.84\pm0.232\times10^5$	ITC	52
111	**H5.22**	**G5.92**	$CHCl_3$: $CH_3OH=1:1$	$2.39\pm0.237\times10^5$	ITC	52
112	**H5.22**	**G5.93**	$CHCl_3$: $CH_3OH=1:1$	$3.13\pm0.210\times10^5$	ITC	52
113	**H5.22**	**G5.94**	$CHCl_3$: $CH_3OH=1:1$	$2.89\pm0.212\times10^5$	ITC	52
114	**H5.22**	**G5.95**	$CHCl_3$: $CH_3OH=1:1$	$3.45\pm0.228\times10^5$	ITC	52
115	**H5.22**	**G5.96**	$CHCl_3$: $CH_3OH=1:1$	$3.06\pm0.232\times10^5$	ITC	52
116	**H5.22**	**G5.97**	$CHCl_3$: $CH_3OH=1:1$	$9.93\pm1.92\times10^3$	ITC	53
117	**H5.22**	**G5.98**	$CHCl_3$: $CH_3OH=1:1$	$5.89\pm1.43\times10^5$	ITC	53
118	**H5.23**	**G5.29**	Acetone-d_6	$6.1\pm0.2\times10^3$	NMR	54
119	**H5.23**	**G5.49**	Acetone-d_6	$2.6\pm0.6\times10^4$	NMR	54
120	**H5.24**	**G5.29**	$CHCl_3$	$1.37\pm0.22\times10^4$	CD	55
121	**H5.25**	**G5.99**	Acetone	$1.25\pm0.08\times10^4$	Fl	56
122	**H5.25**	**G5.100**	Acetone	$8.58\pm0.73\times10^2$	Fl	56
123	**H5.25**	**G5.101**	Acetone	$6.15\pm0.51\times10^2$	Fl	56
124	**H5.25**	**G5.102**	Acetone	$2.84\pm0.19\times10^2$	Fl	56

larger than that observed for the complex between per-methylated pillar[5]arene **H5.2** and 1,5-pentanediamine (**G5.91**, $K=670\pm73.8\ M^{-1}$). The high binding constants of these host–guest complexes were caused by the co-operative electrostatic interactions between the ammonium cations of these guests and the carboxylate anions of **H5.2**, and multiple CH/π interactions between the alkyl chains of alkanediamines and **H5.22**. The diacid **H5.22** also formed a stable host–guest complex with *p*-xylylenediamine [**G5.97**, $K=(5.89\pm1.43)\times10^5\ M^{-1}$, run 117].[53] The K value of the complex between **H5.22** and *p*-phenylenediamine (**G5.98**) was $(9.93\pm1.92)\times10^3\ M^{-1}$ (run 116), which is smaller than that observed for the complex with **G5.97**. This arises from the difference in the pK_a values of these diamines. The pK_a value of **G5.97** (9.46 ± 0.10) is larger than that of **G5.98** (6.17 ± 0.10).

Ma, Li, Jiang and Wen and co-workers synthesized a pillar[5]arene and crown ether-fused bicyclic host (**H5.23**; Figure 5.23).[54]

The pillar[5]arene moiety in **H5.23** formed a stable host–guest complex with 1,4-dicyanobutane (**G5.29**) [$K=(6.1\pm0.2)\times10^3\ M^{-1}$ (run 118)]. In contrast, the crown ether moiety in **H5.23** captured paraquat (**G5.49**) stably [$K=(2.6\pm0.6)\times10^4\ M^{-1}$ (run 119)]. When the two guest molecules were added to **H5.23** at the same time, the selective inclusion of two guest molecules by pillar[5]arene and crown ether rings took place, which was confirmed by single X-ray crystalline analysis of the complex.

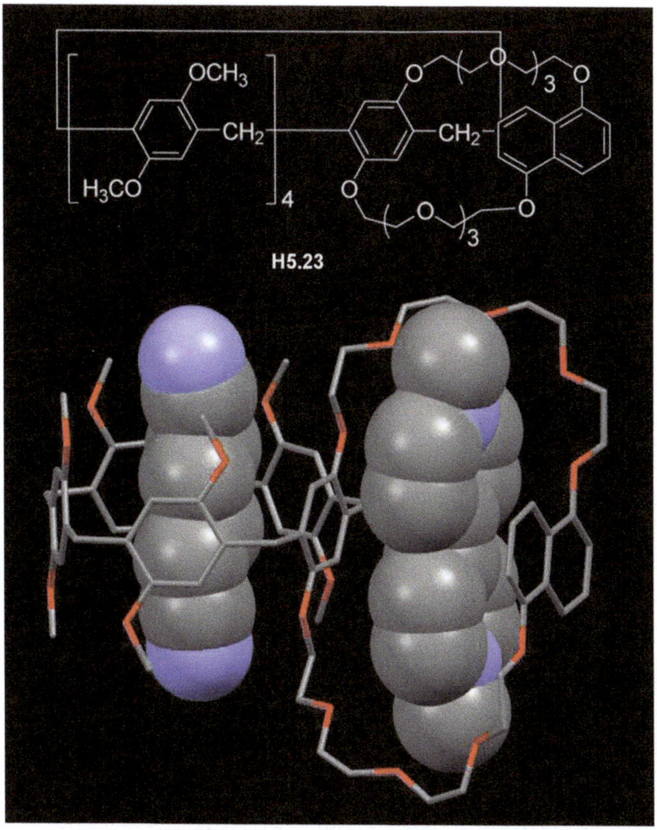

Figure 5.23 X-ray crystal structure of 1,4-dicyanobutane⊂**H5.23**⊃viologen complex. Hydrogen atoms and PF_6^- are omitted for clarity.

To monitor the inclusion of guest molecules, we synthesized a pillar[5]arene containing one π-conjugated unit **H5.24** (Figure 5.24).[55] Direct incorporation of the π-conjugated unit in **H5.24** prevented the rotation of units, thus allowing the isolation of planar chiral *pS* and *pR* enantiomers.

1,4-Dicyanobutane (**G5.29**) was used as an achiral guest molecule. Inclusion of 1,4-dicyanobutane induced not only UV and fluorescence signal changes but also a CD signal change. The *K* value of the complex determined by CD titration was $(1.37 \pm 0.22) \times 10^4$ M^{-1} (run 120).

Huang and co-workers synthesized a novel non-symmetric pillar[5]arene with five triazole moieties on one side of the rims (**H5.25**; Figure 5.25).[56] **H5.25** formed a complex with F^- anions with high selectivity and affinity. The *K* value of the complex was the highest among the anions (runs 121–124). The highly selective complexation with F^- resulted from the co-operative hydrogen bonding interaction between F^- and five triazole rings on one side.

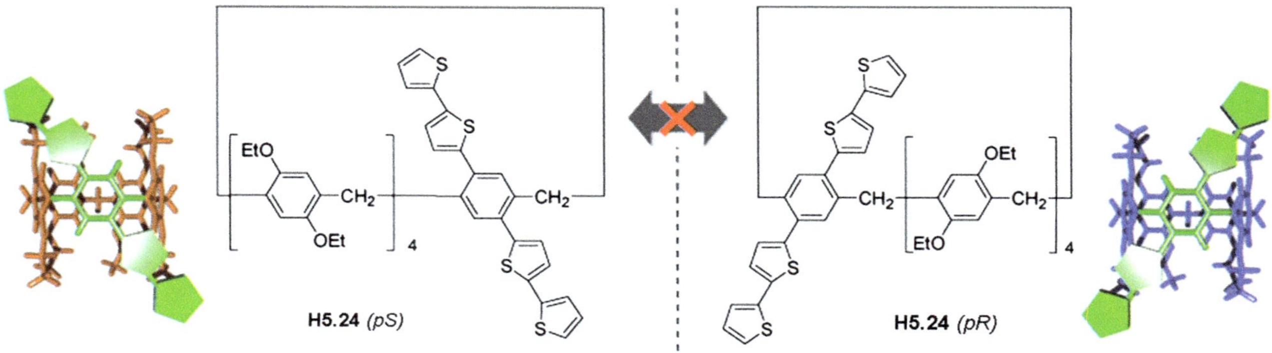

Figure 5.24 Planar chiral pillar[5]arenes (**H5.24**) with *pS* and *pR* forms produced by incorporating one π-conjugated unit. Reproduced from ref. 55 with permission from The Royal Society of Chemistry.

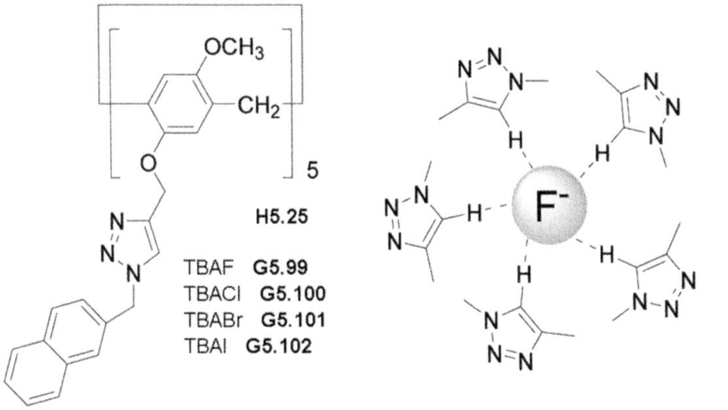

Figure 5.25 Chemical structures of non-symmetrical pillar[5]arene **H5.25** and guests for **H5.25**.

5.3.4.3 Host–Guest Properties of Per-functionalized Pillar[5]arenes

For the extraction of metal ions, Yuan and co-workers synthesized a pillar[5]arene with 10 phosphine oxide groups (**H5.26**) as chelating groups. **H5.26** showed selective recognition of the f-block elements (Figure 5.26a).[57]

Compared with the unit model and classical extractant, tri-*n*-octylphosphine oxide (**H5.26**) showed remarkable selectivity for thorium(IV) and uranyl(VI) caions. The spacer length between the phosphine oxide moiety and the pillar[5]arene also affected the extraction efficiency. As the length of the spacer increased, the extraction efficiency of these cations increased. per-Methylated pillar[5]arene **H5.2** did not extract these cations, therefore the 10 phosphine oxide moieties on the pillar[5]arene rim acted as an extractant. The efficiency was remarkably decreased in the unit model, indicating that the pillar[5]arene cavity is required to extract these cations efficiently. **H5.26** also captured heavy-metal ions, including Co^{2+}, Cu^{2+}, Ni^{2+}, Zn^{2+}, Cd^{2+}, Pb^{2+}, Ag^{2+} and Hg^{2+}. The K value of the complex between **H5.26** and Hg^{2+}, determined by UV titration, was $(4.4 \pm 0.9) \times 10^{3}$ M^{-1} for 1 : 1 stoichiometry. Thus, **H5.26** can be used for the separation and detection of the Hg^{2+} ion from environmental samples.[58]

Yuan and co-workers also synthesized pillar[5]arenes with 10 diglycoamide arms on the rims (**H5.27**; Figure 5.26b).[59] **H5.27** formed complexes with americium(III) and europium(III) cations. The characteristic emission was detected from the complexation.[60]

Li and co-workers synthesized **H5.28** by a CuAAC reaction between a pillar[5]arene containing 10 alkyne moieties and phenyl azide (Figure 5.27).[61]

The K values of the complexes between five diaminoalkane guests and **H5.28** were all in the vicinity of 10^{3} M^{-1} (Table 5.7, runs 125–129), which is much larger than that observed for per-methylated pillar[5]arene

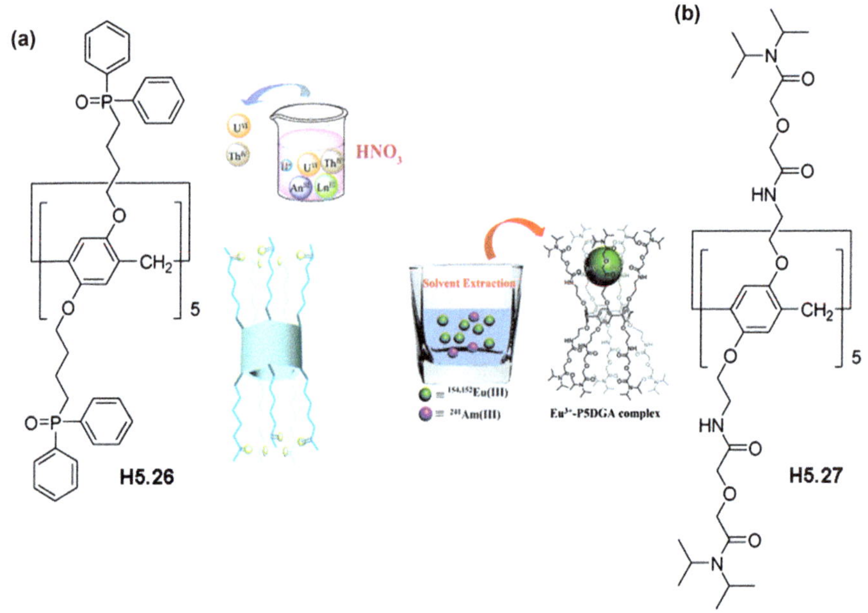

Figure 5.26 Selective extraction of metal cations by pillar[5]arenes with 10 phosphine oxide (**H5.26**) and 10 diglycoamide arms (**H5.27**).
Reproduced from ref. 57 and 59 with permission from The Royal Society of Chemistry.

($K < 10^2\ M^{-1}$). Additional hydrogen bonding, such as CH–N, NH–N, and CH/π interactions, caused by introducing phenyl substituents using a CuAAC reaction, improved the host–guest ability compared with simple per-alkylated pillar[5]arenes.

Huang, Xue and Yao groups reported the host–guest properties of a pillar[5]arene with mono(ethylene oxide) substituents (**H5.29**) and viologen derivatives (**G5.105–G5.109**).[62–64] The *K* values of the complexes between **H5.29** and these viologen derivatives were all in the vicinity of $10^4\ M^{-1}$ (runs 130–132). Formation of such stable host–guest complexes could be attributed to CH/O interactions between O atoms of the mono(ethylene oxide) substituents and the protons of these guests, along with multiple CH/π interactions between benzene moieties of the pillar[5]arene and the protons of these guests.

We synthesized a pillar[5]arene with 10 tetra-alkyl phosphonium bromide groups, and demonstrated substrate-selective oxidation of alkenes with **H5.30** as a phase transfer catalyst (Figure 5.28).[65] Tetra-alkyl phosphonium bromide groups, introduced in the pillar[5]arene arms, are amphiphilic moieties, therefore **H5.30** was soluble in both aqueous and organic media. In addition, the tetra-alkyl phosphonium bromide groups can be used as a phase-transfer catalyst. Thus, we investigated the oxidation of alkenes by $KMnO_4$ in the presence of **H5.30**. Compared with the traditional phase

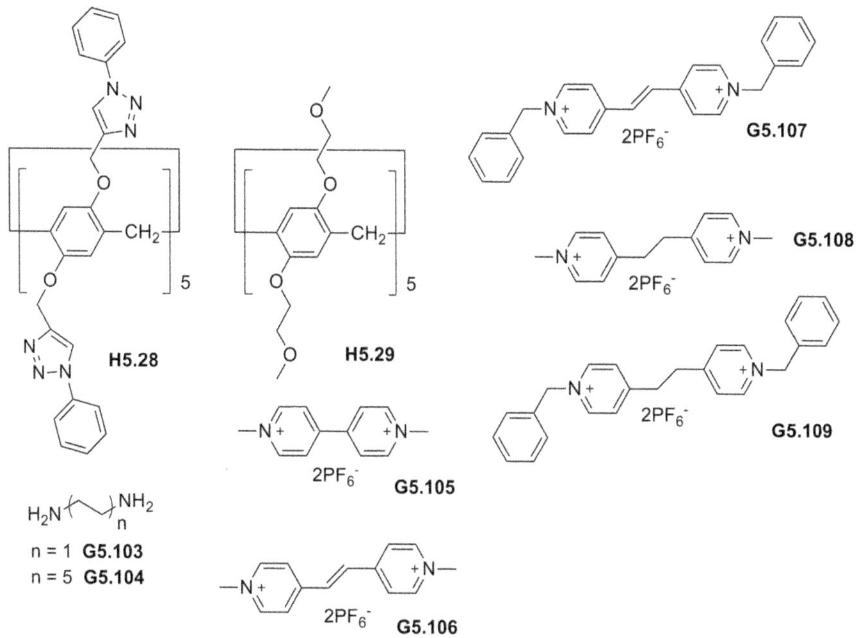

Figure 5.27 Chemical structures of per-functionalizing pillar[5]arenes and their guests.

Table 5.7 Host–guest complexes of per-functionalized pillar[5]arenes.

Run	Host	Guest	Solvent	Association constant (*K*) (M^{-1})	Method	Ref.
125	**H5.28**	**G5.103**	$CDCl_3$	$3.3 \pm 0.5 \times 10^3$	NMR	61
126	**H5.28**	**G5.79**	$CDCl_3$	$9.0 \pm 0.7 \times 10^3$	NMR	61
127	**H5.28**	**G5.81**	$CDCl_3$	$7.1 \pm 0.5 \times 10^3$	NMR	61
128	**H5.28**	**G5.83**	$CDCl_3$	$2.6 \pm 0.4 \times 10^3$	NMR	61
129	**H5.28**	**G5.104**	$CDCl_3$	$2.4 \pm 0.3 \times 10^3$	NMR	61
130	**H5.29**	**G5.105**	CH_3CN	$3.35 \pm 0.22 \times 10^4$	ITC	62
131	**H5.29**	**G5.106**	CH_3CN	$4.7 \pm 0.3 \times 10^4$	ITC	63
132	**H5.29**	**G5.107**	CH_3CN	$1.2 \pm 0.2 \times 10^4$	ITC	63
133	**H5.29**	**G5.108**	CH_3CN	$2.0 \pm 0.5 \times 10^3$	ITC	63
134	**H5.29**	**G5.109**	CH_3CN	$1.5 \pm 0.5 \times 10^3$	ITC	63

transfer catalysts, the oxidation reactions reached completion even with very small amounts of **H5.30**, which indicates that **H5.30** can act as an efficient phase transfer catalyst. Oxidation of the linear alkenes proceeded efficiently, but that of the branched and long chain alkenes did not. The main reason for this observation is that pillar[5]arenes formed stable host–guest complexes with short linear hydrocarbons, but did not with linear long chain and branched hydrocarbons.

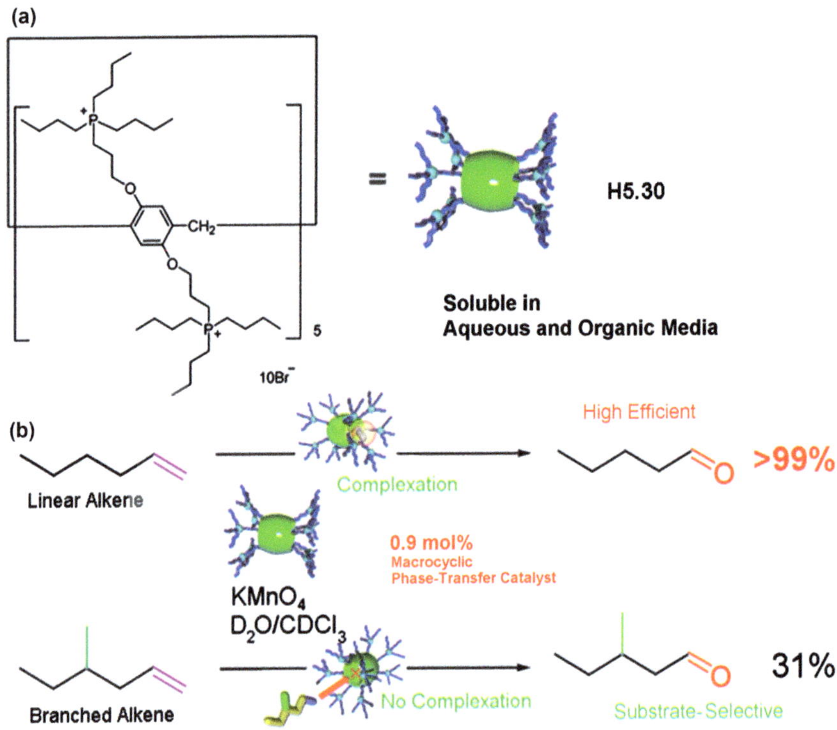

Figure 5.28 The highly efficient and substrate-selective phase transfer catalyst **H5.30**. Reproduced with permission from ref. 65. Copyright 2013 American Chemical Society.

5.4 Host–Guest Properties of Pillar[6]arenes

5.4.1 per-Hydroxylated Pillar[6]arene

As with per-hydroxylated pillar[5]arene **H5.1**, per-hydroxylated pillar[6]arene **H5.31** is also soluble in polar organic solvents, such as acetone, acetonitrile, DMF and DMSO (Figure 5.29).

Thus, the ideal media for the host–guest complexation are these polar organic solvents. The host–guest complexation ability of pillar[6]arenes was first investigated using electron-poor cationic molecules as a guest.[66] **H5.31** formed host–guest complexes with various pyridinium and viologen salts, but the binding ability of **H5.31** toward these cationic guest molecules was weaker [Table 5.8, run 135, $K = (2.2 \pm 0.3) \times 10^2\ M^{-1}$] than that of per-hydroxylated pillar[5]arene **H5.1**.

Our group investigated the host–guest ability of **H5.31** in the crystal state. The units in **H5.31** are arranged in an up-to-down manner by intra-molecular hydrogen bonds. The stabilized hexagonal structure assembled into infinite one-dimensional channels in the solid state (Figure 5.30; detail in Chapter 6).[67]

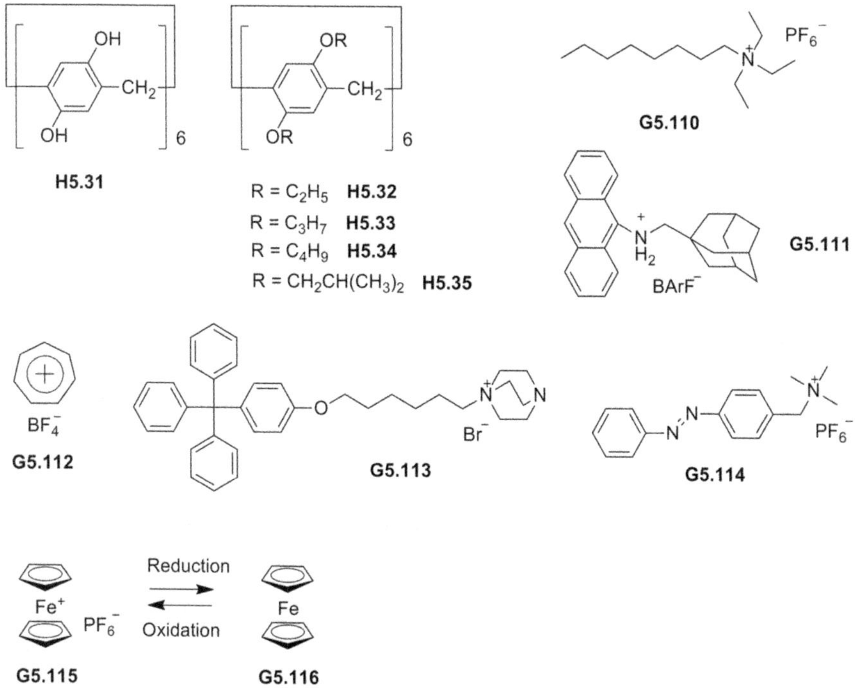

Figure 5.29 Chemical structures of per-hydroxylated and alkylated pillar[6]arenes and their guests.

Table 5.8 Host–guest complexes of per-alkylated pillar[6]arenes.

Run	Host	Guest	Solvent	Association constant (K) (M^{-1})	Method	Ref.
135	**H5.31**	**G5.105**	Acetone-d_6	$2.2 \pm 0.3 \times 10^2$	NMR	66
136	**H5.32**	**G5.25**	$CDCl_3$	$1.91 \pm 0.22 \times 10$	NMR	22
137	**H5.35**	**G5.110**	$CDCl_3$	$3.34 \pm 0.24 \times 10^2$	NMR	68
138	**H5.32**	**G5.111**	$CDCl_3$	$3.4 \pm 0.2 \times 10^3$	NMR	19
139	**H5.32**	**G5.112**	$CDCl_3$	$1.8 \pm 0.3 \times 10^3$	NMR	69
140	**H5.32**	**G5.113**	$CDCl_3$	$5.52 \pm 0.65 \times 10^2$	NMR	20
141	**H5.33**	**G5.114 (*trans*)**	$CDCl_3$	$2.22 \pm 0.34 \times 10^3$	NMR	70
142	**H5.33**	**G5.114 (*cis*)**	$CDCl_3$	$2.64 \pm 0.29 \times 10^2$	NMR	70
143	**H5.34**	**G5.115**	$CHCl_3$: $CH_3CN = 5:1$	$3.1 \pm 0.1 \times 10^4$	Fl	71
144	**H5.34**	**G5.116**	$CDCl_3$: $CD_3CN = 5:1$	$1.8 \pm 0.05 \times 10$	NMR	71

Our group investigated gas and vapor adsorption using the microporous channels constructed from **H5.31** in the crystal state. For activation, acetone molecules contained in the channels were removed by heating. Heating at 120 °C under a reduced pressure was sufficient to remove acetone molecules. The channels were the same as those existing before removal of acetone because the hexagonal structure was stabilized by intra-molecular hydrogen

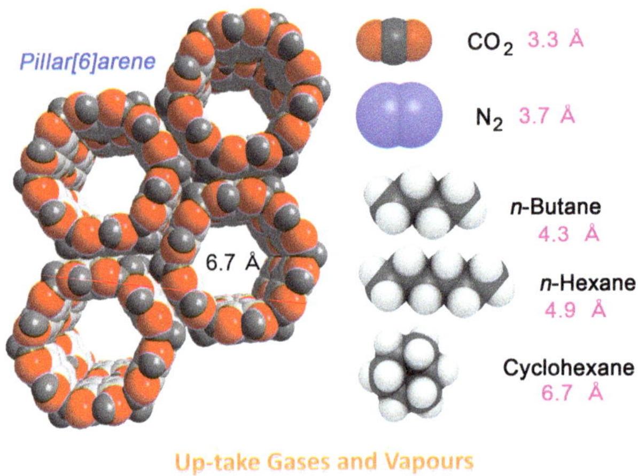

Figure 5.30 Gas and vapor uptake by one-dimensional channels of per-hydroxylated pillar[6]arene **H5.31**.
Reproduced from ref. 67 with permission from The Royal Society of Chemistry.

bonds. The microporous materials can accommodate various gases, such as CO_2 (diameter 3.3 Å), nitrogen (diameter 3.7 Å) and *n*-butane (diameter 4.3 Å) at low pressures. The isotherms were assigned as typical type I, which means they contained micropores. Supramolecular assemblies of **H5.1** took up CO_2, but the amounts of methane and nitrogen adsorbed at low pressures were small.[14] In contrast, the porous one-dimensional channels from **H5.31** can accommodate not only CO_2 but also larger gases, such as nitrogen and *n*-butane. This is because the cavity size of pillar[6]arene is larger than that of pillar[5]arene. Another possible reason is because pillar[6]arene forms precise one-dimensional channels. The porous one-dimensional channels can therefore capture large gases without blocking of the pores, which is observed for nitrogen adsorption at low temperatures. The one-dimensional channels have large pores, therefore they can accommodate large organic vapors, such as *n*-hexane (diameter 4.9 Å) and cyclohexane (diameter 6.7 Å).

5.4.2 per-Alkylated Pillar[6]arenes

Host–guest complexation using per-alkylated pillar[6]arenes was mainly investigated in $CDCl_3$. One of the reasons for this is that per-alkylated pillar[6]arenes are soluble in nonpolar solvents including chloroform, acetone and toluene. The pyridinium cations including **G5.25**, which is a good guest for per-alkylated pillar[5]arenes,[22] also formed host–guest complexes with per-ethylated pillar[6]arene. However, the K value of the complex was 19.1 ± 2.2 M^{-1} (run 136), which is about 400 times smaller than that observed with per-ethylated pillar[5]arene **H5.3** (Table 5.2, run 25) owing to the size mismatch between pyridinium cations and the cavity of pillar[6]arene.

per-Isobutylated pillar[6]arene (**H5.34**) can form a host–guest complex with a cationic *n*-octyltriethyl ammonium hexafluorophosphate (run 137, **G5.110**, $334 \pm 24\ M^{-1}$), but such a complex is not formed with per-isobutylated pillar[5]arene.[68] This is because the triethyl ammonium moiety of **G5.110** fits the cavity size of pillar[6]arene, but does not fit that of pillar[5]arene. Therefore, cations with bulky moieties fit the cavity of pillar[6]arenes, which is a similar observation to the host–guest complexes of β-CD and CB[7], because the cavity size of pillar[6]arenes is similar to that of β-CD and CB[7].[1,8] Cations with bulky hydrocarbons, such as adamantine (run 138, **G5.111**),[19] tropylium (run 139, **G5.112**)[69] and 1,4-diazabicyclo[2.2.2]octane (DABCO) cations (run 140, **G5.113**),[20] form relatively stable host–guest complexes with per-ethylated pillar[6]arene **H5.32** ($K = ca.\ 10^3$–$10^4\ M^{-1}$).

Polyaromatic compounds also form host–guest complexes with pillar[6]arenes. The *trans* form of azobenzene containing mono-cation (run 141, **G5.114**) can complex with per-propylated pillar[6]arene **H5.33** [$K = (2.2 \pm 0.3) \times 10^3\ M^{-1}$], but the *cis* form of the azobenzene guest hardly forms host–guest complexes with **H5.33** [run 142, $K = (2.6 \pm 0.3) \times 10^2\ M^{-1}$].[70] The cavity size of the pillar[6]arene fits to the *trans* form of the azobenzene guest but not to the *cis* form. per-Butylated pillar[6]arene **H5.34** also forms a stable host–guest complex with the oxidized form of the ferrocenium cation [run 143, **G5.115**, $K = (3.1 \pm 0.1) \times 10^4\ M^{-1}$], but hardly forms a complex with the reduced form of ferrocene (run 144, **G5.116**, $K = 18 \pm 0.5\ M^{-1}$).[71] *Trans-* and *cis*-isomerization of azobenzene moieties can be induced by photo-irradiation. Interconversion between ferrocene and the ferrocenium cation can be switched by a redox reaction. Therefore, these photo- and redox-responsive host–guest systems enabled us to build various stimuli-responsive supramolecular assemblies.

5.4.3 Water-soluble Pillar[6]arenes

Water-soluble pillar[6]arenes can be prepared using the same method as the synthesis of water-soluble pillar[5]arenes. The introduction of water-soluble substituents, such as cationic, anionic and nonionic moieties, allows pillar[6]arenes to solubilize in aqueous media. The first water-soluble pillar[6]arene was synthesized by Huang and co-workers, by introducing 12 carboxylate moieties (Figure 5.31, **H5.36**).[72] **H5.36** could capture paraquat (Table 5.9, run 145, **G5.117**) in water with a high association constant [$K = (1.02 \pm 0.10) \times 10^8\ M^{-1}$], which is much higher than that of the anionic water-soluble pillar[5]arene **H5.11** [Table 5.4, run 63, $K = (8.20 \pm 1.70) \times 10^4\ M^{-1}$].[36]

This is because the width of the 4,4′-bipyridinium group (6.3 Å) is larger than the cavity of pillar[5]arene (*ca.* 4.7 Å), but suitable for the cavity of pillar[6]arene (*ca.* 6.7 Å). **H5.36** also formed stable host–guest complexes with cationic molecules such as pyridinium bromide derivatives, with an association constant in the vicinity of $10^5\ M^{-1}$ (runs 146 and 147).[73–75]

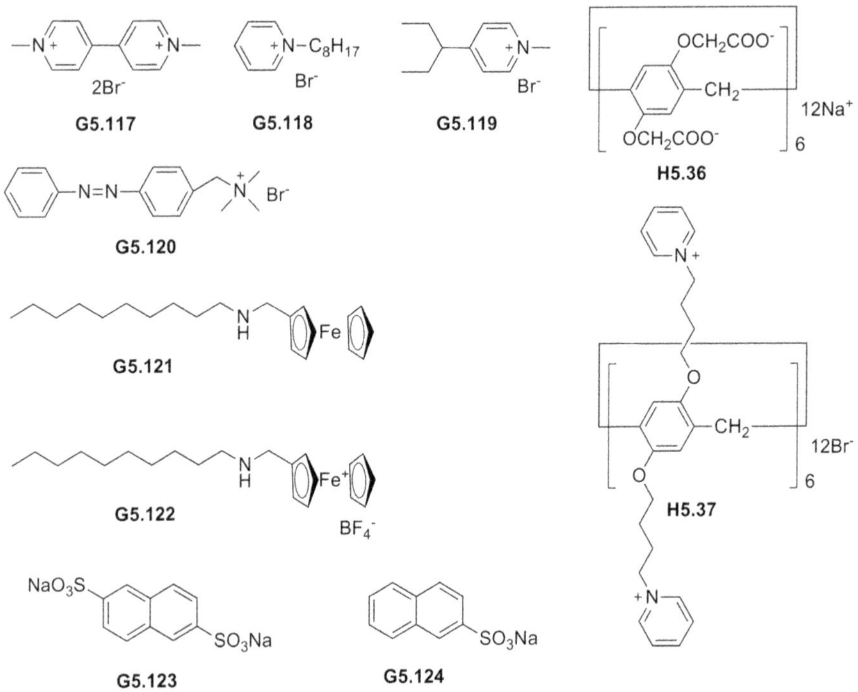

Figure 5.31 Chemical structures of water-soluble pillar[6]arenes and their guests.

The cavity size of pillar[6]arenes is a match for photo-responsive azobenzene and redox-active ferrocene molecules, therefore the complexes can be used as a switch to induce a change of supramolecular structure constructed from the host–guest complexes in organic solvents. The concept can be transferred to aqueous media using water-soluble pillar[6]arenes. Huang and co-workers reported a water-soluble photo-responsive host–guest complex using azobenzene containing a trimethyl ammonium moiety (**G5.120**) as a guest for **H5.36**.[76] **H5.36** formed a stable 1 : 1 host–guest complex with the *trans* form of **G5.120** [run 148, $K = (4.13 \pm 0.19) \times 10^5$ M^{-1}]. Upon irradiation with 365 nm UV light, photo-isomerization from the *trans* to *cis* form took place. At equilibrium, the *trans* : *cis* ratio was 3 : 7. The association constant of the complex between the *cis* form of **G5.120** and **H5.36** was $K = (5.89 \pm 0.33) \times 10^4$ M^{-1} (run 149), indicating that photo-isomerization from the *trans* to *cis* form induces dissociation of the complex. This arises from the size mismatch between the *cis* form of the azobenzene guest and the cavity size of the pillar[6]arene.

Wang and co-workers reported a redox-active host–guest system in aqueous media. *N*-1-decyl-ferrocenylmethylamine (**G5.121**) formed a host–guest complex with water-soluble pillar[6]arene **H5.36**.[77] The association constant of the complex was $(1.27 \pm 0.42) \times 10^5$ M^{-1} (run 150), which is greater than that observed for ferrocene (**G5.34**) with per-alkylated pillar[6]arene in $CDCl_3$ (Table 5.8, run 144, $K = 18 \pm 0.5$ M^{-1}). The hydrophobic interaction

Table 5.9 Host–guest complexes of water-soluble pillar[6]arenes.

Run	Host	Guest	Solvent	Association constant (K) (M^{-1})	Method	Ref.
145	**H5.36**	**G5.117**	H_2O	$1.02 \pm 0.10 \times 10^8$	Fl	72
146	**H5.36**	**G5.118**	H_2O	$3.26 \pm 0.28 \times 10^5$	Fl	74
147	**H5.36**	**G5.119**	D_2O	$1.67 \pm 0.13 \times 10^5$	NMR	75
148	**H5.36**	**G5.120 (*trans*)**	H_2O	$4.13 \pm 0.19 \times 10^5$	ITC	76
149	**H5.36**	**G5.120 (*cis*)**	H_2O	$5.89 \pm 0.33 \times 10^4$	ITC	76
150	**H5.36**	**G5.121**	H_2O	$1.27 \pm 0.42 \times 10^5$	Fl	77
151	**H5.36**	**G5.122**	H_2O	$8.68 \pm 0.72 \times 10^7$	Fl	77
152	**H5.37**	**G5.123**	H_2O	$2.1 \pm 0.3 \times 10^5$	Fl	78
153	**H5.37**	**G5.124**	H_2O	$3.8 \pm 0.2 \times 10^4$	Fl	78
154	**H5.38**	**G5.125**	D_2O	$1.01 \pm 0.0031 \times 10^3$	NMR	79

stabilized the host–guest complexation. In addition, the water-soluble pillar[6]arene can capture ferrocenium tetrafluoroborate salt (**G5.122**), which is generated by oxidation of ferrocene, with a high association constant [run 151, $K = (8.68 \pm 0.72) \times 10^7$ M^{-1}]. The redox-active host–guest system in aqueous media can be applied as a drug delivery system.

A water-soluble positively charged pillar[6]arene (**H5.37**) was synthesized by Li and co-workers.[78] **H5.37** formed stable host–guest complexes with two anionic naphthalenesulfonate derivatives (**G5.123** and **G5.124**). The association constant determined by fluorescence titration was $(2.1 \pm 0.3) \times 10^5$ M^{-1} for **G5.122** with two sulfonate moieties (run 152). The K value for **G5.122** is 5.5 times higher than that for **G5.123** with one sulfonate moiety [run 153, $K = (3.8 \pm 0.2) \times 10^4$ M^{-1}] because co-operative electrostatic interactions between two sulfonate anions of **G5.122** and cationic pyridinium moieties on the pillar[6]arene rims stabilize the complexation.

We reported a water-soluble pillar[6]arene produced by modification of tri(ethylene oxide) chains (**H5.38**).[79] **H5.38** formed a host–guest complex with an azobenzene moiety containing bulky DABCO cations at both ends (**G5.125**). **H5.38** formed the host–guest complex with the *trans*-form azobenzene guest [run 154, $K = (1.01 \pm 0.0031) \times 10^3$ M^{-1}], but not with the *cis*-form azobenzene guest, owing to the size mismatch between the *cis*-form azobenzene guest and the cavity size of the pillar[6]arene. Therefore, the photo-isomerization from the *trans* to *cis* form induced dethreading of the azobenzene guest from the pillar[6]arene cavity. As with the tri(ethylene oxide)-modified pillar[5]arene, **H5.38** showed LCST at 41 °C (Figure 5.32).

The clouding point of the complex between the *trans*-azobenzene guest and **H5.38** was 57 °C. A similar phenomenon was observed in the complex between a pillar[5]arene with 10 tri(ethylene oxide) moieties (**H5.14**) and a viologen guest. Repulsive forces between the complexed cations inhibited aggregation of **H5.38**. When the host–guest complex was irradiated with UV light, the clouding point decreased to 52 °C. Irradiation with visible light then caused an increase of the clouding point to 57 °C. The increases and

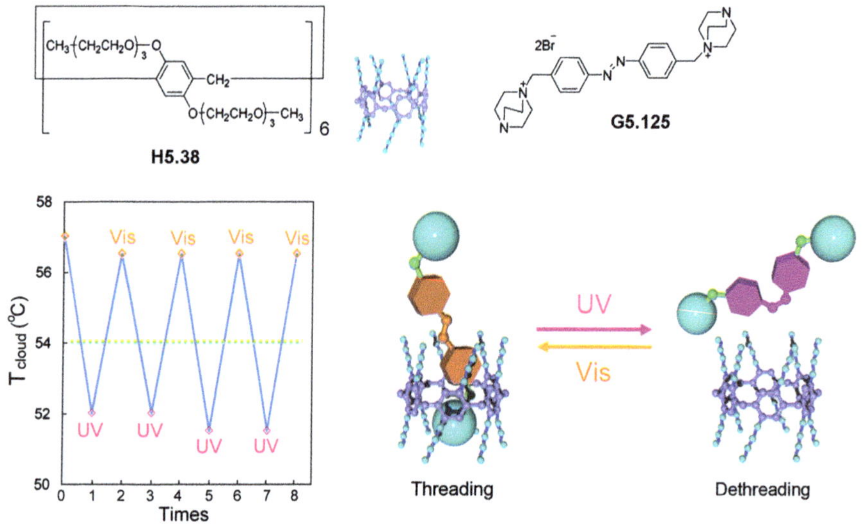

Figure 5.32 Changes of T_{cloud} upon alternating irradiation with UV or visible light, based on host–guest complex between pillar[6]arene with 12 tri(ethylene oxide) chains (**H5.38**) and azobenzene guest (**G5.125**) in aqueous media. Reproduced with permission from ref. 79. Copyright 2012 American Chemical Society.

decreases in the clouding point were reversibly switched by alternating between irradiation with UV and with visible light. Threading and de-threading of the azobenzene guest by photo-irradiation is the trigger for the reversible clouding point change.

5.4.4 Functionalized Pillar[6]arenes

As with pillar[5]arenes, modification of 12 functional substituents on pillar[6]arenes is a useful way to improve their host–guest property. Huang and co-workers synthesized a pillar[5]arene with 12 mono(ethylene oxide) substituents (**H5.39**; Figure 5.33).[80]

per-Hydroxylated pillar[6]arene **H5.31** formed a weak host–guest complex with diquat [**G5.126**, $K = (1.5 \pm 0.4) \times 10\ M^{-1}$], but **H5.39** interacted with diquat (**G5.126**) very strongly [$K = (3.9 \pm 0.1) \times 10^5\ M^{-1}$, which is 10^4 times higher than that observed for the **H5.31**⊃diquat complex]. This is because the oxygen atoms in mono(ethylene oxide) substituents on the rims of **H5.39** form multiple CH–O hydrogen bonds to stabilize the complexation, which is confirmed by X-ray crystalline analysis of the complex.

Huang and co-workers also investigated the host–guest complexation ability of bulky adamantine cationic guest **G5.127** with a pillar[6]arene with one hydroquinone unit (**H5.40**).[81] The *K* value of the **H5.40**⊃**G5.127** complex was $(3.33 \pm 0.99) \times 10^4\ M^{-1}$, which is double that of the per-propylated pillar[6]arene **H5.33**⊃**G5.127** complex [$K = (2.00 \pm 0.63) \times 10^4\ M^{-1}$]. Replacing

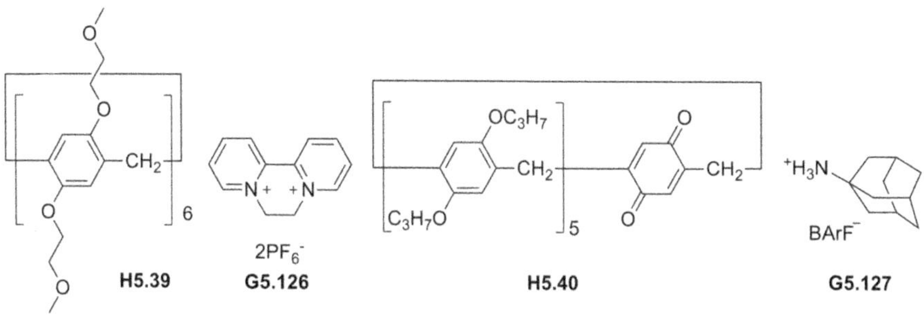

Figure 5.33 Chemical structures of per-functionalized pillar[6]arenes and their guests.

the dialkoxybenzene with a hydroquinone unit reduced the steric hindrance, which contributed to enhancing the host–guest ability toward the bulky adamantine guest.

5.5 Host–Guest Properties of Larger Pillar[*n*]arene Homologs ($n = 7$–14)

Recently, investigation of the host–guest properties of larger pillar[*n*]arene homologs ($n = 7$–14) has been a hot research topic because it has been enabled by the development of synthetic procedures for larger pillar[*n*]arene homologs (details for the methods of synthesis of larger pillar[*n*]arene homologs are discussed in Chapter 3). In the solution state, they can form various conformations due to the flipping of the constituent units and flexible methylene bridges linking *p*-dialkoxybenzene units. Their diameters, calculated by the diameter of the inscribed circle of the polygon, are 0.71 nm (pillar[7]arene), 0.82 nm (pillar[8]arene), 0.93 nm (pillar[9]arene), 1.05 nm (pillar[10]arene), 1.18 nm (pillar[11]arene), 1.27 nm (pillar[12]arene), 1.38 nm (pillar[13]arene) and 1.49 nm (pillar[14]arene). Large pillar[*n*]arene homologs should have the ability to capture large guest molecules, which cannot be captured by pillar[5]- and pillar[6]arenes.

Hou and co-workers reported the host–guest complexation of *n*-octatrimethylammonium hexafluorophosphate (**G5.18**) with pillar[8–10]arenes.[82] **G5.18** is a good guest molecule for pentagonal pillar[5]arene **H5.3** [$K = (2.4 \pm 0.1) \times 10^3$ M^{-1}] and hexagonal pillar[6]arene **H5.32** [$K = (1.2 \pm 0.1) \times 10^3$ M^{-1}]. Therefore, the formation of pentagon- or hexagon-like conformations allows these larger pillar[*n*]arene homologs to form a complex with **G5.18**. However, pillar[8]arene **H5.42** did not form a host–guest complex with **G5.18** (Figure 5.34).

When pillar[8]arene **H5.42** changes its conformation to generate one-hexagon-like cavity and one quadrilateral cavity, **H5.42** can form a 1 : 1 complex with **G5.18**. However, the formation of the conformation may be impossible owing to an increase of ring strain. Pillar[9]arene **H5.43** formed

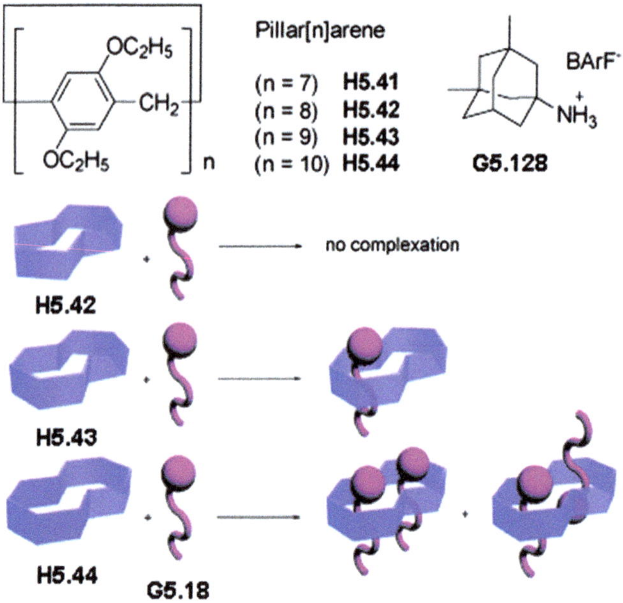

Figure 5.34 Host–guest complexation of large pillar[*n*]arene homologs ($n \geqq 7$) with guests.
Reproduced from ref. 82 with permission from The Royal Society of Chemistry.

a 1:1 host–guest complex with **G5.18** [$K = (3.0 \pm 0.3) \times 10^2\ M^{-1}$] because pillar[9]arene can change its conformation to generate one-hexagon-like cavity without ring strain. Interestingly, pillar[10]arene **H5.44** formed a 1:2 host–guest complex with **G5.18** [$K_1 = (7.2 \pm 0.2) \times 10^2\ M^{-1}$, $K_2 = (5.5 \pm 0.2) \times 10^2\ M^{-1}$]. Pillar[10]arene can change its conformational structure to generate two hexagon-like cavities, therefore it is able to capture two **G5.18** molecules.

Host–guest complexation of per-ethylated pillar[7]arene **H5.41** with an adamantine guest (**G5.128**) was studied by Li and co-workers.[83] The association constant of the complex, determined by ^1H NMR titration, was $(1.2 \pm 0.3) \times 10^5\ M^{-1}$, which is larger than those achieved with per-ethylated pillar[5]arene **H5.3** [$K = (3.1 \pm 0.5) \times 10^3\ M^{-1}$] and per-ethylated pillar[6]arene **H5.32** [$K = (1.9 \pm 0.3) \times 10^4\ M^{-1}$)]. The cavity size of pillar[7]arene **H5.41** is 0.71 nm, which is able to fit the bulky adamantine guest.

Pillar[*n*]arenes ($n \geq 10$) have nano-sized cavities over 1 nm. We investigated the host–guest complexation between these larger pillar[*n*]arene homologs having nano-sized cavities and a nano-size guest, the fullerene C_{60} (Figure 5.35).[84] Among these pillar[*n*]arene homologs, pillar[10]arene **H5.44** selectively formed the most stable host–guest complex with C_{60} [$K = (2.1 \pm 0.2) \times 10^3\ M^{-1}$ in toluene].

This is a result of the cavity size of the pillar[10]arene being 1.05 nm, which fits the C_{60} with a van der Waals diameter of 1.01 nm.

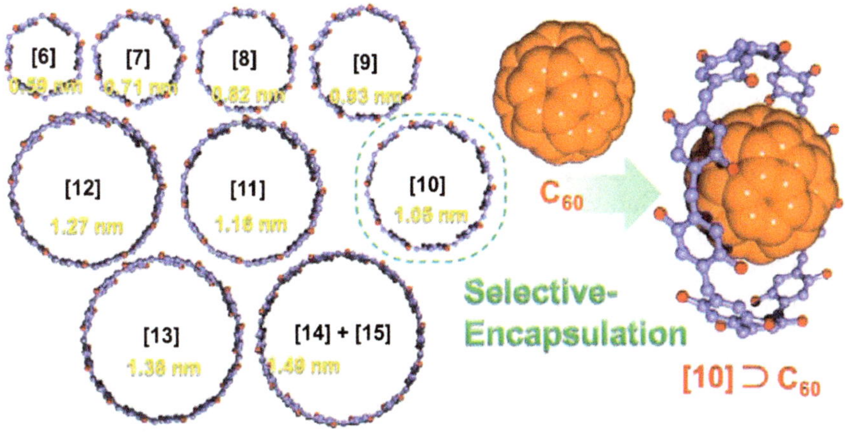

Figure 5.35 Selective complexation of C_{60} with pillar[10]arene. Reproduced with permission from ref. 84. Copyright 2014 American Chemical Society.

5.6 Summary and Outlook

In this chapter, we have discussed the host–guest complexation ability of pillar[*n*]arenes. The host–guest property of pillar[5]arenes has been thoroughly investigated, therefore almost all of their properties have now been determined. Construction of new supramolecular assemblies using the host–guest properties of pillar[5]arenes is the next step. Pillar[6]arenes can form host–guest complexes with photo- and redox-responsive polyaromatic compounds. Therefore, their host–guest properties are very useful for construction of stimuli-responsive materials. Regarding larger pillar[*n*]arene homologs ($n \geq 7$), we have recently been able to investigate their host–guest properties because the synthetic procedures for larger pillar[*n*]arene homologs have been determined. Further investigations of the host–guest properties are necessary because most of the host–guest abilities of these larger pillar[*n*]arene homologs are still unknown, and this will undoubtedly lead to new breakthroughs in pillar[*n*]arene chemistry.

References

1. M. V. Rekharsky and Y. Inoue, *Chem. Rev.*, 1998, **98**, 1875–1918.
2. C. J. Pedersen, *J. Am. Chem. Soc.*, 1967, **89**, 7017–7036.
3. B. Odell, M. V. Reddington, A. M. Z. Slawin, N. Spencer, J. F. Stoddart and D. J. Williams, *Angew. Chem., Int. Ed. Engl.*, 1988, **27**, 1547–1550.
4. A. C. Fahrenbach, Z. Zhu, D. Cao, W. G. Liu, H. Li, S. K. Dey, S. Basu, A. Trabolsi, Y. Y. Botros, W. A. Goddard and J. F. Stoddart, *J. Am. Chem. Soc.*, 2012, **134**, 16275–16288.
5. E. H. Witlicki, S. S. Andersen, S. W. Hansen, J. O. Jeppesen, E. W. Wong, L. Jensen and A. H. Flood, *J. Am. Chem. Soc.*, 2010, **132**, 6099–6107.

6. S. Saha, A. H. Flood, J. F. Stoddart, S. Impellizzeri, S. Silvi, M. Venturi and A. Credi, *J. Am. Chem. Soc.*, 2007, **129**, 12159–12171.
7. T. Ogoshi and T. Yamagishi, *Chem. Commun.*, 2014, **50**, 4776–4787.
8. J. W. Lee, S. Samal, N. Selvapalam, H.-J. Kim and K. Kim, *Acc. Chem. Res.*, 2003, **36**, 621–630.
9. T. Ogoshi, S. Kanai, S. Fujinami, T. A. Yamagishi and Y. Nakamoto, *J. Am. Chem. Soc.*, 2008, **130**, 5022–5023.
10. C. J. Li, Q. Q. Xu, J. Li, F. N. Yao and X. S. Jia, *Org. Biomol. Chem.*, 2010, **8**, 1568–1576.
11. T. Ogoshi, S. Tanaka, T. Yamagishi and Y. Nakamoto, *Chem. Lett.*, 2011, **40**, 96–98.
12. C. J. Li, L. Zhao, J. A. Li, X. Ding, S. H. Chen, Q. L. Zhang, Y. H. Yu and X. S. Jia, *Chem. Commun.*, 2010, **46**, 9016–9018.
13. T. Ogoshi, D. Yamafuji, T. Aoki and T. Yamagishi, *J. Org. Chem.*, 2011, **76**, 9497–9503.
14. L. L. Tan, H. W. Li, Y. C. Tao, S. X. A. Zhang, B. Wang and Y. W. Yang, *Adv. Mater.*, 2014, **26**, 7027–7031.
15. Z. B. Zhang, B. Y. Xia, C. Y. Han, Y. H. Yu and F. H. Huang, *Org. Lett.*, 2010, **12**, 3285–3287.
16. T. Ogoshi, R. Shiga, T. Yamagishi and Y. Nakamoto, *J. Org. Chem.*, 2011, **76**, 618–622.
17. T. Ogoshi, K. Masaki, R. Shiga, K. Kitajima and T. Yamagishi, *Org. Lett.*, 2011, **13**, 1264–1266.
18. C. Han, G. Yu, B. Zheng and F. Huang, *Org. Lett.*, 2012, **14**, 1712–1715.
19. C. Li, X. Shu, J. Li, J. Fan, Z. Chen, L. Weng and X. Jia, *Org. Lett.*, 2012, **14**, 4126–4129.
20. T. Ogoshi, H. Kayama, D. Yamafuji, T. Aoki and T. Yamagishi, *Chem. Sci.*, 2012, **3**, 3221–3226.
21. T. Ogoshi, D. Yamafuji, T. Aoki, K. Kitajima, T. Yamagishi, Y. Hayashi and S. Kawauchi, *Chem. – Eur. J.*, 2012, **18**, 7493–7500.
22. T. Ogoshi, D. Yamafuji, T. Aoki and T. A. Yamagishi, *Chem. Commun.*, 2012, **48**, 6842–6844.
23. S. Y. Dong, B. Zheng, Y. Yao, C. Y. Han, J. Y. Yuan, M. Antonietti and F. H. Huang, *Adv. Mater.*, 2013, **25**, 6864–6867.
24. T. Takata, N. Kihara and Y. Furusho, *Polymer Synthesis*, Springer Berlin, Heidelberg, 2004, vol. 171, pp. 1–75.
25. C. Talotta, N. A. De Simone, C. Gaeta and P. Neri, *Org. Lett.*, 2015, **17**, 1006–1009.
26. G. Gattuso, A. Notti, A. Pappalardo, M. F. Parisi, I. Pisagatti, S. Pappalardo, D. Garozzo, A. Messina, Y. Cohen and S. Slovak, *J. Org. Chem.*, 2008, **73**, 7280–7289.
27. A. Scarso, L. Trembleau and J. Rebek, *J. Am. Chem. Soc.*, 2004, **126**, 13512–13518.
28. D. Ajami and J. Rebek, *Nat. Chem.*, 2009, **1**, 87–90.
29. B. W. Purse and J. J. Rebek, *Chem. Commun.*, 2005, 722–724.

30. N. L. Strutt, R. S. Forgan, J. M. Spruell, Y. Y. Botros and J. F. Stoddart, *J. Am. Chem. Soc.*, 2011, **133**, 5668–5671.
31. X. Shu, S. Chen, J. Li, Z. Chen, L. Weng, X. Jia and C. Li, *Chem. Commun.*, 2012, **48**, 2967–2969.
32. X. Y. Shu, J. Z. Fan, J. Li, X. Y. Wang, W. Chen, X. S. Jia and C. J. Li, *Org. Biomol. Chem.*, 2012, **10**, 3393–3397.
33. X. Y. Shu, W. Chen, D. B. Hou, Q. B. Meng, R. L. Zheng and C. J. Li, *Chem. Commun.*, 2014, **50**, 4820–4823.
34. K. Han, Y. Zhang, J. Li, Y. Yu, X. Jia and C. Li, *Eur. J. Org. Chem.*, 2013, **2013**, 2057–2060.
35. X. S. Hu, H. M. Deng, J. Li, X. S. Jia and C. J. Li, *Chin. Chem. Lett.*, 2013, **24**, 707–709.
36. T. Ogoshi, M. Hashizume, T. A. Yamagishi and Y. Nakamoto, *Chem. Commun.*, 2010, **46**, 3708–3710.
37. C. Li, X. Shu, J. Li, S. Chen, K. Han, M. Xu, B. Hu, Y. Yu and X. Jia, *J. Org. Chem.*, 2011, **76**, 8458–8465.
38. C. J. Li, J. W. Ma, L. Zhao, Y. Y. Zhang, Y. H. Yu, X. Y. Shu, J. Li and X. S. Jia, *Chem. Commun.*, 2013, **49**, 1924–1926.
39. T. Adiri, D. Marciano and Y. Cohen, *Chem. Commun.*, 2013, **49**, 7082–7084.
40. P. Wang, Y. Yao and M. Xue, *Chem. Commun.*, 2014, **50**, 5064–5067.
41. G. C. Yu, J. Yang, D. Y. Xia and Y. Yao, *Rsc Adv.*, 2014, **4**, 18763–18771.
42. Y. J. Ma, X. F. Ji, F. Xiang, X. D. Chi, C. Y. Han, J. M. He, Z. Abliz, W. X. Chen and F. H. Huang, *Chem. Commun.*, 2011, **47**, 12340–12342.
43. Y. J. Ma, M. Xue, Z. B. Zhang, X. D. Chi and F. H. Huang, *Tetrahedron*, 2013, **69**, 4532–4535.
44. X. B. Hu, L. Chen, W. Si, Y. H. Yu and J. L. Hou, *Chem. Commun.*, 2011, **47**, 4694–4696.
45. T. Ogoshi, R. Shiga and T. Yamagishi, *J. Am. Chem. Soc.*, 2012, **134**, 4577–4580.
46. M. F. Ni, Y. F. Guan, L. Wu, C. Deng, X. Y. Hu, J. L. Jiang, C. Lin and L. Y. Wang, *Tetrahedron Lett.*, 2012, **53**, 6409–6413.
47. B. Y. Xia and M. Xue, *Chem. Commun.*, 2014, **50**, 1021–1023.
48. T. Ogoshi, K. Demachi, K. Kitajima and T. Yamagishi, *Chem. Commun.*, 2011, **47**, 10290–10292.
49. C. Li, K. Han, J. Li, H. Zhang, J. Ma, X. Shu, Z. Chen, L. Weng and X. Jia, *Org. Lett.*, 2012, **14**, 42–45.
50. R. Manoni, P. Neviani, P. Franchi, E. Mezzina and M. Lucarini, *Eur. J. Org. Chem.*, 2014, **2014**, 147–151.
51. Q. P. Duan, W. Xia, X. Y. Hu, M. F. Ni, J. L. Jiang, C. Lin, Y. Pan and L. Y. Wang, *Chem. Commun.*, 2012, **48**, 8532–8534.
52. G. C. Yu, B. Hua and C. Y. Han, *Org. Lett.*, 2014, **16**, 2486–2489.
53. B. Hua and G. Yu, *Tetrahedron Lett.*, 2014, **55**, 6274–6276.
54. W. B. Hu, H. M. Yang, W. J. Hu, M. L. Ma, X. L. Zhao, X. Q. Mi, Y. A. Liu, J. S. Li, B. Jiang and K. Wen, *Chem. Commun.*, 2014, **50**, 10460–10463.

55. T. Ogoshi, D. Yamafuji, T. Akutsu, M. Naito and T. Yamagishi, *Chem. Commun.*, 2013, **49**, 8782–8784.
56. G. Yu, Z. Zhang, C. Han, M. Xue, Q. Zhou and F. Huang, *Chem. Commun.*, 2012, **48**, 2958–2960.
57. Y. Fang, L. Wu, J. Liao, L. Chen, Y. Yang, N. Liu, L. He, S. Zou, W. Feng and L. Yuan, *RSC Adv.*, 2013, **3**, 12376–12383.
58. Y. Jia, Y. Fang, Y. Li, L. He, W. Fan, W. Feng, Y. Yang, J. Liao, N. Liu and L. Yuan, *Talanta*, 2014, **125**, 322–328.
59. L. Wu, Y. Y. Fang, Y. M. Jia, Y. Y. Yang, J. L. Liao, N. Liu, X. S. Yang, W. Feng, J. L. Ming and L. H. Yuan, *Dalton Trans.*, 2014, **43**, 3835–3838.
60. A. Sengupta, L. Wu, W. Feng, L. H. Yuan and V. Natarajan, *J. Lumin.*, 2015, **158**, 356–364.
61. H. M. Deng, X. Y. Shu, X. S. Hu, J. Li, X. S. Jia and C. J. Li, *Tetrahedron Lett.*, 2012, **53**, 4609–4612.
62. X. D. Chi, M. Xue, Y. Yao and F. H. Huang, *Org. Lett.*, 2013, **15**, 4722–4725.
63. X. Chi and M. Xue, *RSC Adv.*, 2014, **4**, 365–368.
64. Y. J. Zhou, Z. T. Li, X. D. Chi, C. Thompson and Y. Yao, *Chem. Commun.*, 2014, **50**, 10482–10484.
65. T. Ogoshi, N. Ueshima and T.-a. Yamagishi, *Org. Lett.*, 2013, **15**, 3742–3745.
66. Y. Ma, X. Chi, X. Yan, J. Liu, Y. Yao, W. Chen, F. Huang and J. L. Hou, *Org. Lett.*, 2012, **14**, 1532–1535.
67. T. Ogoshi, R. Sueto, K. Yoshikoshi and T. A. Yamagishi, *Chem. Commun.*, 2014, **50**, 15209–15211.
68. C. Y. Han, F. Y. Ma, Z. B. Zhang, B. Y. Xia, Y. H. Yu and F. H. Huang, *Org. Lett.*, 2010, **12**, 4360–4363.
69. J. Z. Fan, H. M. Deng, J. Li, X. S. Jia and C. J. Li, *Chem. Commun.*, 2013, **49**, 6343–6345.
70. G. Yu, C. Han, Z. Zhang, J. Chen, X. Yan, B. Zheng, S. Liu and F. Huang, *J. Am. Chem. Soc.*, 2012, **134**, 8711–8717.
71. W. Xia, X. Y. Hu, Y. Chen, C. Lin and L. Y. Wang, *Chem. Commun.*, 2013, **49**, 5085–5087.
72. G. C. Yu, X. R. Zhou, Z. B. Zhang, C. Y. Han, Z. W. Mao, C. Y. Gao and F. H. Huang, *J. Am. Chem. Soc.*, 2012, **134**, 19489–19497.
73. P. Wang, X. Z. Yan and F. H. Huang, *Chem. Commun.*, 2014, **50**, 5017–5019.
74. G. Yu, M. Xue, Z. Zhang, J. Li, C. Han and F. Huang, *J. Am. Chem. Soc.*, 2012, **134**, 13248–13251.
75. Y. Cao, X. Y. Hu, Y. Li, X. Zou, S. Xiong, C. Lin, Y. Z. Shen and L. Wang, *J. Am. Chem. Soc.*, 2014, **136**, 10762–10769.
76. D. Y. Xia, G. C. Yu, J. Y. Li and F. H. Huang, *Chem. Commun.*, 2014, **50**, 3606–3608.
77. Q. P. Duan, Y. Cao, Y. Li, X. Y. Hu, T. X. Xiao, C. Lin, Y. Pan and L. Y. Wang, *J. Am. Chem. Soc.*, 2013, **135**, 10542–10549.

78. W. Chen, Y. Zhang, J. Li, X. Lou, Y. Yu, X. Jia and C. Li, *Chem. Commun.*, 2013, **49**, 7956–7958.
79. T. Ogoshi, K. Kida and T. Yamagishi, *J. Am. Chem. Soc.*, 2012, **134**, 20146–20150.
80. X. D. Chi, M. Xue, Y. J. Ma, X. Z. Yan and F. H. Huang, *Chem. Commun.*, 2013, **49**, 8175–8177.
81. C. Y. Han, L. Y. Gao, G. C. Yu, Z. B. Zhang, S. Y. Dong and F. H. Huang, *Eur. J. Org. Chem.*, 2013, 2529–2532.
82. X. B. Hu, Z. X. Chen, L. Chen, L. Zhang, J. L. Hou and Z. T. Li, *Chem. Commun.*, 2012, **48**, 10999–11001.
83. J. Fan, Y. Chen, D. Cao, Y.-W. Yang, X. Jia and C. Li, *RSC Adv.*, 2014, **4**, 4330–4333.
84. T. Ogoshi, N. Ueshima, F. Sakakibara, T. Yamagishi and T. Haino, *Org. Lett.*, 2014, **16**, 2896–2899.

CHAPTER 6

Assembled Structures of Pillar[n]arenes

TOMOKI OGOSHI* AND TADA-AKI YAMAGISHI

Graduate School of Natural Science and Technology, Kanazawa University, Kakuma-machi, Kanazawa 920-1192, Japan
*Email: ogoshi@se.kanazawa-u.ac.jp

6.1 Introduction

Investigating the assembled structures of macrocyclic compounds in the solid state is very important because such studies reveal the optical properties, mechanical and thermal stabilities in macrocyclic-based bulk materials, and the complexation ability of macrocyclic compounds. Assembled structures of macrocyclic compounds, such as cyclodextrins (CDs),[1–4] calix[*n*]arenes[5–8] and cucutbit[*n*]urils (CB[*n*]),[9–12] in the crystal state have been investigated, and their assembled structures were shown to relate closely to the shape of the macrocyclic compounds. Figure 6.1 shows the typical assembled structures of CDs.

Packing of the CD molecules in the crystal lattice occurs in two principal modes: cage-type and channel-type.[1] In the case-type packing structure, adjacent CD molecules are oriented perpendicularly to block the cavity, giving a herringbone structure. In the channel-type packing structure, preformed CD dimers are formed by two neighboring CDs assembled into infinite one-dimensional channels. The types of solvent and guest molecules dominate which structures will be selectively formed. When 1-propanol, methanol, iodine and krypton are used as guests for α-CD, the these guest molecules are small enough to be fully included in the cavity.[1]

Monographs in Supramolecular Chemistry No. 18
Pillararenes
Edited by Tomoki Ogoshi
© The Royal Society of Chemistry 2016
Published by the Royal Society of Chemistry, www.rsc.org

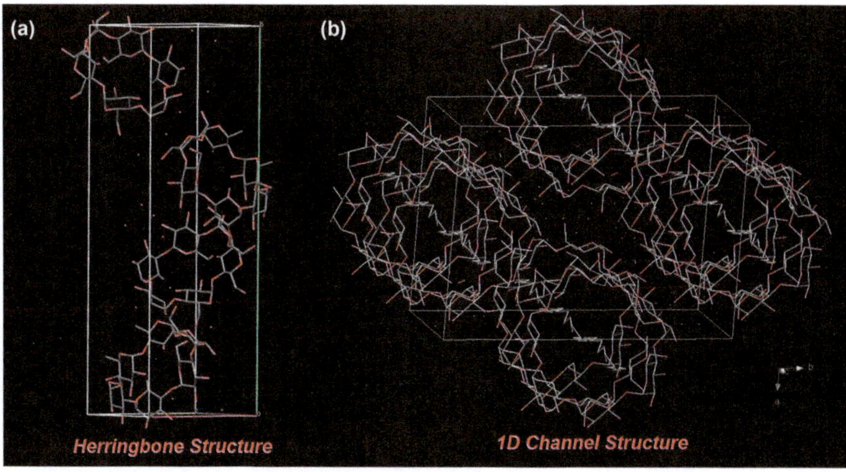

Figure 6.1 Crystal structures of CDs. (a) Herringbone structure of α-CD⊃methanol complexes and (b) one-dimensional channel structure of β-CD⊃poly(trimethylene oxide) complexes.

Therefore, α-CD molecules assemble into a herringbone structure (Figure 6.1a). In contrast, when the guest is too large to be included in the cavity, to cover the un-included part of the guest molecules, the assembled structures from CDs arrange into one-dimensional channels. Harada *et al.* reported the formation of one-dimensional channels from β-CD molecules by inclusion of a polymeric chain. Inclusion of poly(trimethylene oxide) in β-CD cavities triggers the formation of a one-dimensional channel structure (Figure 6.1b).[13] In the crystal structure of CB[*n*]s, as with CDs, CB[*n*] forms herringbone and one-dimensional channel structures. The CB[5] molecules form a herringbone structure, and both ends of the cavity are closed by adjacent CB[5] molecules (Figure 6.2a).[14]

Kim *et al.* reported one-dimensional channels from CB[6] molecules by coordination of Rb cations (Figure 6.2b). Rb cations form inter-molecular connections between CB[6]molecules that result in the formation of one-dimensional channels.[15]

In this chapter, first, the assembled structures of the cyclic pentamers, pillar[5]arenes, and cyclic hexamers, pillar[6]arenes, in the crystal state are discussed. They have been most widely used because they can be obtained in relatively good yields. Pillar[5]- and pillar[6]arenes possess clear cylindrical pillar-shaped structures when compared with other macrocyclic compounds. Therefore, it is envisioned that the assembled structures of pillar[*n*]arenes in the crystal state should be easily analyzed and this pre-organized conformation will facilitate the formation of herringbone, one-dimensional channel and slipped-stacked pillar[*n*]arenes. It has been revealed that the assembled structures of pillar[*n*]arenes largely depend on the ring size, substituents and solvents used to obtain the single crystal. The assembled structures of pillar[*n*]arenes contribute to the gas and organic vapor adsorption properties.

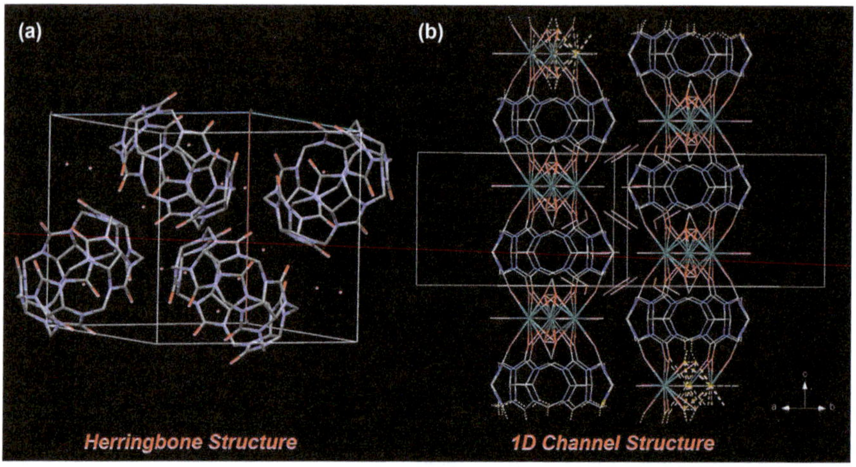

Figure 6.2 Crystal structures of CB[*n*]. (a) Herringbone crystal structure of CB[5] and (b) one-dimensional (1D) channel structure of a complex between CB[6] and Rb cation.

Second, we describe crystal state-assembled structures of pillar[5]arenes, pillar[6]arenes, and larger pillar[*n*]arene homologs ($n = 8$, 9, 10). Furthermore, the bulk materials formed using pillar[*n*]arenes are discussed.

6.2 Assembled Structures of Pillar[5]arenes in Crystal States

6.2.1 per-Alkylated Pillar[5]arenes

Single crystals of per-methylated pillar[5]arene (**6.1**), which was the first pillar[*n*]arene derivative, were obtained by the slow evaporation of acetonitrile solution (Figure 6.3).[16]

The solvent for the crystallization, acetonitrile, was included in the cavity of **6.1**. We proposed that the assembled structure of the **6.1**$\supset$acetonitrile complex would be a one-dimensional channel owing to the highly symmetrical pillar-shaped structure of pillar[*n*]arenes. Surprisingly, the inclusion complexes did not assemble into one-dimensional channels. The neighboring **6.1**$\supset$acetonitrile complexes were arranged perpendicular to each other, giving a herringbone structure (Figure 6.3).

A one-dimensional channel structure, formed by the assembly of pillar[5]arene derivatives, was first reported by Hou *et al.*[17] Pillar[5]arene deca-ester (**6.2**) formed two kinds of one-dimensional channel. Single crystals were obtained by slow evaporation of its solution in both chloroform and a mixture of chloroform and ethylene glycol. When the crystal was grown in chloroform, the molecules of deca-ester **6.2** were aligned to form infinite channels with a central pore size of 6.50 Å, in which the adjacent molecules

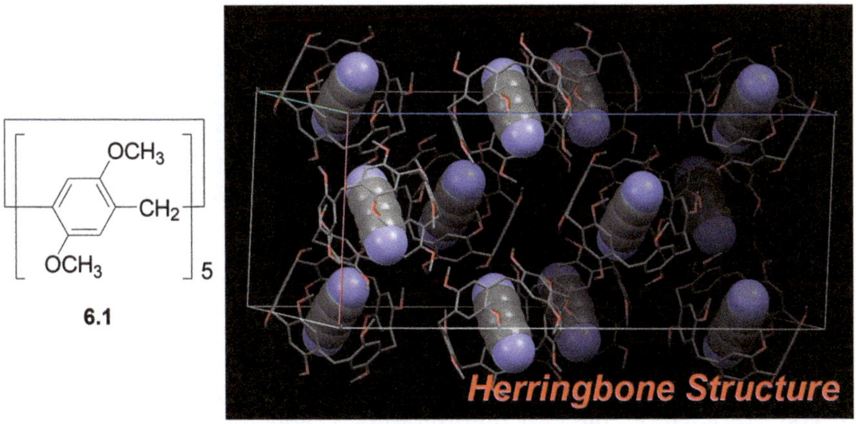

Figure 6.3 Herringbone crystal structures of **6.1** from its acetonitrile solution.

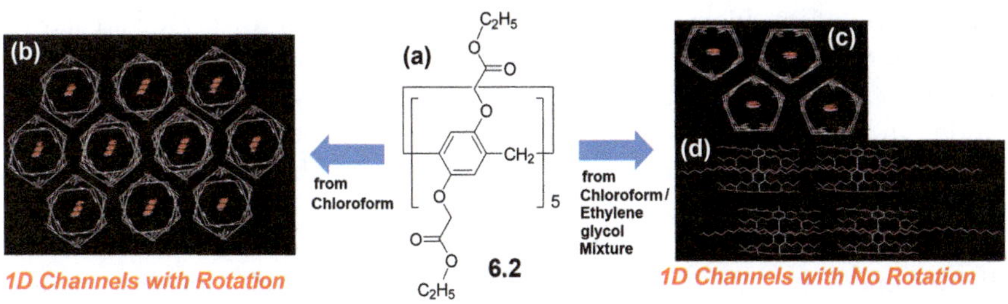

Figure 6.4 Crystal structures of **6.2**. (b) In the 1D channel crystal structure developed from chloroform (a), the adjacent molecules overlapped each other with a rotation angle of 36°. In contrast, molecules stacked in the face-to-face manner form without rotation from a mixture of chloroform and ethylene glycol (c). (d) Partial channels with encapsulated water wire.

overlapped with a rotation angle of 36°. In contrast, when the crystal was grown in a mixture of chloroform and ethylene glycol, another one-dimensional channel was formed. The one-dimensional channels did not stack and no rotation of the neighboring molecules was observed. Water wires were encapsulated in the channels in the crystal. Figure 6.4d shows a partial channel with an encapsulated water wire. Eight water molecules were enclosed per pillar[5]arene module. The distance between the adjacent oxygen atoms of the water molecules was approximately 2.6 Å, indicating that the adjacent water molecules formed inter-molecular hydrogen bonding. When anhydrous chloroform or ethylene glycol was used as a solvent for crystallization, the solids obtained were amorphous, indicating that water molecules are necessary to form the one-dimensional channels.

Simple per-alkylated pillar[5]arenes also formed one-dimensional structures. Single crystals of per-ethylated pillar[5]arene (**6.3**) suitable for X-ray analysis were grown by slow evaporation of its acetone solution (Figure 6.5).[18] In this case, two acetone molecules were included in the cavity

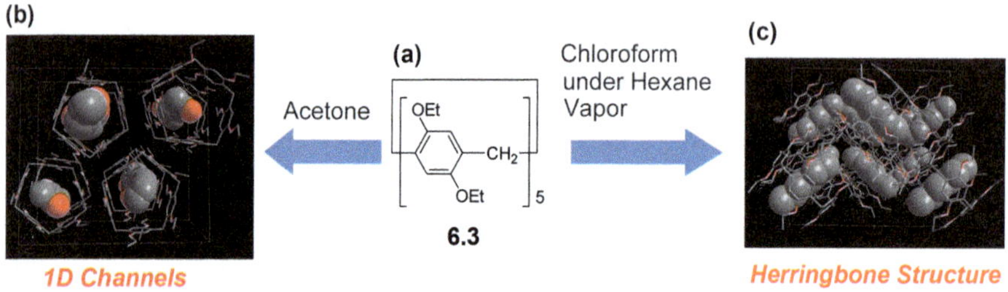

Figure 6.5 Crystal structures of **6.3**. (b) 1D channel crystal structure formed by crystallization from its acetone solution (a). (c) The crystal formed from its chloroform solution under *n*-hexane vapor forms a herringbone structure.

to connect adjacent pillar[5]arene molecules, and thus pillar[5]arene **6.3** assembled into infinite 1D channels along the a-axis. In contrast, **6.3** formed different assembled structures when crystals of **6.3** were grown by slow evaporation of its chloroform under *n*-hexane vapor. In the crystal structure, one *n*-hexane molecule was included in the cavity, which formed a pseudorotaxane. The size of *n*-hexane is sufficiently small to be fully included in the cavity, therefore adjacent **6.3**$\supset$*n*-hexane complexes should be arranged in a herringbone structure, similar to the observed packing modes of CDs.

per-Propylated pillar[5]arene (**6.4**) from its solution in a mixture of dichloromethane and acetonitrile also formed infinite one-dimensional channels (Figure 6.6a).[19] However, in the crystals of per-pentylated pillar[5]arenes (**6.5**) prepared from its acetone solution, the packing mode of the molecules showed slipped-stacked pillar[5]arene molecules.[20] In the case of per-diisobutylated pillar[5]arenes (**6.6**), three **6.6** molecules aligned in one-dimensional channels.[21] Even in the simple per-alkylated pillar[5]arenes, the type of substituents on the rims and the solvents used for crystallization dominate the assembled structures of pillar[5]arenes.

6.2.2 per-Functionalized Pillar[5]arenes

The assembled structure of per-triflated pillar[5]arene (**6.7**) in the crystal state was reported by Yang and co-workers.[22] Pillar[5]arenes **6.7** comprised a close-packed stacked structure and formed one-dimensional channels. However, owing to the bulky substituents of triflate moieties, the diameter of the one-dimensional channels was very narrow (3.15 Å) when compared with those of simple per-alkylated pillar[5]arenes, including per-ethylated (**6.2**) and -propylated pillar[5]arenes (**6.4**).[18,19] The assembled structure of per-carboxylic acid pillar[5]arene (**6.8**) was reported by Danylyuk and Sashuk.[23] Decacarboxylic acid (**6.8**) formed an inclusion complex with ethanol in a 1 : 2 stoichiometry that assembled into one-dimensional channels. Carboxylic acid groups on the rims contributed to the formation of one-dimensional channels (Figure 6.7).

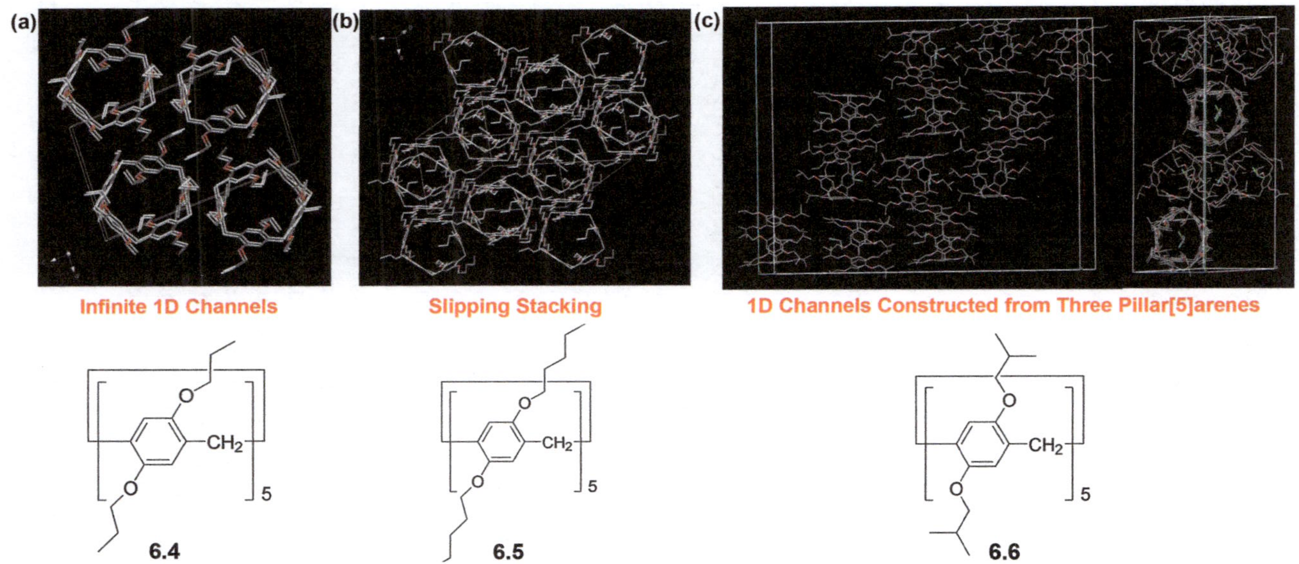

Figure 6.6 Crystal structures of other per-alkylated pillar[5]arenes: **6.4**, **6.5** and **6.6**.

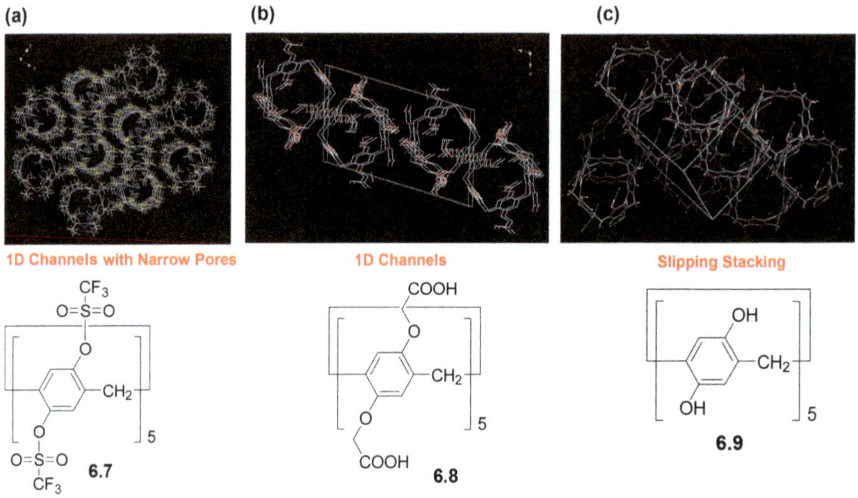

Figure 6.7 Crystal structures of per-functionalized pillar[5]arenes: **6.7**, **6.8** and **6.9**.

As previously mentioned in Chapter 4, per-hydroxylated pillar[5]arenes (**6.9**) form rigid structures owing to the intra-molecular hydrogen bonds. However, because of the odd number of benzene rings in the cyclic pentagonal structure, decanol (**6.9**) cannot form perfect hydrogen bonds in an alternating up-and-down manner. Therefore, **6.9** also forms inter-molecular hydrogen bonds, and the inter-molecular hydrogen bonds cause a slipped-stacked structure for **6.9**.[24]

6.2.3 A1/A2 Pillar[5]arenes

A1/A2 pillar[5]arenes, which have two functional groups in one repeating unit, can adopt various assembled structures. There were two crystal structures of pillar[5]arene derivatives containing one benzoquinone unit. A pillar[5]arene consisting of one benzoquinone and four diethoxybenzene units (**6.10**) assembled into one-dimensional channels by crystallization from its methanol solution.[25] However, the assembled structure of a pillar[5]arene comprising one benzoquinone and four dimethoxybenzene units (**6.11**) in the crystal state was a herringbone formation.[26] The assembled structure of a pillar[5]arene containing one hydroquinone and four 1,4-dimethoxybenzene units (**6.12**) was also a herringbone, while a pillar[5]arene with one π-conjugated unit (**6.13**) formed a 1D channel structure (Figure 6.8).[27]

6.2.4 Pillar[5]arenes with Two and Three Benzoquinone Units

Crystal structures of pillar[5]arenes with two benzoquinones at A/B (**6.14**) and A/C units (**6.15**) were reported by Pan and Xue.[28] The assembled

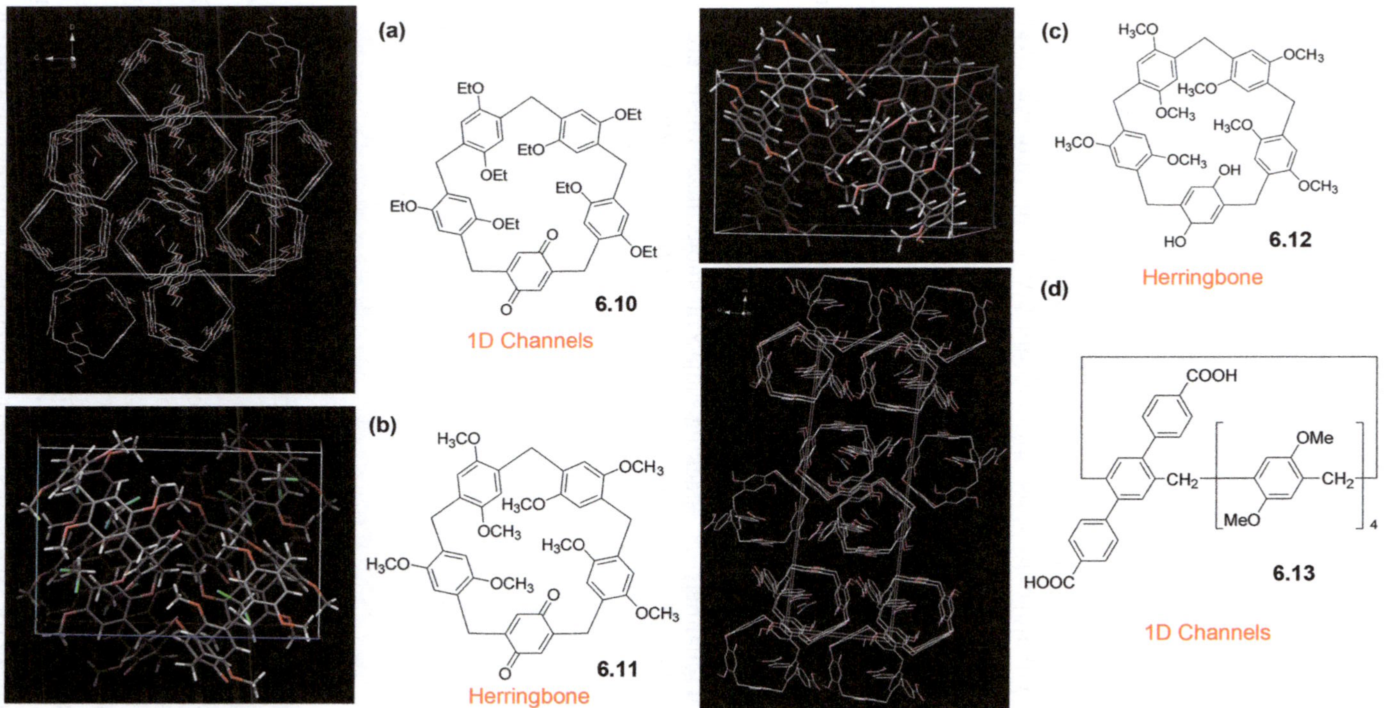

Figure 6.8 Pillar[5]arenes (a) with four 1,4-diethoxybenzene units and one benzoquinone unit (**6.10**), (b) four 1,4-dimethoxybenzene units and one benzoquinone unit (**6.11**), (c) four 1,4-dimethoxybenzene units and one hydroquinone unit (**6.12**) and (d) four 1,4-dimethoxybenzene units and one rigid conjugated unit (**6.13**).

structure of a pillar[5]arene containing two benzoquinones at A/B units obtained from dichloromethane/acetonitrile solution was a one-dimensional tubular structure with slight slipped stacking. In contrast, in the case of a pillar[5]arene with two benzoquinones at A/C units, adjacent two pillar[5]-arene molecules were stacked along the b-axis. The stacking resulted in the formation of a one-dimensional tubular dimer structure. However, these tubular dimers did not assemble to form one-dimensional channels. The position of the benzoquinone units affected the assembled structure of pillar[5]arenes (Figure 6.9).

Pan and Xue also synthesized pillar[5]arenes with three different kinds of substituent unit.[29] Pillar[5]arenes containing one (**6.16**) and two (**6.17**) benzoquinone units assembled into one-dimensional channels by the stacking of pillar[5]arene molecules (Figure 6.10). The channels arranged into honeycomb-like structures. In contrast, two overlapped tubes were observed in pillar[5]arene with three benzoquinone units (**6.18**).

6.2.5 Non-symmetrical Pillar[5]arenes

Four constitutional isomers were obtained by cyclization of 1,4-dialkoxybenzene monomers with different alkoxy substituents. Huang *et al.* reported the crystal structures of four isomers prepared from 1-butoxy-4-methoxybenzene (**6.19**; Figure 6.11).[30] All four isomers formed one-dimensional channels by stacking adjacent molecules. There are two stacking modes. Isomers **1** and **3** formed a tubular structure with no rotation, while isomers **2** and **4** assembled into one-dimensional channels with a rotation of 72°.

Huang *et al.* also reported the assembled structures of an amphiphilic pillar[5]arene having all five ester groups on the same side (**6.20**) in the crystal state.[31] The amphiphilic pillar[5]arene also formed one-dimensional channels with a rotation angle of 36° (Figure 6.12). The trend is the same as for the non-symmetrical pillar[5]arenes having all five alkoxy groups on the same side, and for a pillar[5]arene with 10 ester moieties.

6.2.6 Host–Guest Complexes between Pillar[5]arenes and Guests

Host–guest complexes also form various assembled structures in the solid state. Host–guest complexes between per-ethylated pillar[5]arene **6.3** and 1,4-dichlorobutane were arranged perpendicular to each other (Figure 6.13a, herringbone structure).[32]

The complexes between per-methylated pillar[5]arene (**6.1**) and 1,6-dibromohexane also did not arrange in a row (Figure 6.13b).[32] The molecular stacking mode of the complexes between paraquat and mono(ethylene oxide)-functionalized pillar[5]arene **6.21** was slipped-stacked (Figure 6.13c).[33]

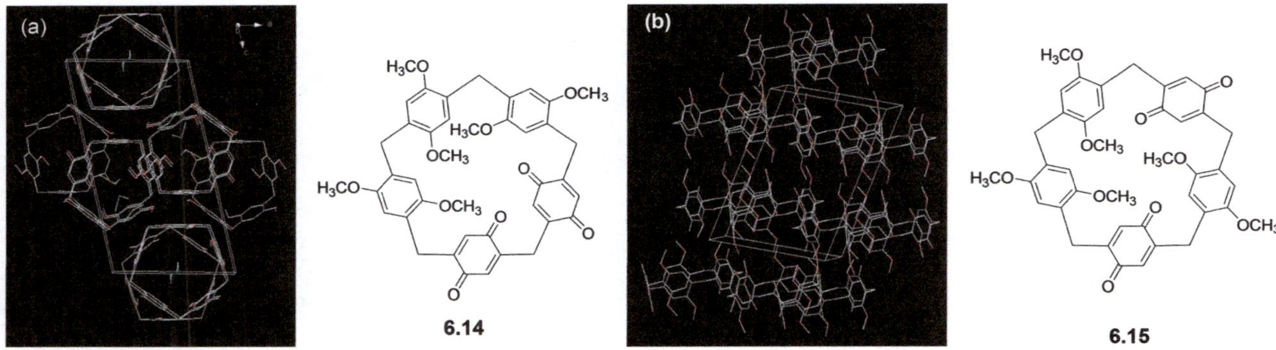

6.14

6.15

Figure 6.9 Crystal structures of pillar[5]arenes with two benzoquinones at (a) A/B (**6.14**) and (b) A/C units (**6.15**).

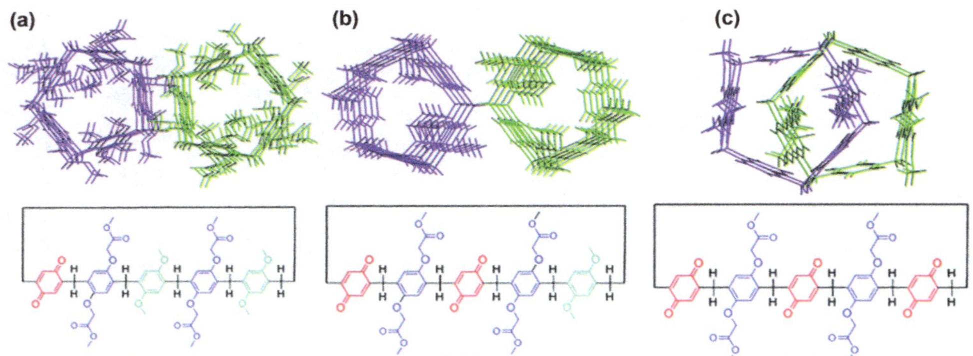

6.16 6.17 6.18

Figure 6.10 Crystal structures of pillar[5]arenes with one (**6.16**), two (**6.17**) and three (**6.18**) benzoquinone units. Reproduced from ref. 29 with permission from The Royal Society of Chemistry.

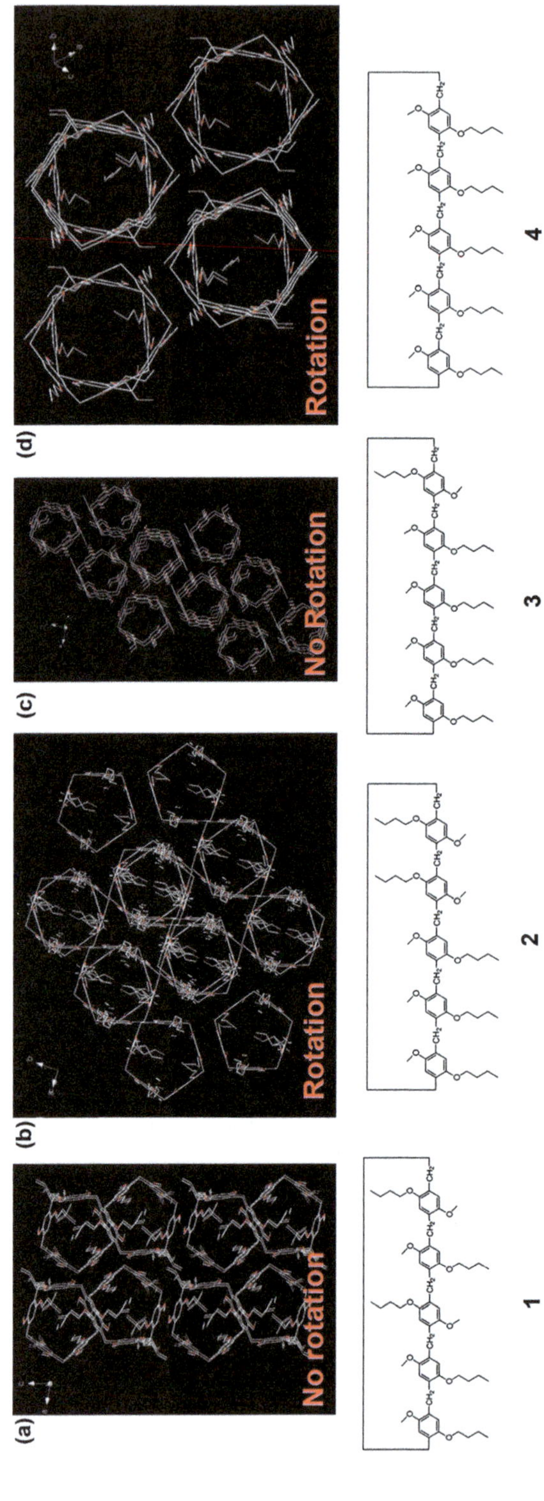

Figure 6.11 Crystal structures of non-symmetirc pillar[5]arene (6.19).

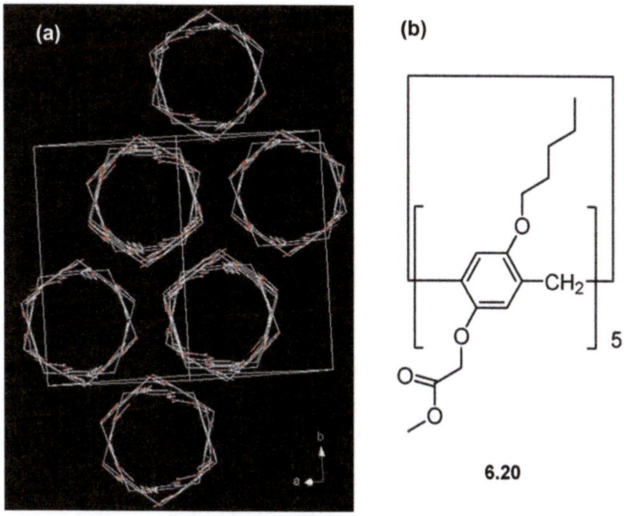

Figure 6.12 Crystal structure of amphiphilic non-symmetrical pillar[5]arene (**6.20**).

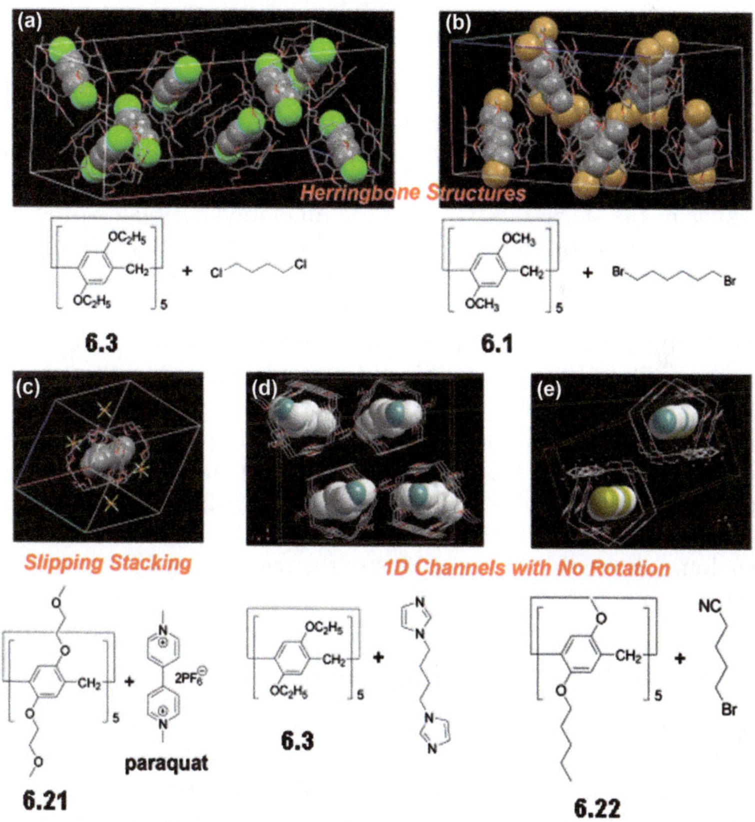

Figure 6.13 Crystal structures of host–guest complexes between pillar[5]arene derivatives and guests.

Formation of one-dimensional channels with no rotation was observed in the complexes between per-ethylated pillar[5]arene **6.3** and 1,4-bis(imidazol-1-yl)butane (Figure 6.13d).[34] The assembly of host–guest complexes of 5-bromovaleronitrile and a non-symmetrical pillar[5]arene with five methyls on one side and five *n*-pentyls on the other (**6.22**) also formed one-dimensional channels (Figure 6.13e).[35] The application incorporating the alignment of these guest molecules in the one-dimensional channels constructed from pillar[5]arenes is a topic for future investigation.

6.3 Pillar[6]arenes

We show examples of the assembled structures of pillar[5]arene molecules in the solid state. They mainly show three assembled structures: one-dimensional channels, slipped-stacked and herringbone structures. Considering the structural difference between cyclic pentagonal pillar[5]arenes and hexagonal pillar[6]arenes, the assembly of pillar[6]arenes is more regular than that of pillar[5]arenes. This is because pillar[6]arenes are highly symmetrical hexagonal structures.

6.3.1 per-Alkylated Pillar[6]arenes

The crystal structures of pillar[6]arenes were first reported by Huang *et al.*[21] They synthesized per-diisobutylated pillar[6]arene (**6.23**), then obtained single crystals of **6.23** by slow evaporation of its solution in a mixture of dichloromethane and acetonitrile (Figure 6.14).

Because of the steric hindrance of the diisobutyl (DIB) substituents, the structure was a distorted hexagon. Nevertheless, the units assembled into one-dimensional channel structures. In the case of the cyclic pentagonal molecule, per-diisobutylated pillar[5]arene **6.6**, the infinite one-dimensional channels did not form. This also supports the idea that hexagonal molecules, pillar[6]arenes, exhibit more regular arrangements than pentagonal molecules, pillar[5]arenes. Huang *et al.* also reported the single X-ray crystal structure of per-pentylated pillar[6]arene (**6.24**).[19] Owing to the low steric hindrance of the pentyl moieties compared with DIB moieties, the pentagonal structure of **6.24** is more of a regular hexagonal structure when compared with **6.23**. Therefore, **6.24** also formed regular one-dimensional channels by the stacking of the regular pentagonal assemblies of **6.24**.

6.3.2 per-Hydroxylated Pillar[6]arene

per-Hydroxylated pillar[6]arene (**6.25**) has 12 OH moieties and an even number of benzene rings, and therefore forms multiple intra-molecular hydrogen bonds in an alternating up-and-down manner, as previously mentioned in Chapter 4. The multiple intra-molecular hydrogen bonds provide **6.25** with a rigid structure. Assembly of the rigid hexagonal

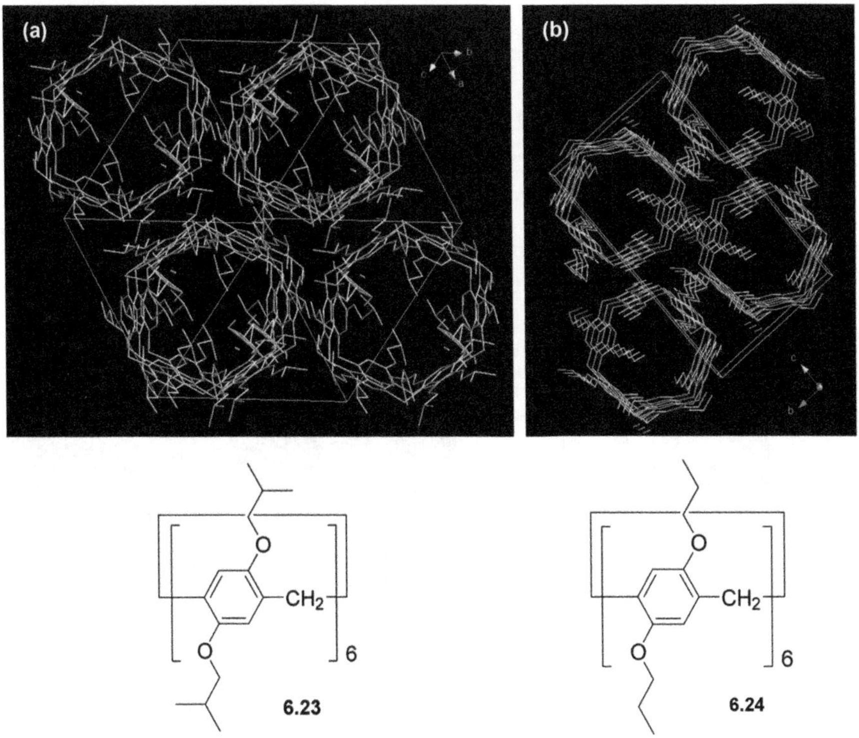

Figure 6.14 Crystal structures of simple per-alkylated pillar[6]arenes: **6.23** and **6.24.**

scaffold forms a hexagonal arrangement and has infinite 1D channels along the c-axis (Figure 6.15), which are derived from the cavity size of pillar[6]arene.[36]

Because the assembled structure was stabilized by hydrogen bonding, the one-dimensional channels were stable even after removing the remaining solvents. The porous organic crystals, consisting of one-dimensional channels, can adsorb gas and organic vapors.[37]

6.3.3 Pillar[6]arene with One Benzoquinone Unit

The assembled structure of pillar[6]arene with one benzoquinone unit (**6.26**; Figure 6.16) was reported by Huang *et al.*[38]

In this case, even with the highly symmetrical hexagonal pillar[6]arene structure, the molecules were slipped-stacked. This is because either the introduction of one benzoquinone unit reduces the symmetry of pillar[6]-arene, or the formation of the inter-molecular charge-transfer complex between the benzoquinone and 1,4-dipropoxybenzene units disturbs the regular hexagonal assembly of pillar[6]arene molecules.

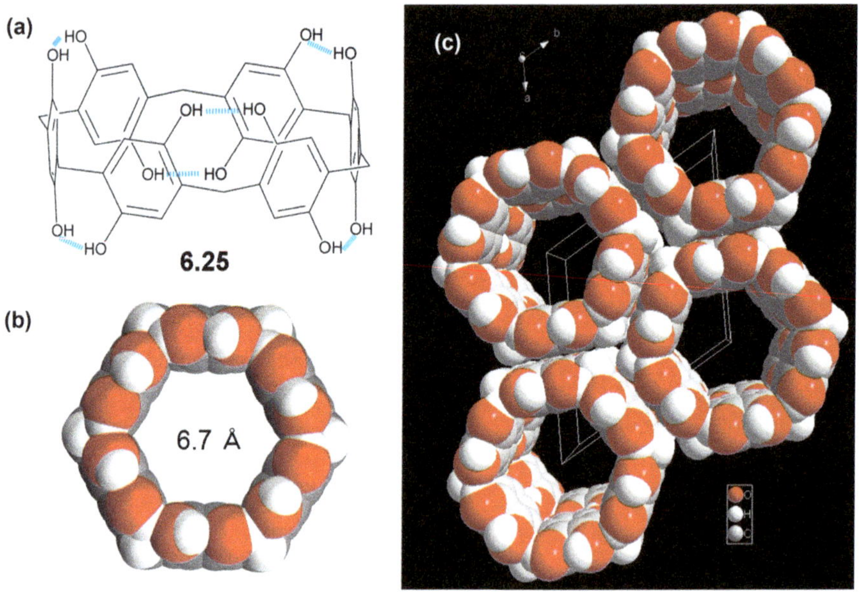

Figure 6.15 (a) Chemical and (b, c) X-ray crystal structures of per-hydroxylated pillar[6]arene **6.25**. The solvent (acetone) in the structure was removed. (c) View along the c axis.
Reproduced from ref. 37 with permission from The Royal Society of Chemistry.

6.3.4 Host–Guest Complexes between Pillar[6]arenes and Guests

There are a few examples of the crystal structures of host–guest complexes between pillar[6]arene and guests. The crystal structure of the 1 : 1 complex between per-hydroxylated pillar[6]arene **6.25** and paraquat was reported by Huang and Hou *et al.*[36]

The complexes were arranged into infinite one-dimensional channels. The 1 : 1 host–guest complexes between mono(ethylene oxide)-functionalized pillar[6]arene (**6.27**) and diquat also formed one-dimensional channels.[39] In both cases, the guest molecules, paraquat and diquat, are aligned in a linear fashion. Pillar[6]arene molecules are predisposed to assemble into one-dimensional channels, therefore they are good candidates to align guest molecules in their one-dimensional channels (Figure 6.17).

6.3.5 Two-dimensional Sheets from per-Hydroxylated Pillar[6]arene

The assembly of hexagonal molecules is a good method for obtaining well-defined supramolecular architectures, and is known as molecular tiling. However, the concept has mainly been applied to planar polyaromatic

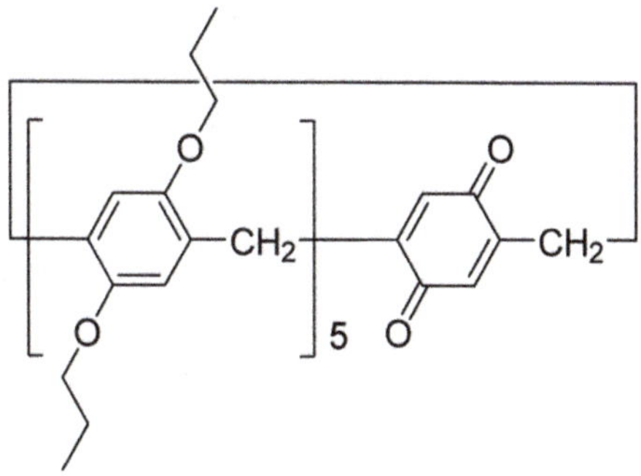

6.26

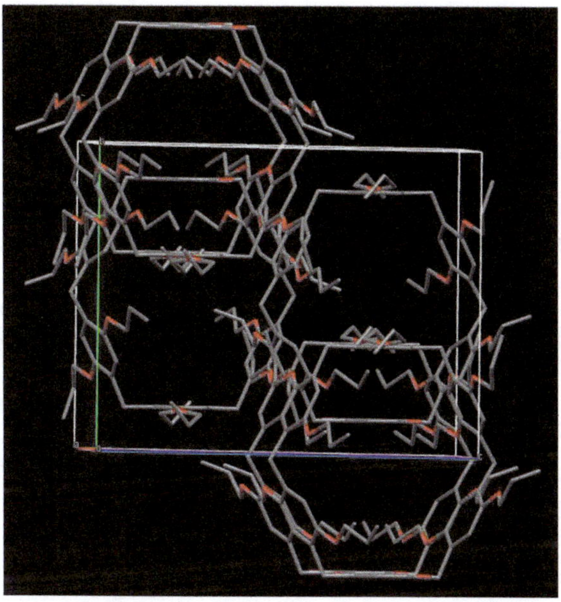

Figure 6.16 Crystal structure of a per-alkylated pillar[6]arene containing one benzoquinone unit (**6.26**).

compounds. We speculated that the assembly of the three-dimensional hexagonal structure of pillar[6]arene would be a potential candidate to construct two-dimensional sheets, because pillar[6]arene molecules likely assemble into a hexagonal close-packed structure in the crystal state. To obtain two-dimensional sheets, two-dimensional supramolecular polymerization was investigated using pillar[6]arene as the monomer.[40] Therefore, we employed charge-transfer complex formation between hydroquinone and benzoquinone (quinhydrone). The formation of intermolecular charge-transfer complexes produced two-dimensional supramolecular polymerization of pillar[6]arene molecules (Figure 6.18b). Owing to the hexagonal structure of **6.25**, assembly from inter-molecular

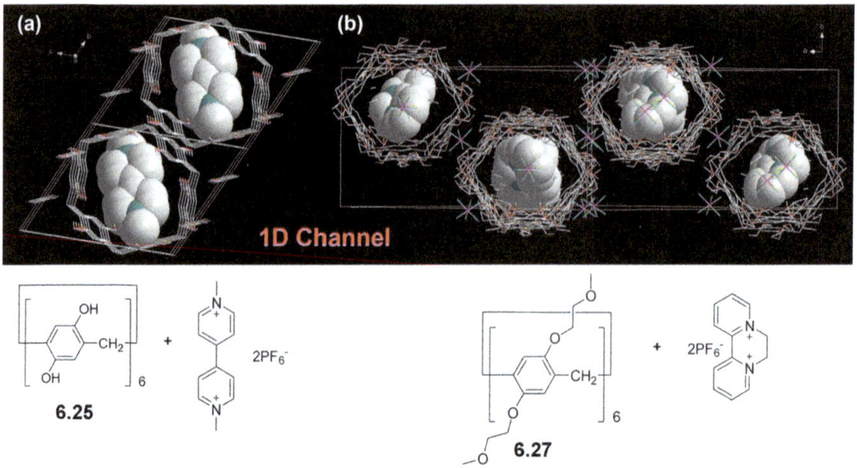

Figure 6.17 X-ray crystal structures of host–guest complexes between pillar[6]arene derivatives and guests.

charge-transfer complex formation, induced by the two-dimensional supramolecular polymerization, gave two-dimensional porous sheets by hexagonal close packing of the charge-transfer complex, and the highly ordered structure of the two-dimensional sheets formed regular fibrous structures.

However, the assembly of pentagonal pillar[5]arene **6.9**, induced by oxidation, afforded an amorphous structure because packing of the pentagonal molecules cannot afford a well-defined structure (Figure 6.18a).

6.4 Liquid and Liquid Crystal Pillar[5,6]arenes

Pillar[5]- and pillar[6]arenes are rigid and highly symmetrical structures, therefore pillar[5]- and pillar[6]arene derivatives are usually in a solid state at room temperature. However, incorporation of functional groups at the rims of pillar[5]- and pillar[6]arene derivatives can change their physical properties from the solid to a liquid state. Ionic liquid moieties are able to influence the state of matter of pillar[5]arenes. For example, the appearances of pillar[5]arenes with 10 imidazolium hexafluorophosphate (**6.28**) and 10 imidazolium bis(trifluoromethanesulfonyl)amide (**6.29**) were a white solid and a slightly yellowish liquid, respectively (Figure 6.19).[41]

The unit model with two imidazolium bis(trifluoromethanesulfonyl)amide moieties was a white solid, indicating that the pillar-shaped macrocyclic structure and modification with 10 imidazolium bis(trifluoromethanesulfonyl)-amide groups contributed to the liquid state at room temperature. The liquid pillar[5]arene **6.29** showed moderate ionic conductivity (6.00×10^{-6} S cm^{-1}) at 324 K. The introduction of soft and flexible tri(ethylene oxide) moieties also

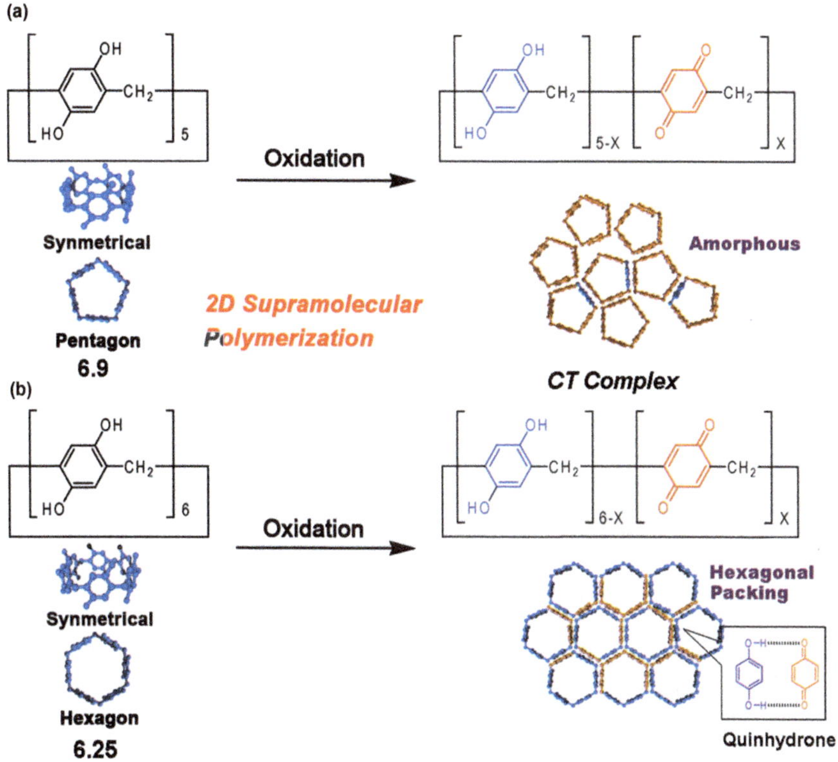

Figure 6.18 (a) Pentagonal and hexagonal structures of per-hydroxylated pillar[5]-arene **6.9** and pillar[6]arene **6.25**. (b) 2D supramolecular polymerization by oxidation of hydroquinone units.
Reproduced with permission from ref. 40. Copyright 2015, Wiley-VCH Verlag Gmbh & Co. KGaA, Weinheim.

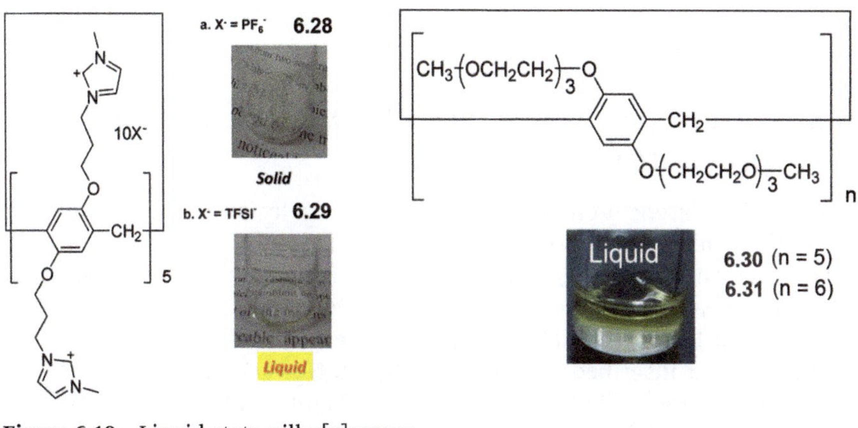

Figure 6.19 Liquid state pillar[*n*]arenes.

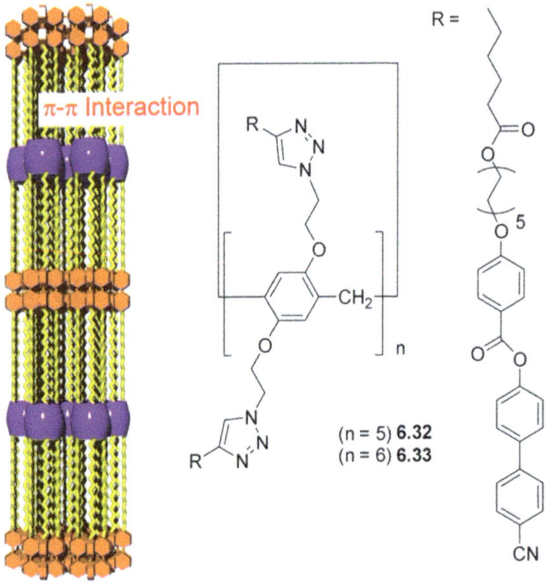

Figure 6.20 Ionic liquid pillar[*n*]arenes.

changes the state of pillar[5]- (**6.30**) and pillar[6]arenes (**6.31**) from solid to liquid.[42,43] They are in the liquid state even at −50 °C, and are stable up to 250 °C. Therefore, they can be used as solvents for chemical reactions over a wide range of temperatures. By mixing guest molecules into these liquid pillar[5]arenes, complexation occurred even in the absence of additional solvents.

Nierengarten *et al.* reported liquid crystalline pillar[5]- and pillar[6]arenes.[44] They synthesized pillar[5]arene **6.32** and pillar[6]arene **6.33** with 10 and 12 mesogenic cyanobiphenyl units, respectively (Figure 6.20). Both compounds showed smectic A phase (focal-conic fan texture and homotropic areas) over a broader temperature range when compared with the model compound. This can be explained as the macrocyclic effect: Inter-molecular π–π interactions between neighboring pillar[*n*]arene cores stabilized the smectic A phase.

6.5 Larger per-Ethylated Pillar[*n*]arene Homologs ($n = 8$–10)

Large pillar[*n*]arene homologs possess large cavity sizes. In the solution state, they can form various conformations because flipping of the units takes place and the C–CH_2–C angle at the methylene linkers becomes larger. Thus, pillar[*n*]arene homologs ($n = 5$–14) have various cavity sizes ranging from 0.47 to 1.49 nm, which roughly calculate their diameters based on the diameter of the inscribed circle of the polygon. Hou *et al.* reported the crystal structures of large pillar[*n*]arene homologs (Figure 6.21) [$n = 8$ (**6.34**), $n = 9$ (**6.35**), $n = 10$ (**6.36**)].[45]

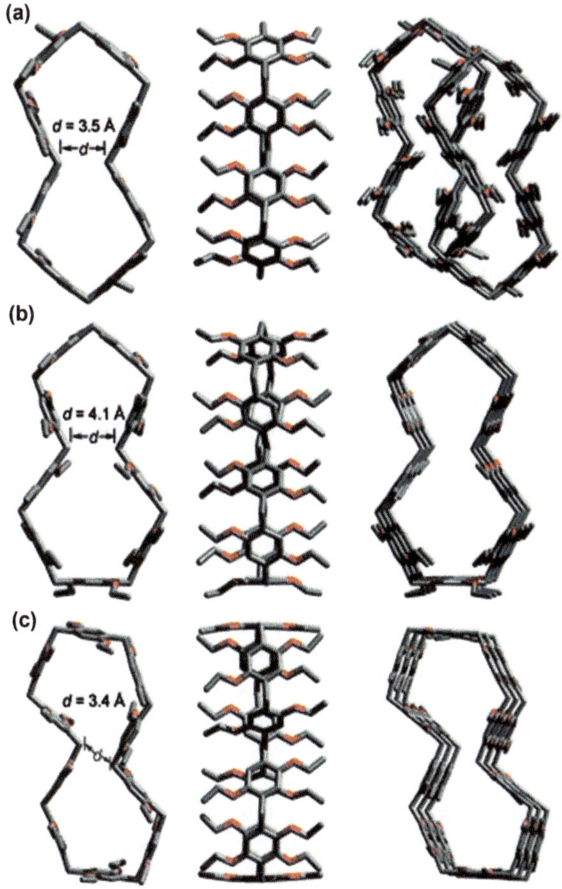

Figure 6.21 X-ray crystal structures of (a) per-ethylated pillar[8]arene **6.34**, (b) pillar[9]arene **6.35** and (c) pillar[10]arene **6.36**. Left: upper view; middle: side view; right: packing model.
Reproduced from ref. 45 with permission from The Royal Society of Chemistry.

In these large pillar[*n*]arene homologs, two of the methylene groups were oriented inwardly, most likely to release the ring strain. This orientation resulted in these large pillar[*n*]arene homologs ($n = 8$, 9, 10) producing two cavities, which is very different from the small pillar[*n*]arenes ($n = 5$ and 6), which have a single cavity. Large pillar[*n*]arene homologs ($n = 8$, 9, 10) have two pentagons, one pentagon and one hexagon, and two hexagons, respectively, reflecting their different numbers of repeating units. The same trend is also observed in calix[*n*]arene chemistry. Owing to the structural stability of tetragonal calix-shaped structures, calix[8]arene has two cavities of tetragonal calix-shaped structures.

The distances between the two inwardly oriented methylene carbons in large pillar[*n*]arene homologs ($n = 8$, 9, 10) were 3.5, 4.1, and 3.4 Å, respectively, which are all shorter than the distances between the two adjacent methylene carbons of pillar[5]arene (5.8 Å) or pillar[6]arene (5.9 Å). Thus, the sizes of the pentagon and hexagon cavities in pillar[8]-, pillar[9]- and pillar[10]arenes are smaller than those of the single pentagonal or hexagonal cavity in pillar[5]- and pillar[6]arenes.

The assembled structures of these large pillar[*n*]arene homologs are shown in the right-hand column of Figure 6.21. Pillar[9]- and pillar[10]arene assembled to form one-dimensional channels, but the packing mode of pillar[8]arene exhibited slipped stacking of pillar[8]arene molecules.

6.6 Conclusion and Outlook

In this chapter, we have discussed the assembled structures of pillar[*n*]arenes in the crystal state. Packing of pillar[*n*]arenes mainly occurs in three principal modes: one-dimensional channels, slipped-stacked and herringbone structures. In the case of pillar[5]arenes, the types of solvent, substituents on the rims, and additional guests dominate their assembled structures. In contrast, pillar[6]arenes form highly regular hexagonal close-packed structures because pillar[6]arenes have a highly symmetrical hexagonal structure, when compared with pillar[5]arenes. The application of the assembled structures of pillar[*n*]arenes should be one of the next challenges for research in the field of pillar[*n*]arene chemistry.

References

1. K. Harata, *Chem. Rev.*, 1998, **98**, 1803–1828.
2. R. A. Smaldone, R. S. Forgan, H. Furukawa, J. J. Gassensmith, A. M. Z. Slawin, O. M. Yaghi and J. F. Stoddart, *Angew. Chem., Int. Ed.*, 2010, **49**, 8630–8634.
3. J. J. Gassensmith, H. Furukawa, R. A. Smaldone, R. S. Forgan, Y. Y. Botros, O. M. Yaghi and J. F. Stoddart, *J. Am. Chem. Soc.*, 2011, **133**, 15312–15315.
4. R. S. Forgan, R. A. Smaldone, J. J. Gassensmith, H. Furukawa, D. B. Cordes, Q. Li, C. E. Wilmer, Y. Y. Botros, R. Q. Snurr, A. M. Z. Slawin and J. F. Stoddart, *J. Am. Chem. Soc.*, 2012, **134**, 406–417.
5. J. L. Atwood, L. J. Barbour and A. Jerga, *Science*, 2002, **296**, 2367–2369.
6. S. J. Dalgarno, G. W. V. Cave and J. L. Atwood, *Angew. Chem., Int. Ed.*, 2006, **45**, 570–574.
7. H. Mansikkamäki, M. Nissinen and K. Rissanen, *Angew. Chem., Int. Ed.*, 2004, **43**, 1243–1246.
8. B. H. Hong, S. C. Bae, C.-W. Lee, S. Jeong and K. S. Kim, *Science*, 2001, **294**, 348–351.
9. E. Lee, J. Kim, J. Heo, D. Whang and K. Kim, *Angew. Chem., Int. Ed.*, 2001, **40**, 399–402.
10. S. Lim, H. Kim, N. Selvapalam, K.-J. Kim, S. J. Cho, G. Seo and K. Kim, *Angew. Chem., Int. Ed.*, 2008, **47**, 3352–3355.

11. H. Kim, Y. Kim, M. Yoon, S. Lim, S. M. Park, G. Seo and K. Kim, *J. Am. Chem. Soc.*, 2010, **132**, 12200–12202.
12. M. Yoon, K. Suh, H. Kim, Y. Kim, N. Selvapalam and K. Kim, *Angew. Chem., Int. Ed.*, 2011, **50**, 7870–7873.
13. S. Kamitori, O. Matsuzaka, S. Kondo, S. Muraoka, K. Okuyama, K. Noguchi, M. Okada and A. Harada, *Macromolecules*, 2000, **33**, 1500–1502.
14. J. Kim, I. S. Jung, S. Y. Kim, E. Lee, J. K. Kang, S. Sakamoto, K. Yamaguchi and K. Kim, *J. Am. Chem. Soc.*, 2000, **122**, 540–541.
15. J. Heo, S. Y. Kim, D. Whang and K. Kim, *Angew. Chem., Int. Ed.*, 1999, **38**, 641–643.
16. T. Ogoshi, S. Kanai, S. Fujinami, T. Yamagishi and Y. Nakamoto, *J. Am. Chem. Soc.*, 2008, **130**, 5022–5023.
17. W. Si, X. B. Hu, X. H. Liu, R. H. Fan, Z. X. Chen, L. H. Weng and J. L. Hou, *Tetrahedron Lett.*, 2011, **52**, 2484–2487.
18. T. Ogoshi, R. Sueto, K. Yoshikoshi, Y. Sakata, S. Akine and T. Yamagishi, *Angew. Chem., Int. Ed.*, 2015, **54**, 9849–9852.
19. G. Yu, C. Han, Z. Zhang, J. Chen, X. Yan, B. Zheng, S. Liu and F. Huang, *J. Am. Chem. Soc.*, 2012, **134**, 8711–8717.
20. T. Ogoshi, K. Kitajima, T. Aoki, S. Fujinami, T. Yamagishi and Y. Nakamoto, *J. Org. Chem.*, 2010, **75**, 3268–3273.
21. C. Y. Han, F. Y. Ma, Z. B. Zhang, B. Y. Xia, Y. H. Yu and F. H. Huang, *Org. Lett.*, 2010, **12**, 4360–4363.
22. L. L. Tan, H. W. Li, Y. C. Tao, S. X. A. Zhang, B. Wang and Y. W. Yang, *Adv. Mater.*, 2014, **26**, 7027–7031.
23. O. Danylyuk and V. Sashuk, *CrystEngComm*, 2015, **17**, 719–722.
24. T. Ogoshi, T. Aoki, K. Kitajima, S. Fujinami, T. Yamagishi and Y. Nakamoto, *J. Org. Chem.*, 2011, **76**, 328–331.
25. T. Ogoshi, D. Yamafuji, D. Kotera, T. Aoki, S. Fujinami and T. Yamagishi, *J. Org. Chem.*, 2012, **77**, 11146–11152.
26. C. Y. Han, Z. B. Zhang, G. C. Yu and F. H. Huang, *Chem. Commun.*, 2012, **48**, 9876–9878.
27. N. L. Strutt, D. Fairen-Jimenez, J. Iehl, M. B. Lalonde, R. Q. Snurr, O. K. Farha, J. T. Hupp and J. F. Stoddart, *J. Am. Chem. Soc.*, 2012, **134**, 17436–17439.
28. M. G. Pan and M. Xue, *Eur. J. Org. Chem.*, 2013, 4787–4793.
29. M. G. Pan and M. Xue, *Rsc Adv.*, 2014, **4**, 260–263.
30. Z. Zhang, Y. Luo, B. Xia, C. Han, Y. Yu, X. Chen and F. Huang, *Chem. Commun.*, 2011, **47**, 2417–2419.
31. Y. Yao, M. Xue, J. Z. Chen, M. M. Zhang and F. H. Huang, *J. Am. Chem. Soc.*, 2012, **134**, 15712–15715.
32. X. Y. Shu, J. Z. Fan, J. Li, X. Y. Wang, W. Chen, X. S. Jia and C. J. Li, *Org. Biomol. Chem.*, 2012, **10**, 3393–3397.
33. X. D. Chi, M. Xue, Y. Yao and F. H. Huang, *Org. Lett.*, 2013, **15**, 4722–4725.
34. C. Li, S. Chen, J. Li, K. Han, M. Xu, B. Hu, Y. Yu and X. Jia, *Chem. Commun.*, 2011, **47**, 11294–11296.

35. X. Y. Shu, W. Chen, D. B. Hou, Q. B. Meng, R. L. Zheng and C. J. Li, *Chem. Commun.*, 2014, **50**, 4820–4823.
36. Y. Ma, X. Chi, X. Yan, J. Liu, Y. Yao, W. Chen, F. Huang and J. L. Hou, *Org. Lett.*, 2012, **14**, 1532–1535.
37. T. Ogoshi, R. Sueto, K. Yoshikoshi and T. Yamagishi, *Chem. Commun.*, 2014, **50**, 15209–15211.
38. C. Y. Han, L. Y. Gao, G. C. Yu, Z. B. Zhang, S. Y. Dong and F. H. Huang, *Eur. J. Org. Chem.*, 2013, 2529–2532.
39. X. D. Chi, M. Xue, Y. J. Ma, X. Z. Yan and F. H. Huang, *Chem. Commun.*, 2013, **49**, 8175–8177.
40. T. Ogoshi, K. Yoshikoshi, R. Sueto, H. Nishihara and T. A. Yamagishi, *Angew. Chem., Int. Ed.*, 2015, **54**, 6466–6469.
41. T. Ogoshi, N. Ueshima, T. Yamagishi, Y. Toyota and N. Matsumi, *Chem. Commun.*, 2012, **48**, 3536–3538.
42. T. Ogoshi, T. Aoki, R. Shiga, R. Iizuka, S. Ueda, K. Demachi, D. Yamafuji, H. Kayama and T. Yamagishi, *J. Am. Chem. Soc.*, 2012, **134**, 20322–20325.
43. T. Ogoshi, K. Kida and T. Yamagishi, *J. Am. Chem. Soc.*, 2012, **134**, 20146–20150.
44. I. Nierengarten, S. Guerra, M. Holler, L. Karmazin-Brelot, J. Barbera, R. Deschenaux and J. F. Nierengarten, *Eur. J. Org. Chem.*, 2013, 3675–3684.
45. X. B. Hu, Z. X. Chen, L. Chen, L. Zhang, J. L. Hou and Z. T. Li, *Chem. Commun.*, 2012, **48**, 10999–11001.

CHAPTER 7

Supramolecular Polymers based on Pillararenes

MENGFEI NI, XIAO-YU HU, JULI JIANG AND LEYONG WANG*

Key Laboratory of Mesoscopic Chemistry of MOE, Collaborative Innovation Center of Chemistry for Life Sciences, School of Chemistry and Chemical Engineering, Nanjing University, Nanjing 210093, China
*Email: lywang@nju.edu.cn

7.1 Introduction

Supramolecular polymers, a fascinating type of polymeric architecture where polymer science and supramolecular chemistry meet, have attracted more and more attention from both polymer scientists and supramolecular chemists. In contrast to traditional polymers, in which monomers are connected by covalent bonds, supramolecular polymers are constructed by directional and reversible noncovalent bonds, such as hydrogen bonding, π–π stacking, hydrophilic and hydrophobic interactions, metal–ligand coordination, and electrostatic interactions.[1,2] Most reported supramolecular polymers are constructed by only one type of noncovalent interaction. Given that each type of noncovalent interaction has its own advantages, the combination of different types of noncovalent interaction in an efficient manner promotes the variety and function of supramolecular polymers.[3–5] Recently, Wang and co-workers published two reviews, mainly discussing supramolecular polymers and other supramolecular complexes constructed by the combination of multiple noncovalent interactions, which is named orthogonal self-assembly.[6,7] Generally, supramolecular polymers

Monographs in Supramolecular Chemistry No. 18
Pillararenes
Edited by Tomoki Ogoshi
© The Royal Society of Chemistry 2016
Published by the Royal Society of Chemistry, www.rsc.org

can be categorized into two types depending on the mechanism of formation of the polymeric backbone: (1) "main chain" supramolecular polymers, whose polymeric backbones are formed by supramolecular polymerization, and (2) "side chain" supramolecular polymers, whose polymeric backbones are formed by covalent bonds while the side chains are formed by noncovalent bonds.[8] Significantly, the overlap of polymer science and supramolecular chemistry not only exists in natural systems such as the folding of DNA and peptides, but also has developed into a field for the assembly of synthetic macromolecules and the fabrication of functional supramolecular materials.[9–11]

Compared with traditional polymers, one advantage of supramolecular polymers is the reversibility of the noncovalent interactions between monomeric units. Such reversibility endows supramolecular polymers with dynamic properties that can be responsive to external stimuli. In recent years, functional supramolecular polymers, serving as a type of stimuli-responsive material, have been applied in many research fields such as biology, medicine and materials science.[12–15] The wide applications of supramolecular polymers have also prompted scientists to fabricate various new functional supramolecular polymers, especially macrocycle-based supramolecular polymers. With the development of macrocyclic chemistry, a great number of supramolecular polymers, based on crown ethers,[16] calixarenes,[17,18] cyclodextrins,[19] and cucurbiturils,[20] with potential for the construction of functional materials have been reported in the past decade. In 2008, a new type of macrocycle named pillar[5]arenes was reported by Ogoshi and co-workers.[21] Later, in 2009, Cao, Meier, and co-workers[22] synthesized pillar[6]arenes, pillar[5]arene homologs, and a pillarquinone, which broadened the research field of pillararenes. Pillararenes have unique symmetrical pillar architectures and π-rich cavities, which endow them with outstanding abilities to bind selectively various types of electron-deficient guest such as viologen derivatives and (bis)-imidazolium cations. Because of their unique structures and host–guest binding properties, pillararenes have received great attention in fabricating various interesting supramolecular systems, including rotaxanes, catenanes, supramolecular polymers, metal–organic frameworks (MOFs), supramolecular nanomaterials, chemosensors and organic nanotubes.[23–30] In this chapter, recent advances in the supramolecular polymers based on pillararenes will be summarized in four sections. In the first section, the self-assembly of mono-functionalized pillararenes, leading to three different assemblies (self-inclusion complexes, cyclic dimers and supramolecular polymers) will be discussed. Other linear "main chain" supramolecular polymers will be summarized in the second section, and supramolecular polymer networks will be discussed in the third section. In the last section, some examples of the "side chain" supramolecular polymers that are constructed on the basis of polymer chains will be described.

7.2 The Self-assembly of Mono-functionalized Pillararenes

Mono-functionalized pillararenes, with only one guest unit attached to the rim of a pillararene skeleton, are the most exciting molecules in pillararene chemistry. There are two facile routes to synthesize mono-functionalized pillararenes: co-oligomerization[31,32] and mono-dealkylation.[33,34] It is worth noting that, because of their special structures, mono-functionalized pillararenes can form variable arrays *via* different threading modes of the guest units into the cavities of the pillararenes. The assemblies will be self-inclusion complexes when intramolecular behaviors occur, otherwise intermolecular assembly will result in the formation of oligomers, especially cyclic dimers, and supramolecular polymers (at high concentration), which are also named daisy chains (Figure 7.1).[35,36] These different assemblies can usually be transformed by changing the solution temperature, solvent, and concentration because the equilibrium between the assemblies in solution is sensitive to these external stimuli.

7.2.1 The Self-inclusion of Mono-functionalized Pillararenes

Pillar[5]arenes possess column-shaped cavities which fit well with the linear guests: alkyl chains substituted by electron-poor groups such as ammonium groups.[37,38] Ogoshi and co-workers[33] introduced an ammonium group into the preparation of a mono-guest-functionalized pillar[5]arene (**7.1**) (Figure 7.2a). It was demonstrated that **7.1** predominantly formed a self-inclusion complex in chloroform by variable concentration ^{1}H NMR spectroscopy, ESI-MS and diffusion-ordered ^{1}H NMR spectroscopy (DOSY) measurements. However, the host–guest interaction between the guest moiety and the cavity was weakened in acetone, which led to the de-threading of the guest moiety (Figure 7.2b).

Subsequently, Cao and co-workers[39] reported the self-inclusion behaviors of a series of monoester co-pillar[5]arenes (**7.2a**–**7.2d**; Figure 7.3a) in chloroform at both high and low concentrations. The solid state of **7.2a** also confirmed the formation of a self-inclusion complex (Figure 7.3b, c). However, **7.2e** (Figure 7.3a) bearing a flexible alkyl chain as the linkage did not have the self-inclusion behaviors in solution. Therefore, the length of the linkage between the aromatic unit and the ester group is crucial for the formation of self-inclusion complexes. Moreover, the stable self-inclusion complexes formed by **7.2a**–**7.2d** could selectively recognize linear guests (1,4-dihalobutanes) better than the uncomplexed pillar[5]arene **7.2e**.

Intramolecular hydrogen bonding plays an important role in the conformation of molecules, which probably affects the kinetics of reactions. Recently, Hou and co-workers[40] investigated the aminolysis of a mono-functionalized pillar[5]arene **7.3** (Figure 7.4a) and found that the reactions were accelerated dramatically. This was probably because the intramolecular hydrogen bonding (Figure 7.4b and c) stabilized the introverted structure of the product and the intermediate of the reaction as well.

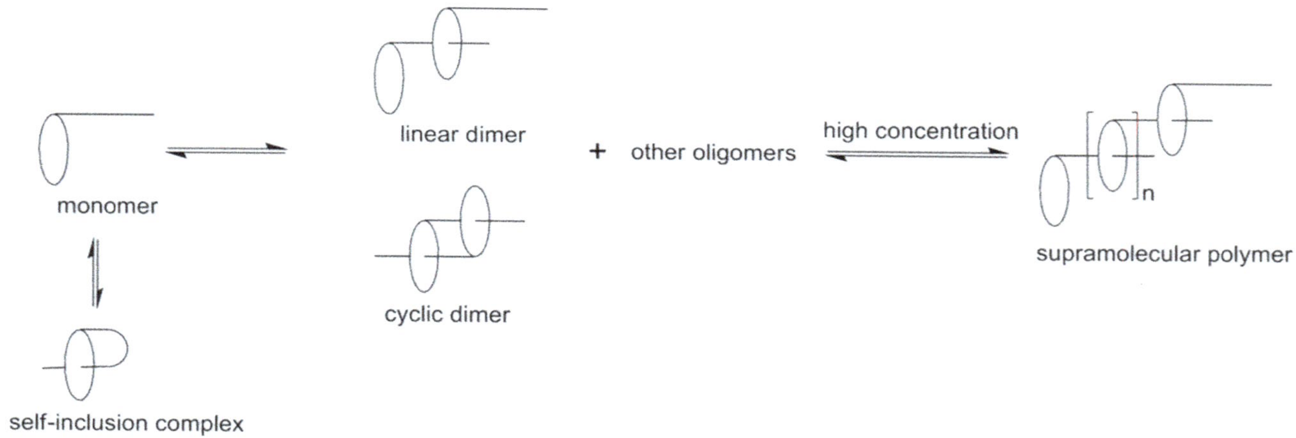

Figure 7.1 The equilibrium of the self-assembly of mono-functionalized pillararenes.

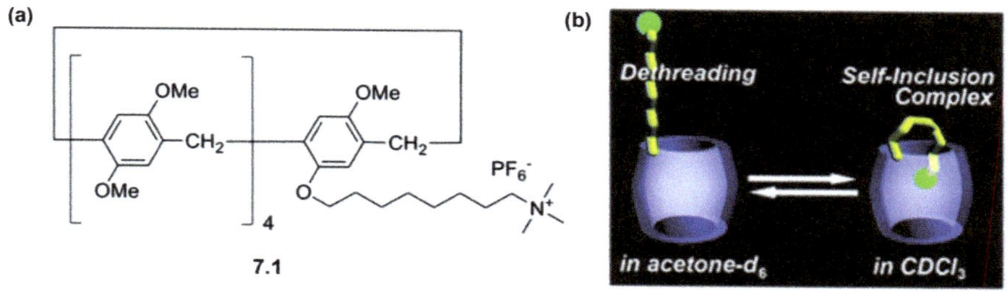

Figure 7.2 The structure of pillar[5]arene **7.1** (a), and the solvent-tuned self-inclusion complex formed by **7.1** (b). Reproduced from ref. 33 with permission from The Royal Society of Chemistry.

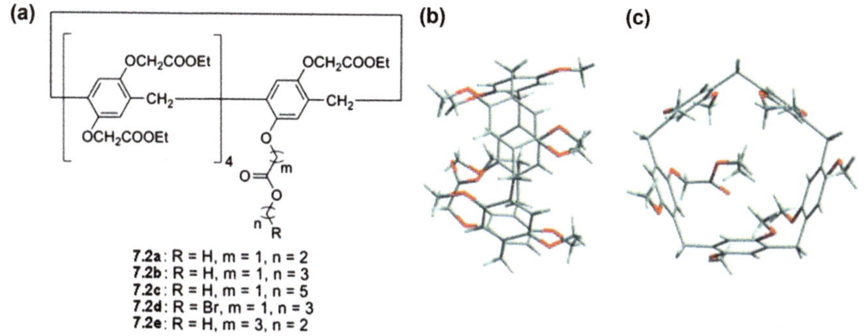

Figure 7.3 The structures of **7.2a–7.2e** (a), and the crystal structure of **7.2a**: side view (b); top view (c).
Reproduced with permission from ref. 39. Copyright 2013 Wiley Publishers.

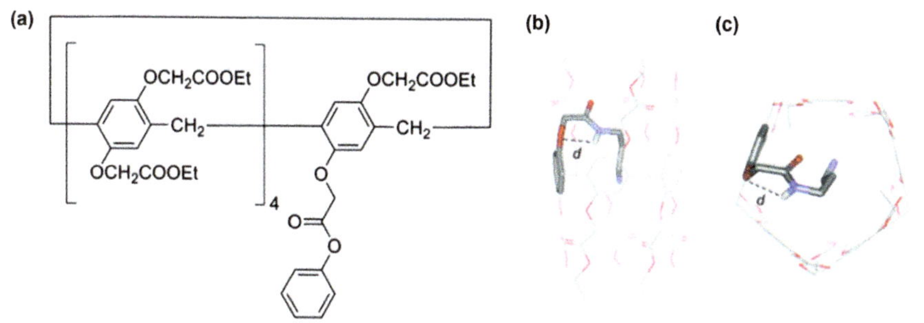

7.3

Figure 7.4 The structures of **7.3** (a), and the crystal structure of the product with intramolecular hydrogen bonding (dotted line): side view (b); top view (c).
Reproduced from ref. 40 with permission from The Royal Society of Chemistry.

It is well known that ureas tend to aggregate *via* N–H$\cdots$O=C hydrogen bonding, and such an aggregation can be disturbed by adding anions because of the more stable hydrogen bonding between ureas and anions.[41–43] Recently, Wang and co-workers[44] developed two mono-urea-functionalized pillar[5]arenes (**7.4a** and **7.4b**; Figure 7.5a) which showed abnormal urea behaviors due to the formation of self-inclusion complexes. Although the solid state of **7.4a** exhibited a cyclic dimer structure in which ureas packed in the C=O$\cdots$O=C form, ^{1}H NMR spectroscopy, ESI-MS, DOSY experiments and theoretical calculations (Figure 7.5b and c) indicated that **7.4a** and **7.4b** formed self-inclusion complexes in chloroform. Because of the inclusion of the ureido group in the cavities of pillar[5]arenes, urea aggregation was disturbed and meanwhile the ureas showed poor anion binding abilities. In contrast, the mono-urea-functionalized pillar[5]arene **7.4c** (Figure 7.5a) could not form a threading structure, probably because of the steric effect of the bulky aromatic unit, and therefore showed a good ability for anion recognition. The same group also studied the self-inclusion behavior of another urea-modified pillar[5]arene (7.7) in the polar solvent DMSO; this

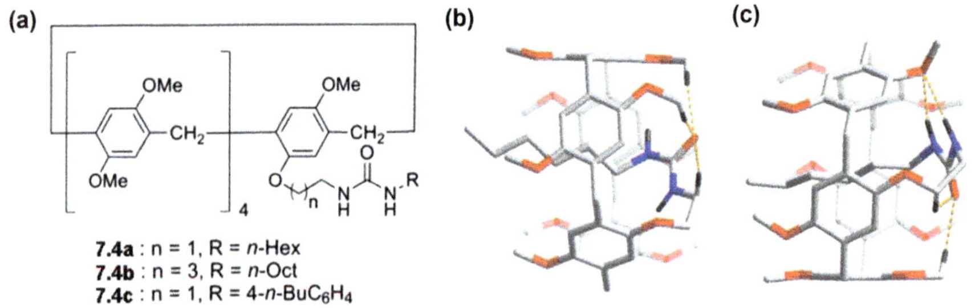

Figure 7.5 The structures of **7.4a–7.4c** (a), and the optimized self-inclusion structures of **7.4a** in two conformations, with two N–H bonds of the ureas oriented in the opposite direction (b) and the same direction (c). Reproduced from ref. 44 with permission from The Royal Society of Chemistry.

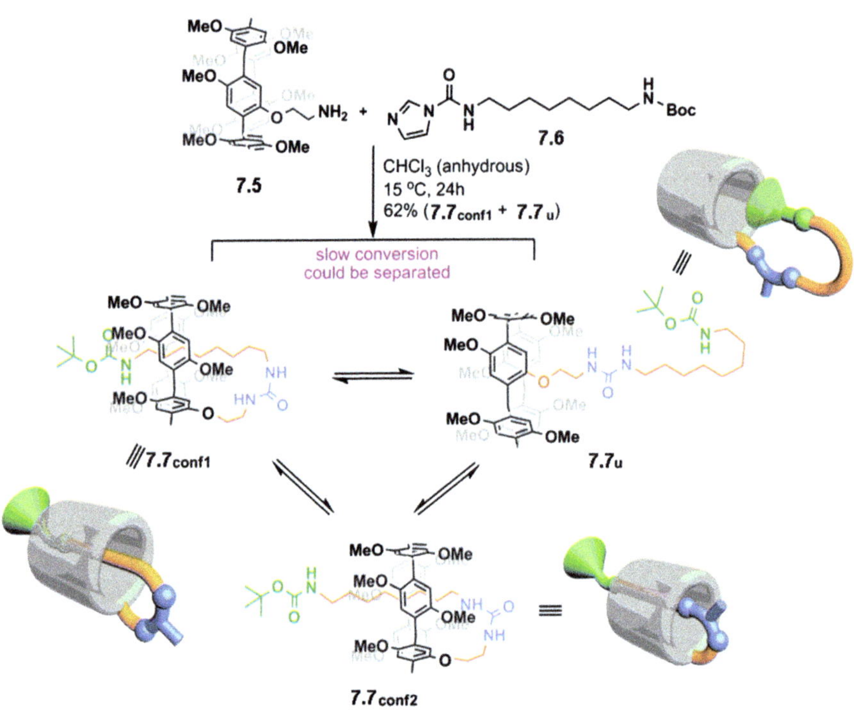

Figure 7.6 Schematic representation of the synthesis of 7.7, and the transformations between the three conformations.
Reproduced from ref. 45 with permission from The Royal Society of Chemistry.

had a large-sized N–Boc stopper.[45] It was surprising that there were three possible stable conformations for 7.7: $7.7_{\mathbf{conf1}}$, $7.7_{\mathbf{conf2}}$, and $7.7_{\mathbf{u}}$ (Figure 7.6), because the host–guest complexes between pillar[5]arenes and neutral guests are normally unstable in polar solvents.[44,46] Moreover, the main products of the reaction between **7.5** and **7.6**, $\mathbf{7.7_{conf1}}$ and $\mathbf{7.7_{u}}$ could be separated temporarily by column chromatography (Figure 7.6). However, pure $\mathbf{7.7_{conf1}}$ and $\mathbf{7.7_{u}}$ each spontaneously and slowly transformed to the mixture of $\mathbf{7.7_{conf1}}$, $\mathbf{7.7_{conf2}}$, and $\mathbf{7.7_{u}}$, reaching an equilibrium after several months at 20 °C. Such a phenomenon, which is very different from the behavior of **7.4a** and **7.4b**, is probably attributable to the presence of multiple hydrogen bonds and the steric effect of the N–Boc moiety.

7.2.2 Cyclic Dimers Formed by Mono-functionalized Pillararenes

Cyclic dimers, also named [c2]daisy chains,[36] are a special kind of mechanically interlocked molecule. Cyclic dimers have been widely studied because of their characteristic topological structures and potential applications in nanotechnology, materials science, and medicine.[35,47]

The first pillararene-based cyclic dimer was reported by Huang and co-workers.[48] They found that the alkyl chains of two molecules of co-pillar[5]arene **7.8** (Figure 7.7a) in different conformations threaded into the cavity of another one, resulting in a mirror image cyclic dimer. The crystal structure of **7.8** indicated that multiple interactions, especially van der Waals forces, made the cyclic dimer structure stable in solid state (Figure 7.7b). Besides, co-pillar[5]arene **7.8** could change from self-inclusion complexes to cyclic dimers in chloroform as the concentration increased from 0.25 mM to 48 mM. Based on this work, the group further developed a solvent-driven molecular spring.[49] Functionalizing the stoppers is a general strategy to prepare bi-stable mechanically interlocked molecules from cyclic dimer precursors.[50,51] Similar to **7.8**, an amino-modified co-pillar[5]arene (**7.9**) could also form cyclic dimers in nonpolar solvents (Figure 7.8). By reacting **7.9** with 3,5-bis(trifluoromethyl)phenyl isocyanate in chloroform at −35 °C, the stopper was successfully attached to the cyclic dimer precursor, together with creation of a second station (urea). The solvent-stimulated contractions/extensions of the molecular spring **7.10** could be controlled by changing the polarity of the solvent. As the ratio of DMSO in the mixture with chloroform increased, the molecular spring **7.10** stretched gradually, and contracted in turn when the polarity of the solvent was decreased, which acted as a muscle-like contraction–extension (Figure 7.8).

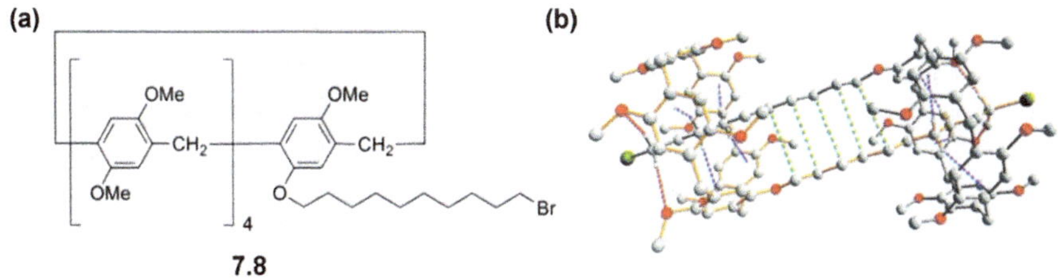

Figure 7.7 The structure of **7.8** (a), and the crystal structure of **7.8** (b). Reproduced with permission from ref. 48. Copyright 2011 American Chemical Society.

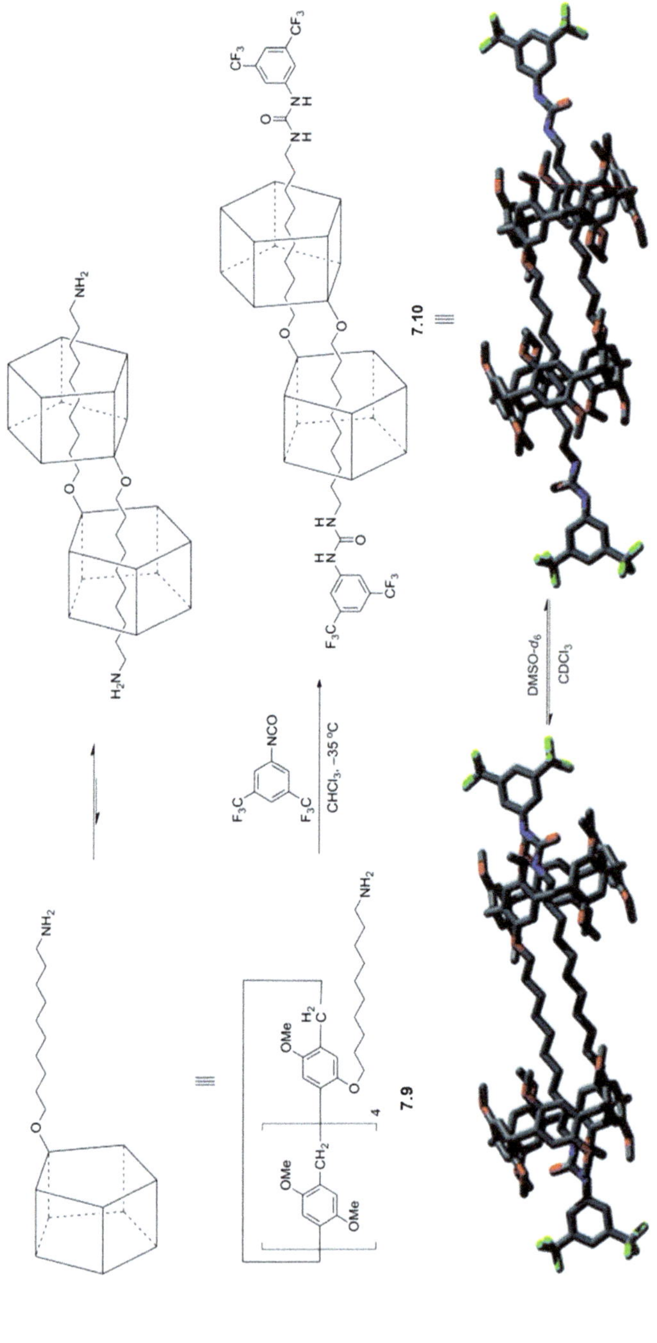

Figure 7.8 A solvent-driven molecular spring (**7.10**). Reproduced from ref. 49 with permission from The Royal Society of Chemistry.

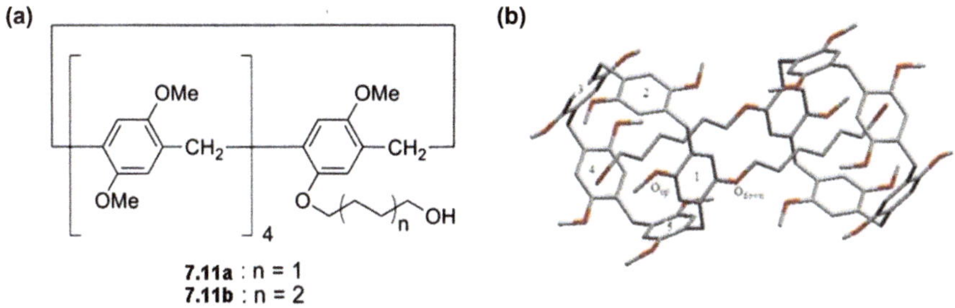

Figure 7.9 The structures of **7.11a** and **7.11b** (a), and the crystal structure of **7.11b** (b). Reproduced with permission from ref. 52. Copyright 2012 American Chemical Society.

Similar dimerization of a ω-hydroxyalkoxy group substituted pillar[5]arene **7.11b** (Figure 7.9a) was also reported by Cao's group.[52] The formation of cyclic dimers from **7.11b** was demonstrated both in solution and in solid state (Figure 7.9b). It was found that the self-assembly of **7.11b** in chloroform was concentration dependent and that the cyclic dimer was formed only at relatively high concentration. Consequently, the formation of cyclic dimer by **7.11b** was reversible and controllable, which could be realized by changing the concentration, the polarity of the solvent, and the temperature. However, **7.11a** (Figure 7.9a) bearing a shorter alkyl chain could not form a cyclic dimer or other threading structures. On the other hand, both **7.11a** and **7.11b** could bind HCOOH, yielding stable 1 : 1 ratio complexes; these were able to be isolated as colorless crystals.

7.2.3 Supramolecular Polymers Formed by Mono-functionalized Pillararenes

Among the assemblies formed by mono-functionalized pillararenes, supramolecular polymers differ from other types of assembly (self-inclusion complexes, cyclic dimmers, *etc.*) in the much higher aggregation numbers in microscopic view. With much longer linear arrays, the formation of supramolecular polymers has to overcome an entropically unfavorable process. Generally, there are two strategies to overcome such disadvantages: (1) to increase the concentration of the monomer, and (2) to strengthen the binding affinity between the macrocycle and the guest.

Inspired by their previous work in which hexanes were included in the cavities of pillar[5]arenes in solid state,[53] Huang and co-workers[54] constructed the first supramolecular polymers driven by C–H···π interactions, formed by a mono-functionalized pillar[5]arene (**7.12a**) with a single octyloxy group (Figure 7.10a). It was found that **7.12a** formed self-inclusion complexes at low concentrations and transformed to supramolecular polymers at high concentrations (in the range of 1–768 mM) in chloroform. The crystal structure of **7.12a** also demonstrated the formation of linear supramolecular polymers (Figure 7.10b). It was presumed that the high concentration of **7.12a** in chloroform and the multiple C–H···π interactions, coupled with the

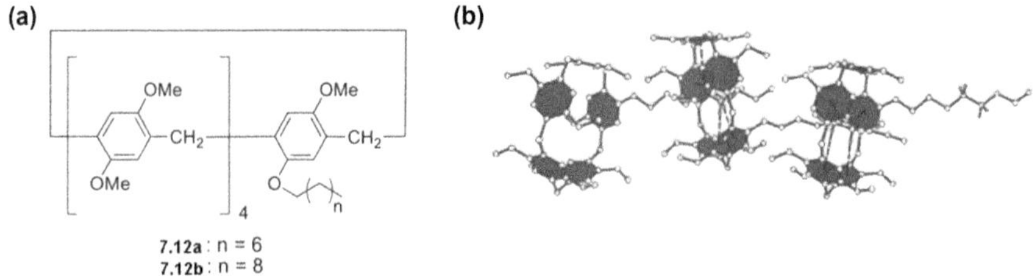

Figure 7.10 The structures of **7.12a** and **7.12b** (a), and the crystal structure of **7.12a** (b). Reproduced with permission from ref. 54. Copyright 2011 Wiley Publishers.

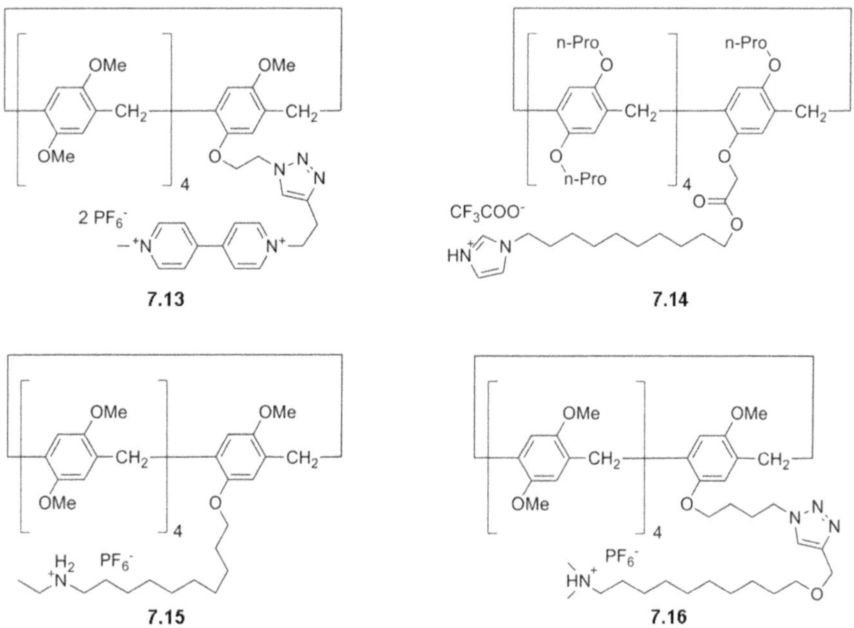

Scheme 7.1 The structures of **7.13**, **7.14**, **7.15**, and **7.16**.

entropic effect of desolvation of the pillar[5]arene cavity, were probably several important reasons for the formation of supramolecular polymers, although C–H···π interactions are the weakest hydrogen bonds (1.5–2.5 kcal mol^{-1}).[55–57] Similarly, **7.12b**, with a longer alkyl chain, could also form linear supramolecular polymers at high concentrations (Figure 7.10a).[48]

By using the "click" reaction, Stoddart and co-workers[58] connected a viologen unit to the pillar[5]arene moiety to obtain a mono-functionalized pillar[5]arene (**7.13**; Scheme 7.1). As the concentration increased in the range of 0.1–100 mM in dichloromethane, the assemblies of **7.13** changed from self-inclusion complexes to linear supramolecular polymers. In addition, gels could be formed after a sealed solution of **7.13** at concentrations above 25 mM in dichloromethane had been left to stand for 12 h.

The construction of stimuli-responsive supramolecular polymers is always attractive to supramolecular chemists. An imidazolium-modified *co*-pillar[5]-arene **7.14** (Scheme 7.1) could form a pH-responsive supramolecular polymer in chloroform on the basis of the pillar[5]arene/imidazolium cation binding motif, and the degree of polymerization decreased sharply after neutralization.[59] On the other hand, the polymerization of an ammonium-modified *co*-pillar[5]arene **7.15** (Scheme 7.1) in chloroform was anion-responsive. It was found that the supramolecular polymers formed by **7.15** would disassemble on adding Cl^-, probably owing to the formation of an intimate ion pair between Cl^- and the ammonium cation, which is too large to fit into the cavity of the pillar[5]arene.[60] Similarly, on the basis of the pillar[5]arene/ammonium recognition motif, Yang and co-workers[61] developed a multiple stimuli-responsive supramolecular polymer formed by another ammonium-modified *co*-pillar[5]arene (**7.16**; Scheme 7.1). Increasing the pH, adding counter ions or increasing the polarity of the solvent could make the supramolecular polymer disassemble, and the reversible assembly could be realized by decreasing the pH, adding AgOTf or decreasing the polarity of the solvent, respectively. Moreover, these authors further fabricated supramolecular polymer nanofibers based on **7.16** by applying the electrospinning technique.

Recently, a photo-responsive supramolecular polymer based on stiff stilbene was reported by Yang's group.[62] They found that a mono-functionalized pillar[5]arene (**7.17**; *Z*) was favorable for the formation of self-inclusion complexes at low concentrations and cyclic dimers at relatively high concentrations in chloroform, owing to the *Z* configuration of stiff stilbene (Scheme 7.2). Irradiation of **7.17** (*Z*) with UV light at 387 nm generated **7.17** (*E*) (97%), which was favorable for the formation of supramolecular polymers. Moreover, **7.17** (*E*) was responsive to acid–base stimuli, and the supramolecular polymers formed by the charged form **7.18** (*E*) had a higher degree of polymerization.

Self-sorting exists extensively in live systems, making different functional architectures operate independently and effectively. Learning from nature, supramolecular chemists are also pursuing the construction of artificial self-sorting systems.[63,64] Based on the highly selective multiple host–guest recognition system that depends on the differences among the cavity sizes of pillar[5]arenes and pillar[6]arenes, Ogoshi and co-workers[34] ingeniously developed a self-sorting supramolecular polymer system (Figure 7.11). They found that the 1,4-diazabicyclo[2.2.2]octane (DABCO) unit fits pillar[6]arenes well while the pyridinium unit fits pillar[5]arenes well. Consequently, the pyridinium-modified pillar[6]arene **7.20** individually formed self-inclusion complexes and oligomers, while no complexation phenomena were observed for the DABCO-modified pillar[5]arene **7.19**. However, when **7.19** and **7.20** were mixed at a 1 : 1 ratio, the highly selective host–guest recognition motifs of pillar[5]arene–pyridinium complexes and pillar[6]arene–DABCO complexes were formed simultaneously, leading to the formation of supramolecular polymers with alternating host–guest recognition units (Figure 7.11). This work provided a new route for the construction of sophisticated supramolecular systems based on pillararenes.

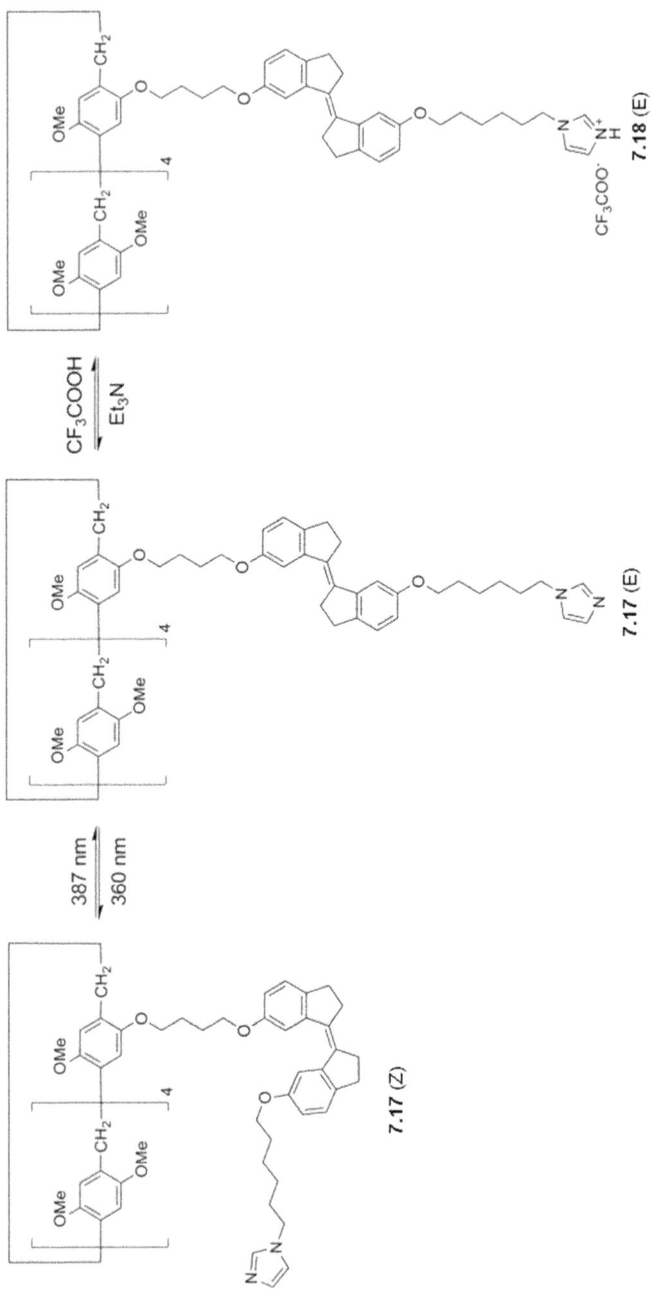

Scheme 7.2 The transformation of **7.17** (*Z*), **7.17** (*E*), and **7.18** (*E*).

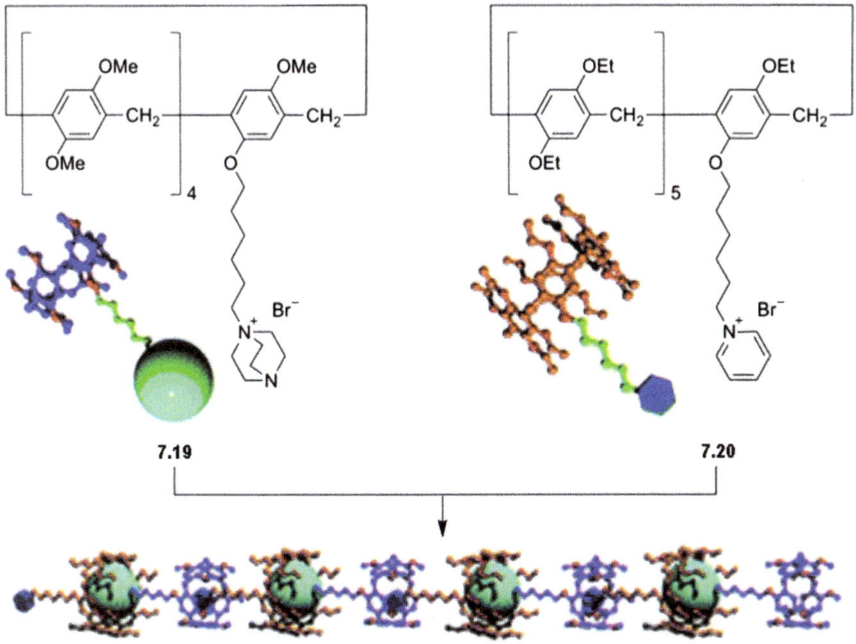

Figure 7.11 The formation of supramolecular polymers by alternating pillar[5]arene and pillar[6]arene units.
Reproduced from ref. 34 with permission from The Royal Society of Chemistry.

7.3 Linear Supramolecular Polymers based on Pillararenes

Orthogonal self-assembly is also an effective strategy for construction of supramolecular polymers. "Orthogonal" means that different types of supramolecular interaction do not affect each other.[65] The combination of various noncovalent interactions provides the diversity of the design and construction of supramolecular polymers.[6]

Recently, Wang and co-workers[66] combined host–guest interaction and hydrogen bonding to develop a pillar[5]arene-based supramolecular polymer (Figure 7.12). Owing to the strong dimerization of the ureidopyrimidinone (UPy) units, the UPy-modified pillar[5]arene **7.21** tended to form dimers in solution (chloroform/acetonitrile = 1/1, v/v). With the addition of the bis-paraquat guest **7.22**, the pyridinium units threaded into the cavities of pillar[5]arenes and connected the pillar[5]arene dimers to form linear supramolecular polymers (Figure 7.12). Subsequently, Li and co-workers[67] developed a supramolecular polymer by orthogonal assembly of two types of host–guest interactions. It was found that co-pillar[5]arene **7.23** bearing complementary recognition sites could form cyclic dimers in chloroform,

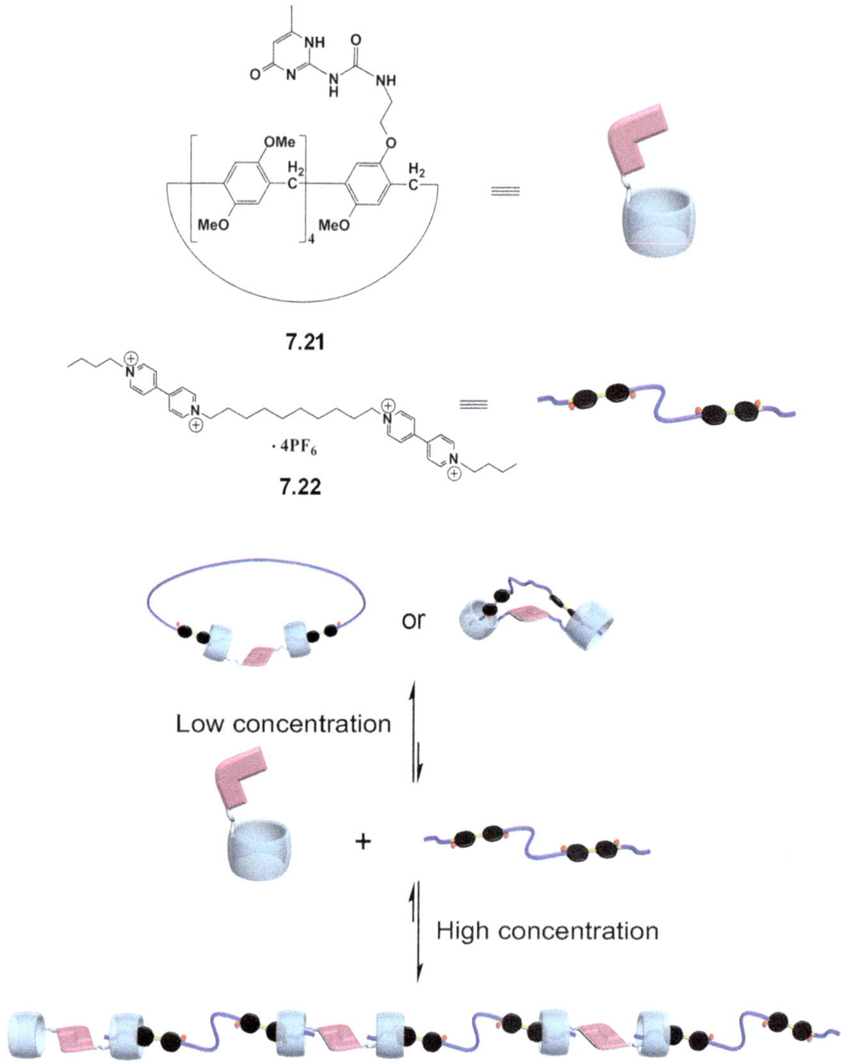

Figure 7.12 Illustration of the formation of supramolecular polymers by **7.21** and **7.22**. Reproduced from ref. 66 with permission from The Royal Society of Chemistry.

which was also confirmed by X-ray crystallographic analysis (Figure 7.13). Subsequently, with the addition of the pillar[5]arene dimer **7.24**, the cyclic dimer units formed by **7.23** could be connected by host–guest interactions, leading to the formation of an alternating supramolecular polymer at high concentrations.

Poly(pseudo)rotaxanes are special categories of supramolecular polymers, which actually are a string of (pseudo)rotaxanes linked by noncovalent bonds. Wang and co-workers[68,69] creatively introduced ureidopyrimidinone

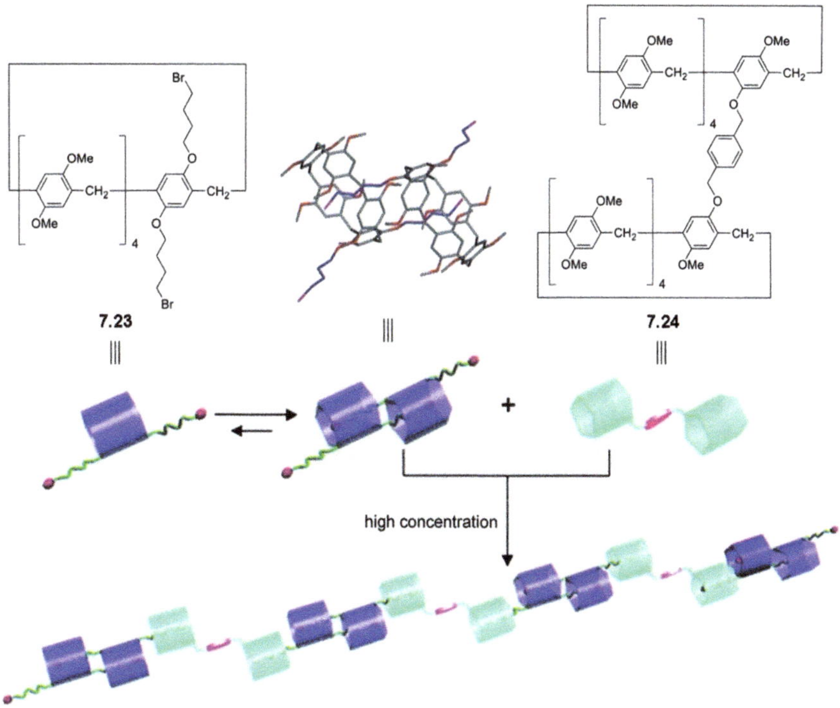

Figure 7.13 The structures of **7.23** and **7.24**, the crystal structure of the cyclic dimer formed by **7.23**, and the graphical representation of the orthogonal assembly.
Reproduced from ref. 67 with permission from The Royal Society of Chemistry.

(UPy) units into the fabrication of pillararene-based poly(pseudo)rotaxanes by orthogonal assembly. The bi-UPy-functionalized pillar[5]arene **7.25** (Figure 7.14) could form linear supramolecular polymers due to the dimerization of UPy units. Subsequently the combination of **7.25** and the diamine guest **7.26** resulted in the formation of polypseudorotaxanes (Figure 7.14).[68] Moreover, utilizing the pillar[5]arene/diamine binding motifs (pseudorotaxane) as the precursors to react with the stoppers, UPy units, the group fabricated a series of rotaxanes in high yields, which could further aggregate to polyrotaxanes by quadruple hydrogen bonding at high concentrations (Figure 7.15).[69]

One advantage of the supramolecular polymers constructed by orthogonal assembly is their enhanced responsiveness to multiple external stimuli. Recently, a thermo- and photo-responsive supramolecular polymer based on pillar[5]arene was reported by Yang's group,[70] where the anthracene group was used as the photo-responsive unit, and the pillar[5]arene/imidazole binding motif was employed as the thermo-responsive unit. When an anthracene-modified pillararene (**7.27**) was irradiated with UV light (λ >360 nm), the anthracene unit dimerized and the pillar[5]arene dimer

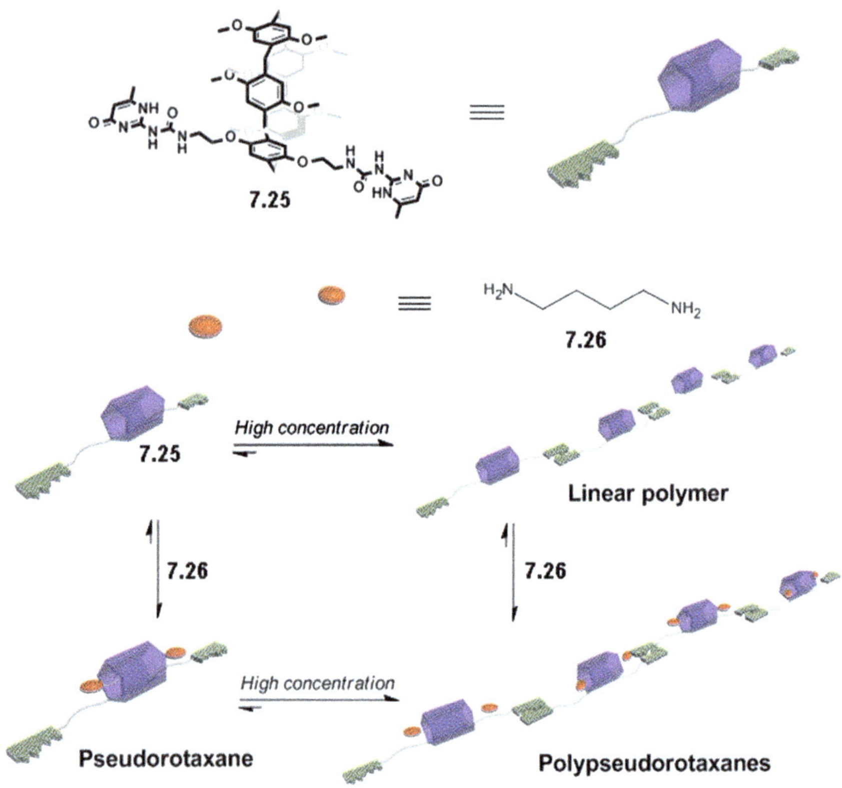

Figure 7.14 Graphical representation of the construction of polypseudorotaxanes from **7.25** and **7.26**.
Reproduced from ref. 68 with permission from The Royal Society of Chemistry.

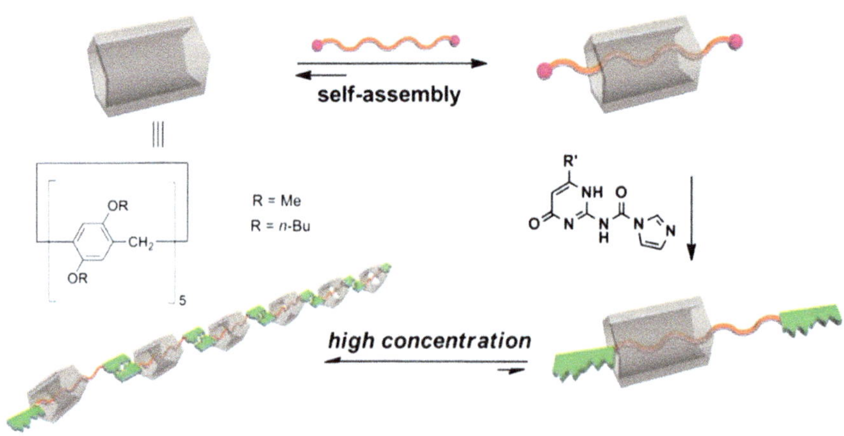

Figure 7.15 Graphical representation of the construction of polyrotaxanes.
Reproduced with permission from ref. 69. Copyright 2012 American Chemical Society.

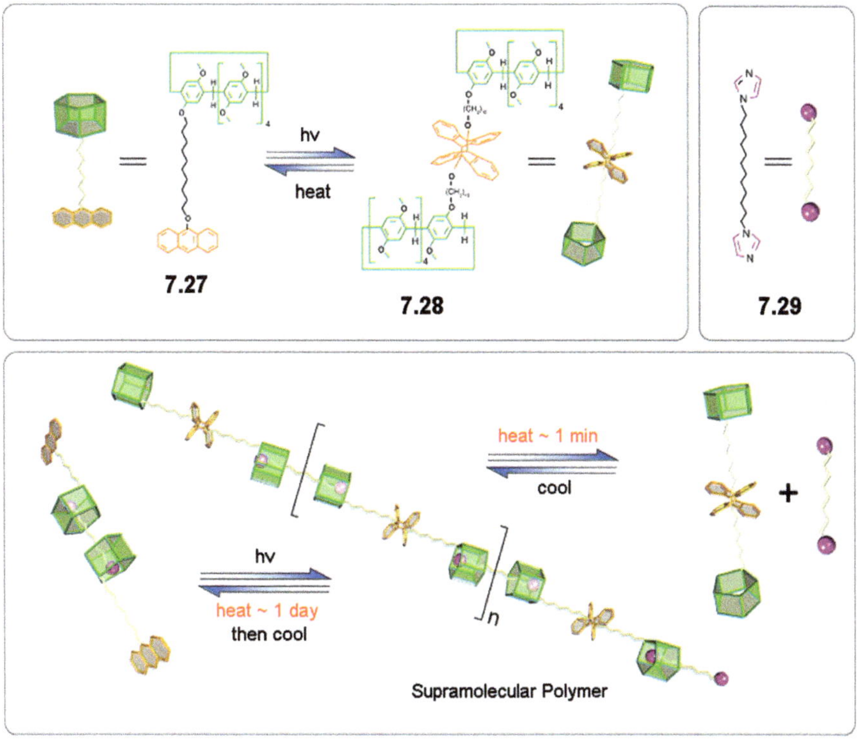

Figure 7.16 Graphical representation of the formation and response of supramolecular polymers.
Reproduced with permission from ref. 70. Copyright 2013 American Chemical Society.

7.28 was generated (Figure 7.16). Then the combination of **7.28** and a bisimidazole alkane guest (**7.29**) led to the formation of supramolecular polymers, and gels could be formed when the heated solution of the mixture cooled to room temperature. Further investigations suggested that such supramolecular gels showed good responsiveness to multiple external stimuli. Heating the gels for 20 s resulted in the de-threading of the imidazole moieties, and the anthracene dimers dissociated when heated for 20 h, both leading to the depolymerization of the supramolecular polymers. However, the gels could be reformed by cooling or irradiation with UV light (λ >360 nm), respectively (Figure 7.16). This work suggested the advantage of the orthogonal assembly in response to multiple external stimuli, and the potential in the exploration of smart materials with high tunability and reversibility.

7.4 Supramolecular Polymer Networks based on Pillararenes

On the basis of previous work on the construction of linear supramolecular polymers *via* the combination of quadruple hydrogen bonding and pillar[5]arene-based host–guest recognition,[66,68,69] Wang and co-workers[71] further developed a supramolecular polymer network. The bi-UPy-functionalized pillar[5]arene **7.25** had been proved to form linear supramolecular polymers in solution.[68] The addition of a bisparaquat guest (**7.30**) further cross-linked the linear polymeric structures *via* the formation of pillar[5]-arene/bisparaquat binding motifs (Figure 7.17). Dark gray spherical aggregates could also be observed by transmission electron microscopy (TEM) measurements, and might originate from further assembly of the supramolecular network. Furthermore, a translucent film could be created by using the supramolecular polymer networks with polyethylene glycol (PEG)-2000 as the polymer matrix.

Using a bi-functionalized pillar[5]arene (**7.31**), Li and co-workers[72] constructed a supramolecular polymer network in chloroform *via* strong

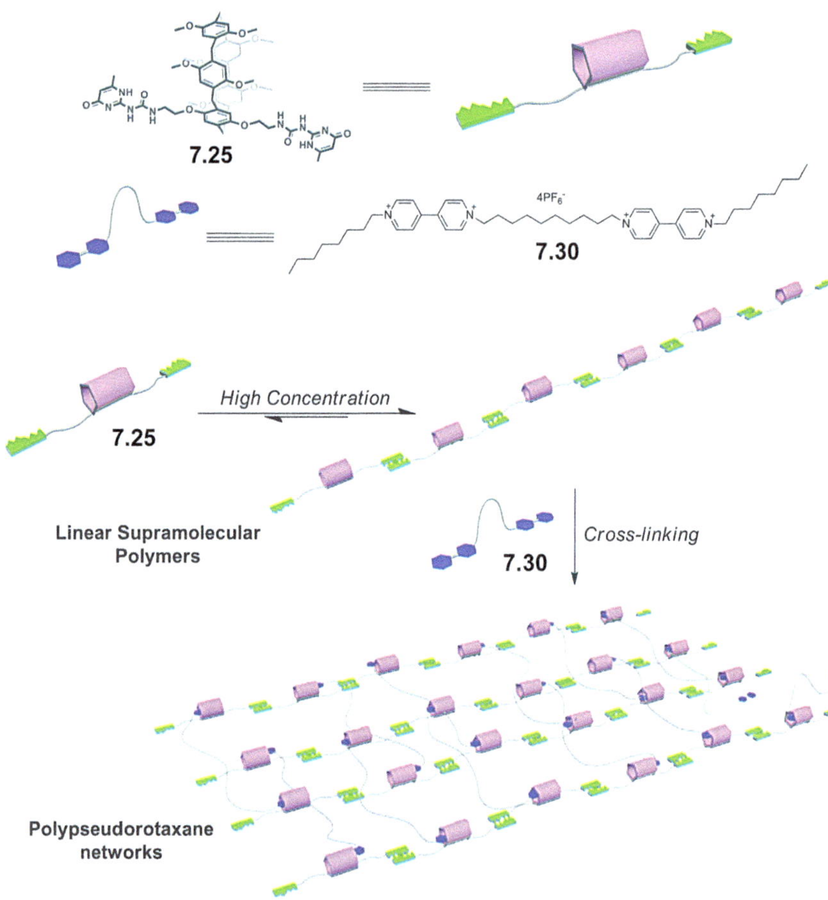

Figure 7.17 Graphical representation of the formation of a supramolecular polymer network from **7.25** and **7.30**.
Reproduced from ref. 71 with permission from The Royal Society of Chemistry.

host–guest recognition motifs, which could also be effectively depolymerized by the addition of a competitive butanedinitrile guest (Figure 7.18). Moreover, based on such strong host–guest recognition motifs, they also fabricated another supramolecular polymer system.[73] The combination of the pillar[5]arene dimer **7.32** and the ditopic guest **7.33** resulted in the formation of a linear supramolecular polymer with a calculated molecular weight of 98 kDa, while use of the tritopic guest **7.34** as a linker led to the generation of a supramolecular polymer network with a higher calculated molecular weight (127 kDa) (Figure 7.19).

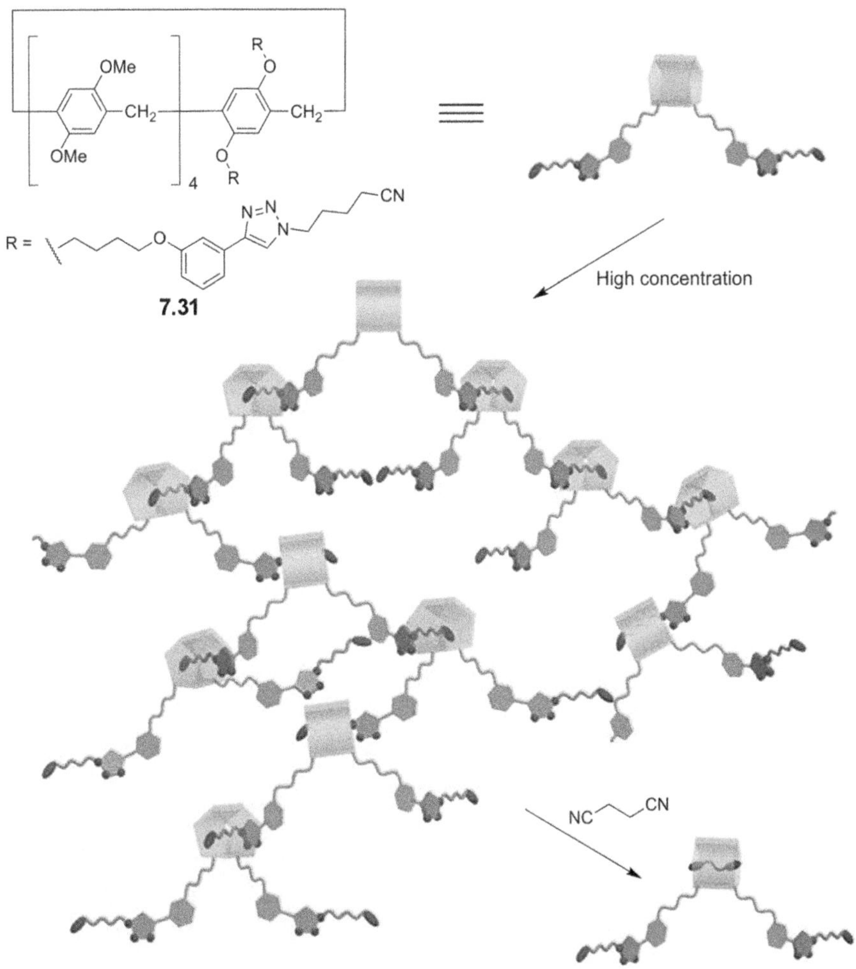

Figure 7.18 Graphical representation of the formation of a supramolecular polymer network from **7.31**.
Reproduced with permission from ref. 72. Copyright 2013 Wiley Publishers.

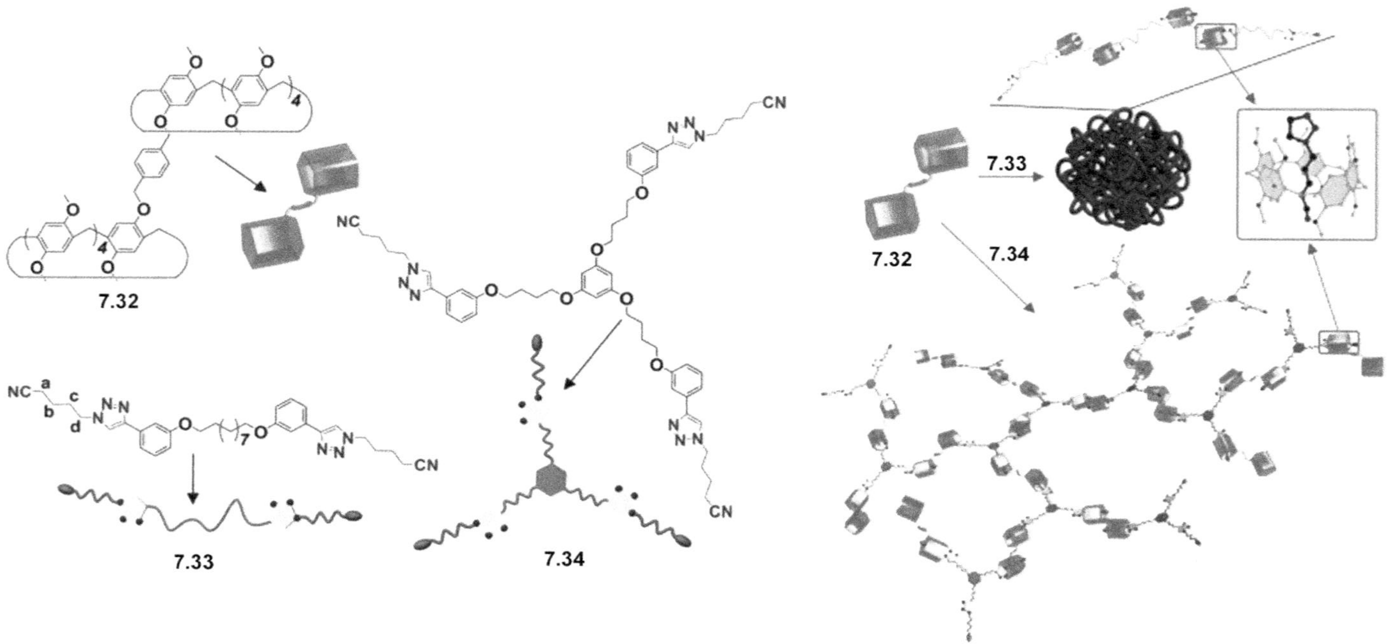

Figure 7.19 Graphical representation of the formation of supramolecular polymers. Reproduced with permission from ref. 73. Copyright 2013 Wiley Publishers.

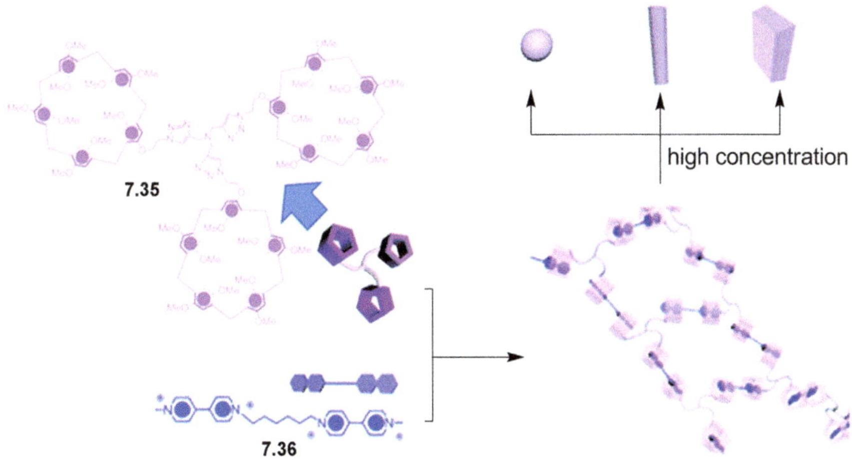

Figure 7.20 Graphical representation of the formation of supramolecular polymer networks and multi-dimensional assemblies.
Reproduced from ref. 74 with permission from The Royal Society of Chemistry.

Recently, Zhao and co-workers[74] constructed a supramolecular polymer network by mixing the pillar[5]arene trimer **7.35** with a bisparaquat linker (**7.36**). In particular, as the concentration of the mixture increased, the supramolecular assemblies obtained transformed from vesicular structures, to tubular objects, layers and stacked layers (Figure 7.20).

7.5 Pillararene-based Poly(pseudo)rotaxanes on Polymer Chains

The symmetrical column-shaped cavities of pillar[5]arenes make it possible for the macrocycles to thread, move and rotate freely on a polymer chain. Based on the host–guest complexation between viologen derivatives and perhydroxylated pillar[5]arenes, Ogoshi and co-workers[75] developed the first pillararene-based polypseudorotaxanes *via* the threading of perhydroxylated pillar[5]arenes on the viologen polymers (Figure 7.21). However, perhydroxylated pillar[5]arenes and viologen polymers with adamantly groups at both ends could not form polypseudorotaxanes owing to the steric effect of the bulky adamantly stoppers. Inspired by this work, the group prepared further polyrotaxanes in an extremely high yield (93%) by attaching adamantly stoppers to the above polypseudorotaxanes (Figure 7.21).[76] They also found that the hydrogen bonding between the hydroxyl groups of pillar[5]arenes reduced the solubility of the polyrotaxanes, and the color of the polyrotaxanes became darker on heating. This was probably because the hydrogen bonding was weakened and the macrocycles shuttled faster along the axle when the temperature rose, leading to an efficient electron transfer between pillar[5]-arenes and viologen units and the stabilization of the radical cation species.

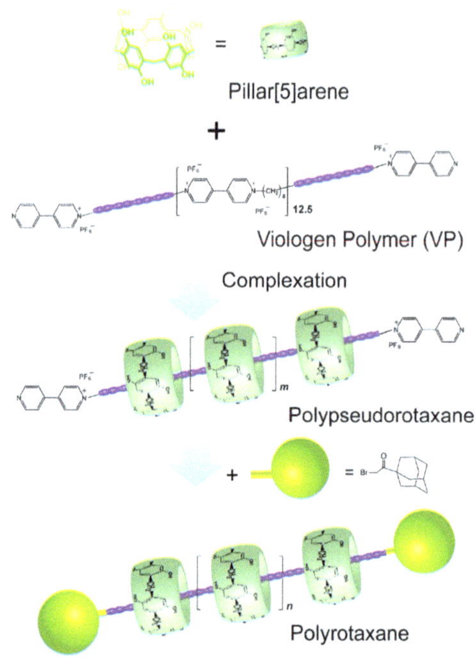

Figure 7.21 The preparation of pillar[5]arene-based polyrotaxanes. Reproduced with permission from ref. 76. Copyright 2010 American Chemical Society.

Recently, Ogoshi and co-workers[77] developed a polypseudorotaxane which was applied in the emeraldine-base polyaniline reduction (Figure 7.22). When emeraldine-base polyaniline was mixed with the perhydroxylated pillar[5]arene built from five hydroquinone units, the solution changed from blue to completely colorless after 180 min (Figure 7.22). Control experiments suggested that the blue solution color hardly changed when emeraldine-base polyaniline was mixed with hydroquinone, although hydroquinone is widely used as a reducing agent.[78] The stronger reducing ability of perhydroxylated pillar[5]arene probably resulted from the formation of polypseudorotaxane by polyaniline and the pillar[5]arene, which enhanced the oxidation–reduction (Figure 7.22). Furthermore, the group constructed a polypseudorotaxane based on the threading of pillar[5]arene **37** (Scheme 7.3) on the polyethylene (PE) chain, which drove extension of the PE chain and consequently increased the melting point of PE.[79]

Stimuli-responsive fluorescent conjugated polymer chemosensors are a type of smart molecular device. They also have a very bright future owing to their unique optical and electronic properties.[80,81] Wang and co-workers[82,83] synthesized a pillar[5]arene-modified conjugated polymer (**7.38**) [polydispersity index (PDI) = 1.58; degree of polymerization (DP) ≈ 8] that could serve as a stimuli-responsive fluorescent chemosensor for the selective detection of halogen ions (Figure 7.23). Following the formation of

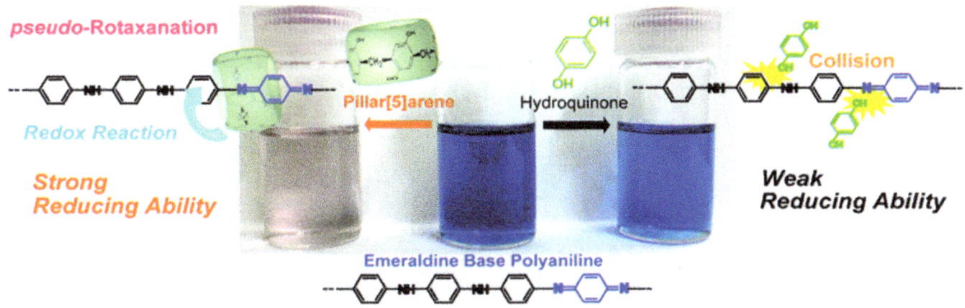

Figure 7.22 The reduction of polyaniline using pillar[5]arenes or hydroquinone. Reproduced with permission from ref. 77. Copyright 2011 American Chemical Society.

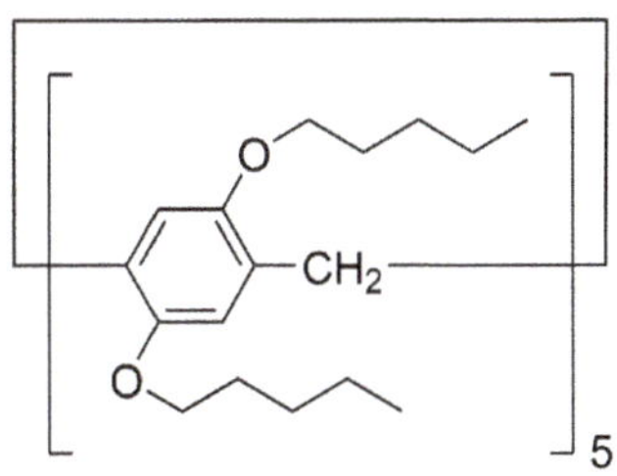

7.37

Scheme 7.3 The structure of pillar[5]arene **7.37**.

polypseudorotaxanes by **7.38** and the *n*-octylpyrazinium guest **7.39**, the strong fluorescence of **7.38** was quenched as a result of the efficient electron transfer from the conjugated backbone to the cationic guest. However, recovery of the fluorescence could be realized *via* the disassembly of the polypseudorotaxanes by adding halogen ions, for which the order of fluorescence enhancement was $I^- < Br^- < Cl^-$, attributed to the different ion-pairing between the three halide anions and the *n*-octylpyrazinium cation moiety.

A lower critical solution temperature (LCST) is an important property of polymer materials. Supramolecular chemists are also trying to control the LCST of polymer materials in a supramolecular way.[84,85] Recently, the pH-responsive supramolecular control of the LCST of a copolymer was reported by Huang's group.[86] A copolymer **7.40** (Figure 7.24) of *N*-isopropylacrylamide (NIPAAM) and styrene, modified with paraquat units as the side chains, was prepared by radical polymerization and subsequent modification. It was found that the turbidity point of **7.40** was 40.5 °C, higher than that of poly(NIPAAM). However, the LCST value of **7.40** increased gradually and reached 53.4 °C and 50.5 °C, respectively after the addition of 1 equiv. of a water-soluble pillar[6]arene (**7.41**) or a water-soluble pillar[5]arene (**7.42**). This was attributed to the formation of polypseudorotaxanes on the side chains of **7.40**, leading to the restraint of the interpolymer aggregation

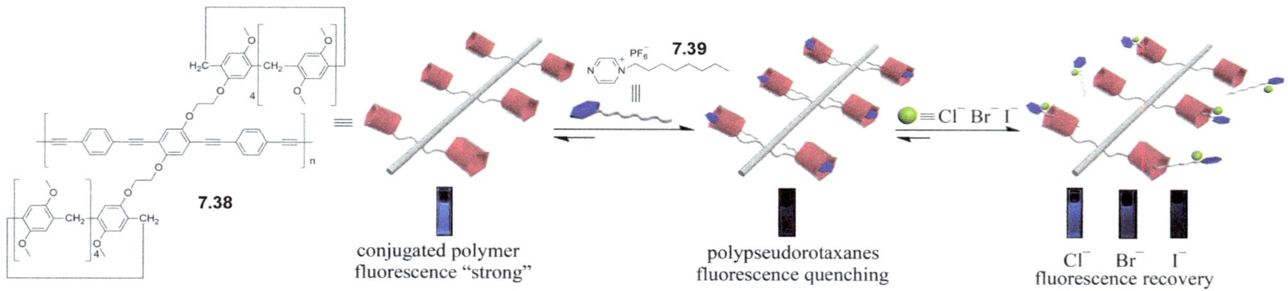

Figure 7.23 Graphical representation of the formation of polypseudorotaxanes and their disassembly induced by halogen ions. Reproduced from ref. 82 with permission from The Royal Society of Chemistry.

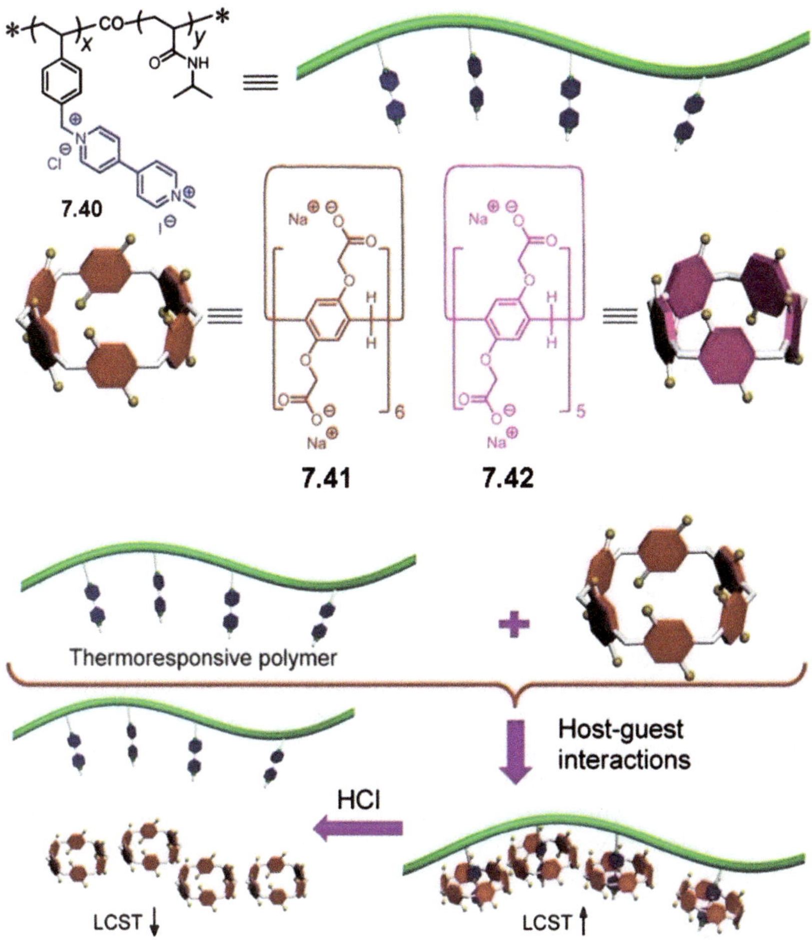

Figure 7.24 The structures of **7.40**, **7.41** and **7.42**, and graphical representation of the formation of polypseudorotaxanes from the polymer chain and the pillararene.
Reproduced with permission from ref. 86. Copyright 2014 American Chemical Society.

(Figure 7.24). In particular, the pillar[6]arene **7.41** was more efficient in increasing the LCST of **7.40**, probably due to the presence of a large cavity and more water-soluble groups when compared with pillar[5]arene **7.42**. Moreover, the recovery of the LCST could also be realized by treatment with acid, which contributed to the disassembly of the polypseudorotaxanes.

7.6 Conclusions and Outlook

With the development of supramolecular chemistry, supramolecular polymers are expanding rapidly. Given that supramolecular polymers are derived from the combination of supramolecular chemistry and polymer science, the development of supramolecular polymers has been integrated into many research fields covering biology, medicine, materials and so on. The intrinsic reversibility of supramolecular polymers provides a platform for the construction of stimuli-responsive materials, which can be applied for drug delivery, photovoltaic devices, porous membranes, and self-healing materials.

Macrocycles play an important role in the development of supramolecular chemistry. Based on the traditional macrocycles such as crown ethers, calixarenes, cyclodextrins, and cucurbiturils, a great many supramolecular polymers as well as their applications have been reported. Because pillararenes have only had 6 years of development (from 2008), the research on pillararene-based supramolecular polymers is at a preliminary stage. However, all of the efforts made and achievements that have been reached suggest that pillararenes as well as pillararene-based supramolecular polymers will have a bright future.

In this chapter, we have tried our best to summarize different examples of the pillararene-based supramolecular polymers shown in the published journals so far. It is worth mentioning that many of these examples have been beyond the simple construction of supramolecular polymers. These outstanding researches are focusing on stimuli-responsive polymers, supramolecular gels, molecular devices, and the modification of polymeric materials. However, more functions and applications of pillararene-based supramolecular polymers need to be explored in the future. On the other hand, most supramolecular polymers discussed in this chapter are based on pillar[5]arenes, which is attributed to the facile synthesis of pillar[5]arenes when compared with other pillar[*n*]arenes. Therefore, the research on supramolecular polymers based on other pillar[*n*]arenes, especially pillar[6]arenes, will also be a challenge.

Acknowledgements

The authors wish to thank all the motivated and talented researchers who have contributed to our own work discussed in the present chapter. The authors would like also to express their acknowledgment to Prof. Yi Pan, Prof. Guigen Li, Prof. Jinglin Zuo, Prof. Weiping Ding, and Prof. Dongzhong Chen from Nanjing University, for their continuous support during the course of study. Finally, we would like to pay special homage to Prof. Hongwen Hu, who initiated the field of Host–Guest Chemistry in Nanjing University.

References

1. L. Brunsveld, B. J. B. Folmer, E. W. Meijer and R. P. Sijbesma, *Chem. Rev.*, 2001, **101**, 4071–4098.
2. T. F. A. De Greef, M. M. J. Smulders, M. Wolffs, A. P. H. J. Schenning, R. P. Sijbesma and E. W. Meijer, *Chem. Rev.*, 2009, **109**, 5687–5754.
3. H. Hofmeier and U. S. Schubert, *Chem. Commun.*, 2005, 2423–2432.

4. C.-H. Wong and S. C. Zimmerman, *Chem. Commun.*, 2013, **49**, 1679–1695.
5. E. Elacqua, D. S. Lye and M. Weck, *Acc. Chem. Res.*, 2014, **47**, 2405–2416.
6. S.-L. Li, T. Xiao, C. Lin and L. Wang, *Chem. Soc. Rev.*, 2012, **41**, 5950–5968.
7. X.-Y. Hu, T. Xiao, C. Lin, F. Huang and L. Wang, *Acc. Chem. Res.*, 2014, **47**, 2041–2051.
8. *Supramolecular Polymers*, ed. A. Ciferri, Taylor & Francis, London, U.K., 2nd edn, 2005.
9. *Supramolecular Polymer Chemistry*, ed. A. Harada, Wiley-VCH, Weinheim, Germany, 2012.
10. F. Huang, C. Zhai, B. Zheng and S. Li, *Supramolecular Polymers*, Zhejiang University, Hangzhou, Zhejiang, 2012.
11. X. Yan, F. Wang, B. Zheng and F. Huang, *Chem. Soc. Rev.*, 2012, **41**, 6042–6065.
12. C. de las Heras Alarcon, S. Pennadam and C. Alexander, *Chem. Soc. Rev.*, 2005, **34**, 276–285.
13. R. Yerushalmi, A. Scherz, M. E. van der Boom and H.-B. Kraatz, *J. Mater. Chem.*, 2005, **15**, 4480–4487.
14. W. Weng, J. B. Beck, A. M. Jamieson and S. J. Rowan, *J. Am. Chem. Soc.*, 2006, **128**, 11663–11672.
15. T. Aida, E. W. Meijer and S. I. Stupp, *Science*, 2012, **335**, 813–817.
16. B. Zheng, F. Wang, S. Dong and F. Huang, *Chem. Soc. Rev.*, 2012, **41**, 1621–1636.
17. D.-S. Guo and Y. Liu, *Chem. Soc. Rev.*, 2012, **41**, 5907–5921.
18. D.-S. Guo and Y. Liu, *Acc. Chem. Res.*, 2014, **47**, 1925–1934.
19. A. Harada, Y. Takashima and H. Yamaguchi, *Chem. Soc. Rev.*, 2009, **38**, 875–882.
20. Y. Liu, H. Yang, Z. Wang and X. Zhang, *Chem. – Asian J.*, 2013, **8**, 1626–1632.
21. T. Ogoshi, S. Kanai, S. Fujinami, T. Yamagishi and Y. Nakamoto, *J. Am. Chem. Soc.*, 2008, **130**, 5022–5023.
22. D. Cao, Y. Kou, J. Liang, Z. Chen, L. Wang and H. Meier, *Angew. Chem., Int. Ed.*, 2009, **48**, 9721–9723.
23. M. Xue, Y. Yang, X. Chi, Z. Zhang and F. Huang, *Acc. Chem. Res.*, 2012, **45**, 1294–1308.
24. N. L. Strutt, D. Fairen-Jimenez, J. Iehl, M. B. Lalonde, R. Q. Snurr, O. K. Farha, J. T. Hupp and J. F. Stoddart, *J. Am. Chem. Soc.*, 2012, **134**, 17436–17439.
25. Q. Duan, Y. Cao, Y. Li, X. Hu, T. Xiao, C. Lin, Y. Pan and L. Wang, *J. Am. Chem. Soc.*, 2013, **135**, 10542–10549.
26. T. Ogoshi and T. Yamagishi, *Eur. J. Org. Chem.*, 2013, 2961–2975.
27. X. Wu, Q. Duan, M. Ni, X. Hu and L. Wang, *Chin. J. Org. Chem.*, 2014, **34**, 437–449.
28. Y. Cao, X.-Y. Hu, Y. Li, X. Zou, S. Xiong, C. Lin, Y.-Z. Shen and L. Wang, *J. Am. Chem. Soc.*, 2014, **136**, 10762–10769.

29. N. L. Strutt, H. Zhang, S. T. Schneebeli and J. F. Stoddart, *Acc. Chem. Res.*, 2014, **47**, 2631–2642.
30. C. Li, *Chem. Commun.*, 2014, **50**, 12420–12433.
31. N. L. Strutt, R. S. Forgan, J. M. Spruell, Y. Y. Botros and J. F. Stoddart, *J. Am. Chem. Soc.*, 2011, **133**, 5668–5671.
32. L. Liu, D. Cao, Y. Jin, H. Tao, Y. Kou and H. Meier, *Org. Biomol. Chem.*, 2011, **9**, 7007–7010.
33. T. Ogoshi, K. Demachi, K. Kitajima and T. Yamagishi, *Chem. Commun.*, 2011, **47**, 7164–7166.
34. T. Ogoshi, H. Kayama, D. Yamafuji, T. Aoki and T. Yamagishi, *Chem. Sci.*, 2012, **3**, 3221–3226.
35. L. Fang, M. A. Olson, D. Benitez, E. Tkatchouk, W. A. Goddard Iii and J. F. Stoddart, *Chem. Soc. Rev.*, 2010, **39**, 17–29.
36. J. Rotzler and M. Mayor, *Chem. Soc. Rev.*, 2013, **42**, 44–62.
37. C. Han, G. Yu, B. Zheng and F. Huang, *Org. Lett.*, 2012, **14**, 1712–1715.
38. M. Ni, Y. Guan, L. Wu, C. Deng, X. Hu, J. Jiang, C. Lin and L. Wang, *Tetrahedron Lett.*, 2012, **53**, 6409–6413.
39. Y. Chen, D. Cao, L. Wang, M. He, L. Zhou, D. Schollmeyer and H. Meier, *Chem. – Eur. J.*, 2013, **19**, 7064–7070.
40. L. Chen, Z. Li, Z. Chen and J.-L. Hou, *Org. Biomol. Chem.*, 2013, **11**, 248–251.
41. J. W. Steed, *Chem. Soc. Rev.*, 2010, **39**, 3686–3699.
42. V. Amendola, L. Fabbrizzi and L. Mosca, *Chem. Soc. Rev.*, 2010, **39**, 3889–3915.
43. M. Wenzel, J. R. Hiscock and P. A. Gale, *Chem. Soc. Rev.*, 2012, **41**, 480–520.
44. M. Ni, X.-Y. Hu, J. Jiang and L. Wang, *Chem. Commun.*, 2014, **50**, 1317–1319.
45. Y. Guan, P. Liu, C. Deng, M. Ni, S. Xiong, C. Lin, X.-Y. Hu, J. Ma and L. Wang, *Org. Biomol. Chem.*, 2014, **12**, 1079–1089.
46. C. Li, S. Chen, J. Li, K. Han, M. Xu, B. Hu, Y. Yu and X. Jia, *Chem. Commun.*, 2011, **47**, 11294–11296.
47. M. Hmadeh, L. Fang, A. Trabolsi, M. Elhabiri, A.-M. Albrecht-Gary and J. F. Stoddart, *J. Mater. Chem.*, 2010, **20**, 3422–3430.
48. Z. Zhang, G. Yu, C. Han, J. Liu, X. Ding, Y. Yu and F. Huang, *Org. Lett.*, 2011, **13**, 4818–4821.
49. Z. Zhang, C. Han, G. Yu and F. Huang, *Chem. Sci.*, 2012, **3**, 3026–3031.
50. J. Wu, K. C.-F. Leung, D. Benítez, J.-Y. Han, S. J. Cantrill, L. Fang and J. F. Stoddart, *Angew. Chem., Int. Ed.*, 2008, **47**, 7470–7474.
51. L. Fang, M. Hmadeh, J. Wu, M. A. Olson, J. M. Spruell, A. Trabolsi, Y.-W. Yang, M. Elhabiri, A.-M. Albrecht-Gary and J. F. Stoddart, *J. Am. Chem. Soc.*, 2009, **131**, 7126–7134.
52. L. Liu, L. Wang, C. Liu, Z. Fu, H. Meier and D. Cao, *J. Org. Chem.*, 2012, 77, 9413–9417.
53. Z. Zhang, B. Xia, C. Han, Y. Yu and F. Huang, *Org. Lett.*, 2010, **12**, 3285–3287.

54. Z. Zhang, Y. Luo, J. Chen, S. Dong, Y. Yu, Z. Ma and F. Huang, *Angew. Chem., Int. Ed.*, 2011, **50**, 1397–1401.
55. H. Suezawa, T. Yoshida, M. Hirota, H. Takahashi, Y. Umezawa, K. Honda, S. Tsuboyama and M. Nishio, *J. Chem. Soc., Perkin Trans. 2*, 2001, 2053–2058.
56. M. Nishio, Y. Umezawa, K. Honda, S. Tsuboyama and H. Suezawa, *CrystEngComm*, 2009, **11**, 1757–1788.
57. O. Takahashi, Y. Kohno and M. Nishio, *Chem. Rev.*, 2010, **110**, 6049–6076.
58. N. L. Strutt, H. Zhang, M. A. Giesener, J. Lei and J. F. Stoddart, *Chem. Commun.*, 2012, **48**, 1647–1649.
59. B. Xia, B. Zheng, C. Han, S. Dong, M. Zhang, B. Hu, Y. Yu and F. Huang, *Polym. Chem.*, 2013, **4**, 2019–2024.
60. C. Han, B. Xia, J. Chen, G. Yu, Z. Zhang, S. Dong, B. Hu, Y. Yu and M. Xue, *RSC Adv.*, 2013, **3**, 16089–16094.
61. K. Wang, C.-Y. Wang, Y. Wang, H. Li, C.-Y. Bao, J.-Y. Liu, S. X.-A. Zhang and Y.-W. Yang, *Chem. Commun.*, 2013, **49**, 10528–10530.
62. Y. Wang, J.-F. Xu, Y.-Z. Chen, L.-Y. Niu, L.-Z. Wu, C.-H. Tung and Q.-Z. Yang, *Chem. Commun.*, 2014, **50**, 7001–7003.
63. A. Wu and L. Isaacs, *J. Am. Chem. Soc.*, 2003, **125**, 4831–4835.
64. M. M. Safont-Sempere, G. Fernández and F. Würthner, *Chem. Rev.*, 2011, **111**, 5784–5814.
65. J. M. Pollino, L. P. Stubbs and M. Weck, *J. Am. Chem. Soc.*, 2004, **126**, 563–567.
66. Y. Guan, M. Ni, X. Hu, T. Xiao, S. Xiong, C. Lin and L. Wang, *Chem. Commun.*, 2012, **48**, 8529–8531.
67. X. Wang, K. Han, J. Li, X. Jia and C. Li, *Polym. Chem.*, 2013, **4**, 3998–4003.
68. X.-Y. Hu, P. Zhang, X. Wu, W. Xia, T. Xiao, J. Jiang, C. Lin and L. Wang, *Polym. Chem.*, 2012, **3**, 3060–3063.
69. X.-Y. Hu, X. Wu, Q. Duan, T. Xiao, C. Lin and L. Wang, *Org. Lett.*, 2012, **14**, 4826–4829.
70. J.-F. Xu, Y.-Z. Chen, L.-Z. Wu, C.-H. Tung and Q.-Z. Yang, *Org. Lett.*, 2013, **15**, 6148–6151.
71. X.-Y. Hu, X. Wu, S. Wang, D. Chen, W. Xia, C. Lin, Y. Pan and L. Wang, *Polym. Chem.*, 2013, **4**, 4292–4297.
72. X. Wang, H. Deng, J. Li, K. Zheng, X. Jia and C. Li, *Macromol. Rapid Commun.*, 2013, **34**, 1856–1862.
73. C. Li, K. Han, J. Li, Y. Zhang, W. Chen, Y. Yu and X. Jia, *Chem. – Eur. J.*, 2013, **19**, 11892–11897.
74. H. Zhang, K. T. Nguyen, X. Ma, H. Yan, J. Guo, L. Zhu and Y. Zhao, *Org. Biomol. Chem.*, 2013, **11**, 2070–2074.
75. T. Ogoshi, Y. Nishida, T. Yamagishi and Y. Nakamoto, *Macromolecules*, 2010, **43**, 3145–3147.
76. T. Ogoshi, Y. Nishida, T. Yamagishi and Y. Nakamoto, *Macromolecules*, 2010, **43**, 7068–7072.
77. T. Ogoshi, Y. Hasegawa, T. Aoki, Y. Ishimori, S. Inagi and T. Yamagishi, *Macromolecules*, 2011, **44**, 7639–7644.

78. S. D. Perrault and W. C. W. Chan, *J. Am. Chem. Soc.*, 2009, **131**, 17042–17043.
79. T. Ogoshi, H. Kayama, T. Aoki, T. Yamagishi, R. Ohashi and M. Mizuno, *Polym. J.*, 2014, **46**, 77–81.
80. T. Ogoshi, Y. Takashima, H. Yamaguchi and A. Harada, *Chem. Commun.*, 2006, 3702–3704.
81. X. Ji, Y. Yao, J. Li, X. Yan and F. Huang, *J. Am. Chem. Soc.*, 2013, **135**, 74–77.
82. S. Sun, X.-Y. Hu, D. Chen, J. Shi, Y. Dong, C. Lin, Y. Pan and L. Wang, *Polym. Chem.*, 2013, **4**, 2224–2229.
83. S. Sun, J.-B. Shi, Y.-P. Dong, C. Lin, X.-Y. Hu and L.-Y. Wang, *Chin. Chem. Lett.*, 2013, **24**, 987–992.
84. C. Koopmans and H. Ritter, *J. Am. Chem. Soc.*, 2007, **129**, 3502–3503.
85. L. Sambe, V. R. de La Rosa, K. Belal, F. Stoffelbach, J. Lyskawa, F. Delattre, M. Bria, G. Cooke, R. Hoogenboom and P. Woisel, *Angew. Chem., Int. Ed.*, 2014, **53**, 5044–5048.
86. X. Ji, J. Chen, X. Chi and F. Huang, *ACS Macro Lett.*, 2014, **3**, 110–113.

CHAPTER 8

Mechanically Interlocked Molecules Based on Pillar[n]arenes

TOMOKI OGOSHI* AND TADA-AKI YAMAGISHI

Graduate School of Natural Science and Technology, Kanazawa University, Kakuma-machi, Kanazawa 920-1192, Japan
*Email: ogoshi@se.kanazawa-u.ac.jp

8.1 Introduction

Mechanically interlocked molecules (MIMs), including rotaxanes,[1–7] catenanes,[1,5–8] polyrotaxanes[2,9] and polycatenanes,[10,11] are constructed from macrocyclic hosts, such as cyclodextrins, crown ethers, calix[*n*]arenes, blue boxes, and cucurbit[*n*]urils. For the preparation of MIMs, it is necessary to use an interlocking chemical reaction while maintaining host–guest complexation. The complexation medium for host–guest complexation of guest molecules with cyclodextrins and cucurbit[*n*]urils is mainly an aqueous medium, hence the formation of MIMs using cyclodextrins and cucurbit[*n*]urils was carried out in an aqueous medium. Therefore, when cyclodextrins and cucurbit[*n*]urils were used as ring components, the limitations of chemical reactions in aqueous media were the main problem for the synthesis of MIMs. However, blue boxes and crown ethers mainly form host–guest complexes in organic solvents. The complexation in organic solvents allows the use of various organic reactions in organic media to form MIMs, therefore various blue box and crown ether-based MIMs have been constructed using various organic reactions. Organo- and water-soluble

Monographs in Supramolecular Chemistry No. 18
Pillararenes
Edited by Tomoki Ogoshi
© The Royal Society of Chemistry 2016
Published by the Royal Society of Chemistry, www.rsc.org

pillar[*n*]arenes can be synthesized by selecting the substituents, and they form stable host–guest complexes in both aqueous and organic media. Therefore, pillar[*n*]arenes should be good wheel segments for the construction of MIMs. Recently, various pillar[*n*]arene–bead MIMs have been reported.[12–22] The copper(I)-catalyzed alkyne–azide cycloaddition (CuAAC) reaction is very useful in the formation of pillar[*n*]arene-based MIMs because the CuAAC reaction between terminal alkynes and azides is high-yielding, functional-group tolerant, and compatible with a wide range of substrates. Furthermore, the *n*-butyl linker with two 1-substituted 1,2,3-triazole moieties, which can be easily synthesized by the CuAAC reaction, is a good station for pillar[5]arenes.[23] Using the CuAAC reaction, pillar[6]arene-based [2]rotaxane and hetero-rotaxane consisting of pillar[*n*]arene and cucurbit[*n*]uril wheels have been synthesized.[16–18] We provide a historical account of the synthesis and properties of pillar[*n*]arene-based MIMs.

8.2 Pillar[5]arene-based Rotaxanes

8.2.1 Pillar[5]arene-based [*n*]Rotaxanes

Rotaxanes are an important class of MIM that have attracted tremendous attention because of their unique structural features and wide potential applications, for example molecular machines and devices, logic gates and molecular actuators. There are typically four methods used to synthesize rotaxanes: end-capping, clipping, slipping and snapping. The first pillar[5]-arene-based polyrotaxane was prepared by Stoddart *et al.* by the end-capping method using the imidation of amine and aldehyde compounds (Scheme 8.1).[12]

End-capping of 1,8-diaminobutane (axle **8.1**) with bulky 3,2-di-*tert*-butyl-benzaldehyde (stopper **8.2**) in the presence of the per-methylated pillar[5]-arene wheel **8.3** afforded a [2]rotaxane (**8.4**). The ^{1}H NMR signals from the central alkyl groups of [2]rotaxane **8.4** showed dramatic upfield shifts compared with those of the free dumbbell, indicating the pillar[5]arene wheels located on the central alkyl groups of the axle. However, its yield was low (7%) because the host–guest complex between 1,8-diaminobutane **8.1** and per-methylated pillar[5]arene **8.3** in $CDCl_3$ is an unfavorable combination ($K = 70 \pm 10$ M^{-1}). The other [2]rotaxanes were synthesized by Huang *et al.* through the inclusion of bulky stoppers by way of ester and urethane bonds ([2]rotaxane **8.5**, Figure 8.1).[24,25] However, in both cases, because of the low association constants between axles and wheel pillar[5]arenes, their yields were also low and their properties were not investigated. We reported the synthesis of [2]- and [3]rotaxanes in good yield that were constructed from a per-ethylated pillar[5]arene wheel (**8.6**) and a pyridinium axle containing an alkyne end (**8.7**) using the CuAAC reaction (Scheme 8.2).[15]

The solubility of per-ethylated pillar[5]arene **8.6** is higher than that of per-methylated pillar[5]arene **8.3** owing to the lower crystallinity of **8.6**, therefore we chose **8.6** as a wheel. Owing to the cationic/π interaction, the

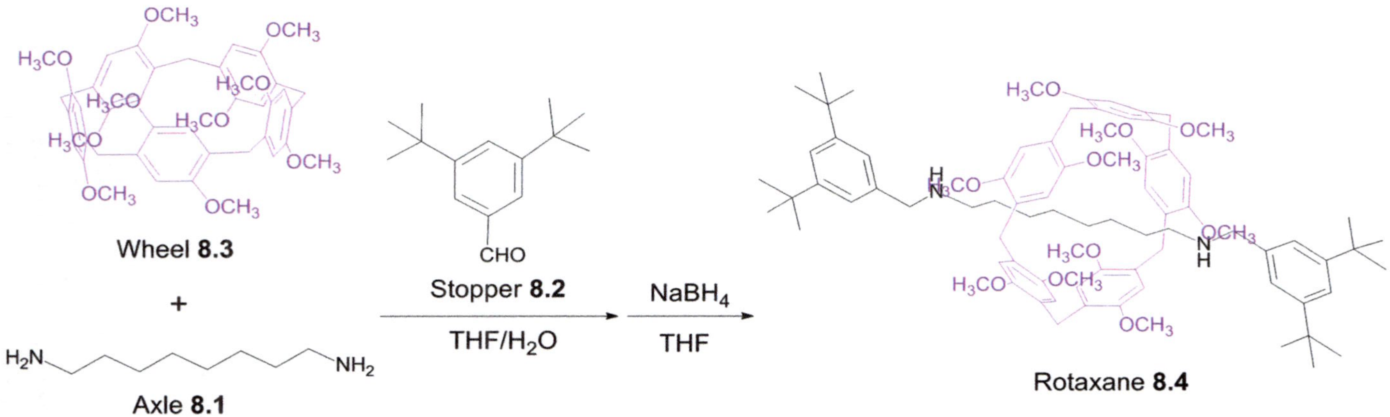

Scheme 8.1 The first pillar[*n*]arene-based [2]rotaxane (**8.4**) prepared by the end-capping method using the imidation between a diamine axle (**8.1**) and aldehyde stopper (**8.2**).

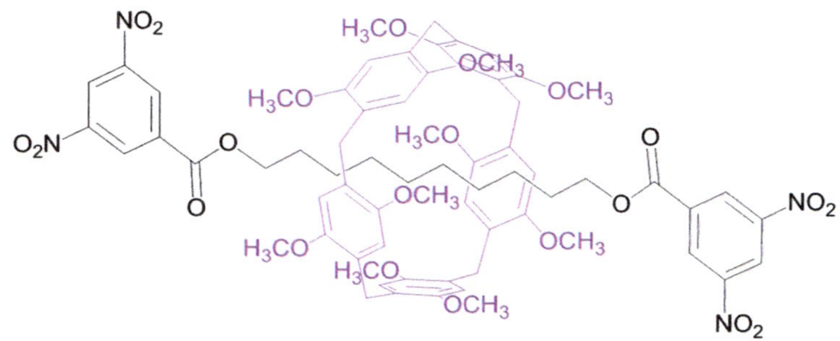

[2]Rotaxane **8.5**

Figure 8.1 [2]Rotaxane (**8.5**) formed by end-capping with a bulky stopper *via* ester formation.

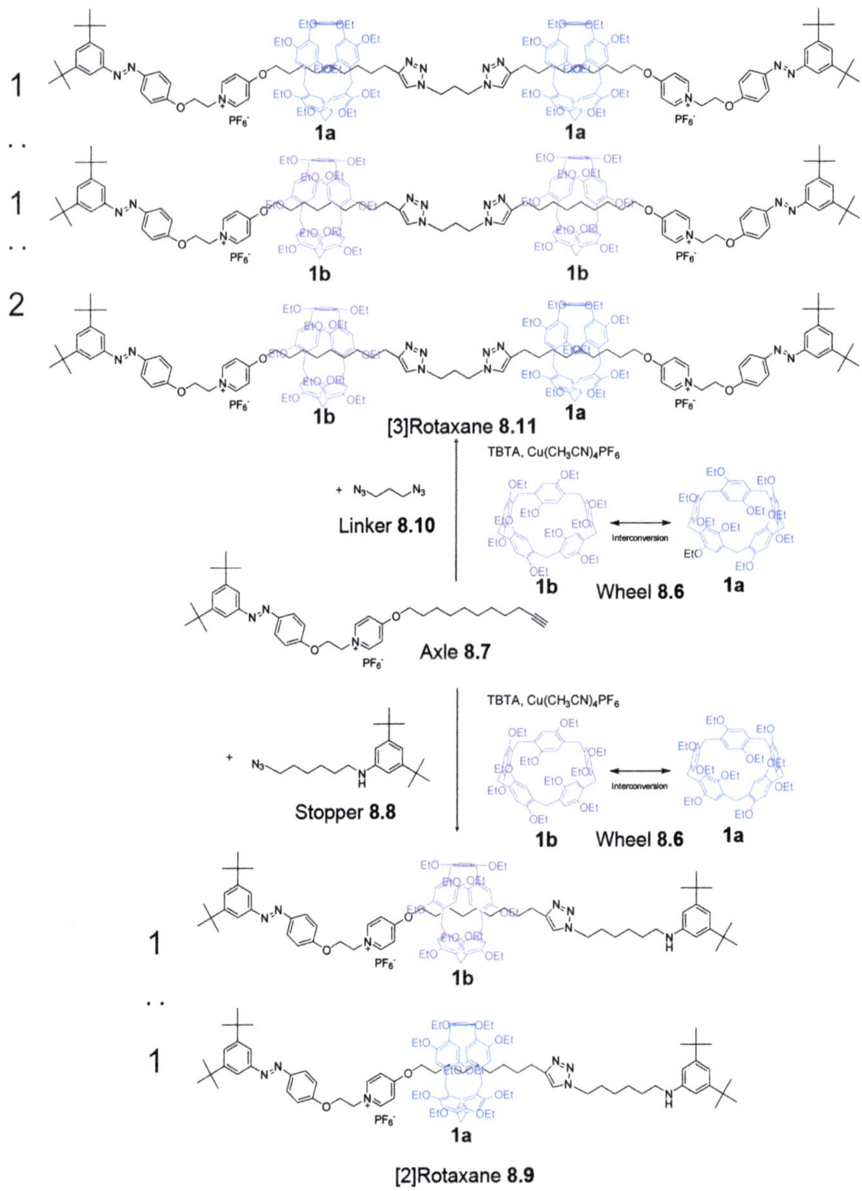

Scheme 8.2 Planar-chiral [2]- (**8.9**) and [3]rotaxanes (**8.11**) prepared by CuAAC reaction.

host–guest complex between axle **8.7** and wheel **8.6** is very stable [$K = (7.6 \pm 1.1) \times 10^4$ M^{-1}], which is favorable for formation of the rotaxane. We used snapping, in which the axle containing an end group allows threading by the cyclic bead (yielding a pseudorotaxane), and then bears another end group. The CuAAC reaction between azide-terminated stopper **8.8** and alkyne-terminated axle **8.7** in the presence of excess wheel **8.6** afforded a [2]rotaxane (**8.9**) in high yield (75%). A [3]rotaxane (**8.11**) was also synthesized in moderate yield (45%) by connecting two pseudo[2]rotaxanes having an alkyne end by the diazide linker **8.10**. We can obtain these rotaxanes in good yields, therefore the planar chirality of these rotaxanes was investigated. On chiral column chromatography of the [2]rotaxane **8.9**, two peaks with equal area (50.8 : 49.2) were observed, indicating that the [2]rotaxane **8.9** was a mixture of two enantiomers. We corrected these two fractions, and obtained circular dichroism (CD) spectra of each fraction. The CD spectra of the fractions were complete mirror images, indicating isolation of enantiopure [2]rotaxane. Inclusion of the axle in the pillar[5]arene cavity inhibited the rotation of pillar[5]arene units, and led to the separation of these enantiomers. Pillar[5]arene has four conformational isomers (see Chapter 4 for details). From the ^{1}H NMR spectrum of [2]rotaxane, the proton signals of the aromatic protons from the pillar[5]arene wheel in [2]rotaxane showed equivalent splitting patterns, indicating that the conformational structure of the pillar[5]arene wheel in [2]rotaxane was C_5 symmetrical **1a** and **1b**. The other [2]rotaxanes, consisting of other possible conformational wheels, were not detected, indicating that the formation of [2]rotaxane is diastereoselective. This probably arises from the fact that the **1a** and **1b** forms should be of lower energy and possess a larger cavity, which allows inclusion of the axle segment. As with [2]rotaxane **8.9**, the pillar[5]arene wheels in [3]rotaxane **8.11** are either C_5 symmetrical **1a** or **1b** conformers, and the formation of [3]rotaxane **8.11** is also diastereoselective. From chiral column chromatography and CD measurements, [3]rotaxane **8.11** was found to be a 1 : 2 : 1 mixture with one axle threaded through two **1a** wheels, one **1a** and one **1b** wheel (*meso* form), and two **1b** wheels.

We recently reported a one-pot synthesis of pillar[5]arene-based [2]rotaxanes containing one and two stations, using a stepwise CuAAC reaction (Scheme 8.3).[18] Pillar[5]arenes form very stable host–guest complexes with *n*-butylene with two 1-substituted 1,2,3-triazole moieties at its end [$K = (1.6 \pm 0.3) \times 10^4\ M^{-1}$]. Triazole moieties can be synthesized easily by the CuAAC reaction, therefore we synthesized [2]rotaxanes using a stepwise CuAAC reaction. First, a CuAAC reaction of an excess of diyne with diazide afforded an axle containing a good station for pillar[5]arenes and reactive alkyne moieties at both ends (axle **8.12**). Second, the CuAAC reaction was used to synthesize a [2]rotaxane containing one wheel and one axle: a CuAAC reaction between axle **8.12** and azide-terminated stopper **8.13** in the presence of an excess of per-ethylated pillar[5]arene **8.6** afforded [2]rotaxane **8.14** (yield 54%). Various axles can be designed by choosing appropriate diyne and diazide derivatives in the synthetic method, therefore the procedure is useful for synthesis of various pillar[5]arene-based [2]rotaxanes. A [2]rotaxane containing two stable stations with equal energy levels can also be prepared by changing the combination of diyne and diazide. We first synthesized an axle with azide-ends (axle **8.15**) by CuAAC reaction of an excess of diazide with diyne. Second, a CuAAC reaction of axle with azide-ends **8.15** and alkyne-terminated stopper **8.16** in the presence of excess

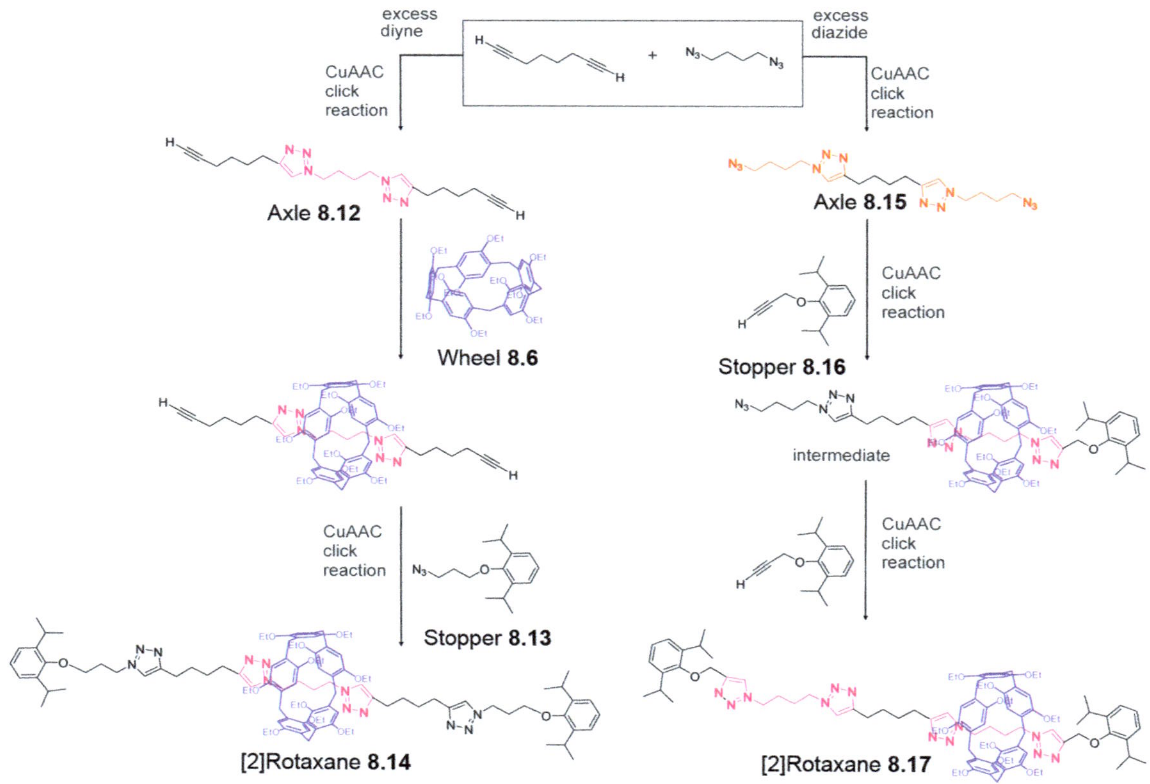

Scheme 8.3 [2]Rotaxanes containing one and two stable stations formed by stepwise CuAAC reaction.

per-ethylated pillar[5]arene **8.6** afforded a [2]rotaxane containing two stable stations (**8.17**). The mechanism for the formation of the [2]rotaxane **8.17** involves an *in situ* stepwise formation of the stations. The CuAAC reaction between the azide-terminated axle **8.15** and alkyne-terminated stopper **8.16** generates an intermediate containing one station. Thus, *in situ* formation of a pseudo[2]rotaxane between the intermediate and wheel **8.6** takes place. Subsequently, another CuAAC reaction between the pseudo[2]rotaxane and alkyne-end stopper **8.16** affords [2]rotaxane **8.17**. The second CuAAC reaction also generates the second station. However, the wheel (**8.6**) cannot slip over the bulky stopper ends (**8.16**). Therefore, a stepwise CuAAC reaction affords [2]rotaxane containing two stations (**8.17**). For the synthesis of [2]rotaxanes containing two stations, the introduction of a protecting group at one station site is required in many cases to inhibit the formation of [3]rotaxanes with two wheels on two stations.[4,26] Therefore, changing the synthesis method is a useful way to produce pillar[5]arene-based [2]rotaxanes with two stations. The [2]rotaxane **8.17** has two stations with equal energy levels, therefore the wheel on the axle shuttles between the two stations. We investigated the shuttling behavior of the wheel between the two stations by variable temperature ^{1}H NMR measurements. The rate constant of the pillar[5]arene wheel shuttling in the [2]rotaxane **8.17** at 25 °C (k) in DMSO-d_6 was 11.0 s^{-1}, which is approximately 22 times faster than that observed in toluene-d_8 (0.5 s^{-1}). This arises from the more efficient solvation between the polar solvent (DMSO-d_6) and the station when compared with that between the nonpolar solvent (toluene-d_8) and the station. Thus, the simple method used to produce [2]rotaxanes with two stations can also be used in the formation of various pillar[5]arene-based [2]rotaxanes that show wheel shuttling between two stations.

Stoddart *et al.* reported the high-yield synthesis of highly ordered hetero-rotaxanes consisting of pillar[5]arene and cucurbit[6]uril wheels (Scheme 8.4).[16] The co-operative hydrogen-bonding interactions between urea moieties of cucurbit[6]urils and phenol moieties of pillar[5]arenes led to a high level of pre-organization for the rotaxanation, and yielded two [4]rotaxanes and one [5]rotaxane in >90% yields. In the case where short spacers were used between the viologen and azide moieties **8.18** ($n=1$) and **8.19** ($n=2$), [4]rotaxanes (**8.23** and **8.24**), consisting of two cucurbit[6]uril and one pillar[5]arene (**8.22**) wheels, were obtained in >95% yields. In contrast, with the use of a long spacer (**8.20**) ($n=3$), a [5]rotaxane (**8.25**) consisting of two cucurbit[6]urils and two pillar[5]arene (**8.22**) wheels was obtained in >90% yields. This is because the space is sufficient to include two pillar[5]arene wheels in the case of long alkyl spacers. In the case of a longer spacer (**8.21**) ($n=4$), a hetero[5]rotaxane (**8.26**) was obtained but its yield was not high (30%). From these results, the length between the cucurbit[6]uril and the pillar[5]arene (**8.22**) is very important to the formation of inter-molecular hydrogen bonds, which induces pre-organization of the rotaxane structure. The pre-organization also accelerates the reaction rate of 1,3-dipolar cycloadditions between azides and alkynes [*ca.* 10^5-fold

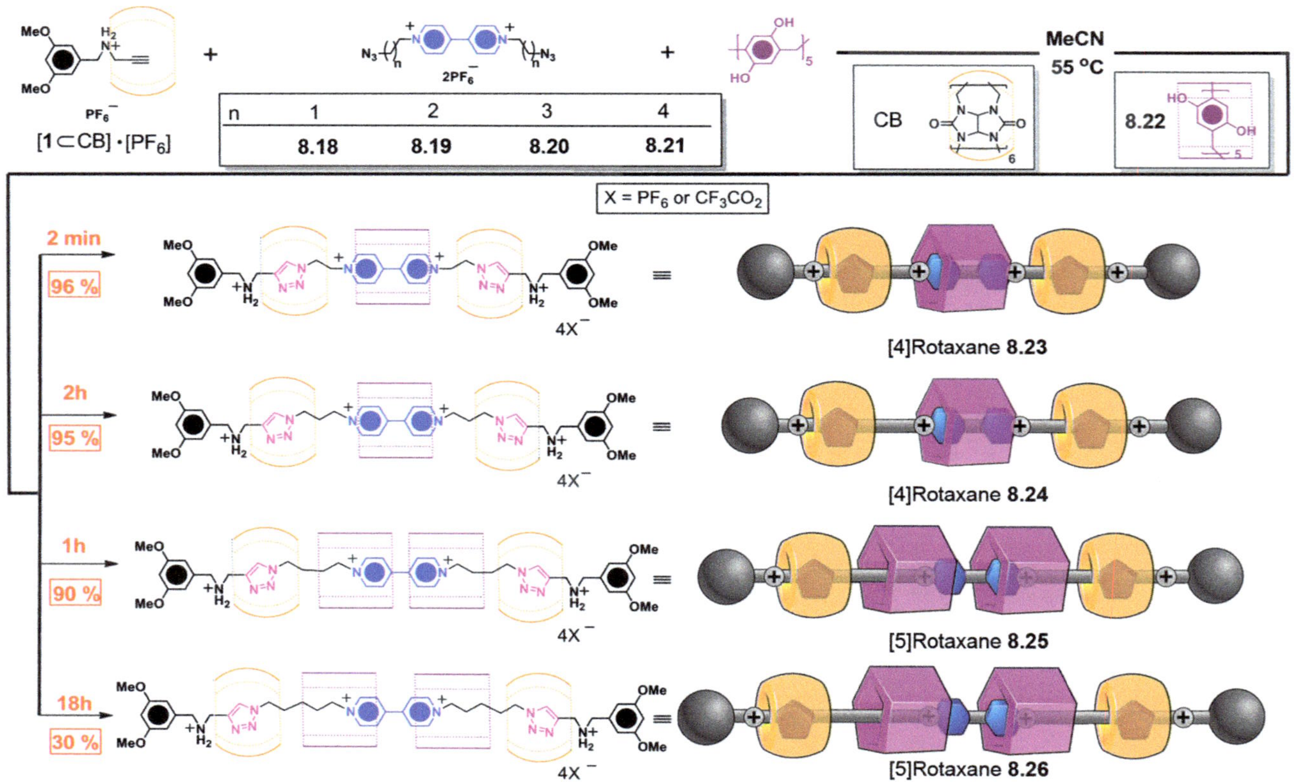

Scheme 8.4 Synthesis of the hetero[4]- and hetero[5]rotaxanes consisting of pillar[5]arene and cucurbit[6]uril wheels by copper-free AAC reaction.
Reproduced with permission from ref. 16. Copyright 2013 American Chemical Society.

compared with the Cu(I) catalyzed reaction], therefore the Cu(I) catalyst is not required for the formation of these rotaxanes. The conformation of the pillar[5]arene wheels in the rotaxanes was mainly C_5 symmetrical **1a/b** forms, but they also contained **2b/a** and **3a/b** forms. Conformational isomers **4a/b**, which are the most stable conformation in per-hydroxylated pillar[5]arene **8.22**, were not detected.[27] The **1a/b**, **2a/b** and **3a/b** forms allow the majority of phenolic groups to participate in the inter-molecular hydrogen bond network with cucurbit[6]urils because **1a/b**, **2a/b** and **3a/b** forms have more than six phenolic groups with no intra-molecular hydrogen bonds. However, **4a/b** conformers only have two phenolic groups with no intra-molecular hydrogen bonds, therefore the majority of the phenolic groups cannot participate in the inter-molecular hydrogen bond network with cucurbit[6]urils.

We have reported the high-yield synthesis of [2]rotaxanes using liquid state pillar[5]arene.[13] The pillar[5]arene carrying 10 tri(ethylene oxide) chains (**8.27**) is a liquid at room temperature because tri(ethylene oxide) chains are soft and contain flexible segments (Scheme 8.5). We used this

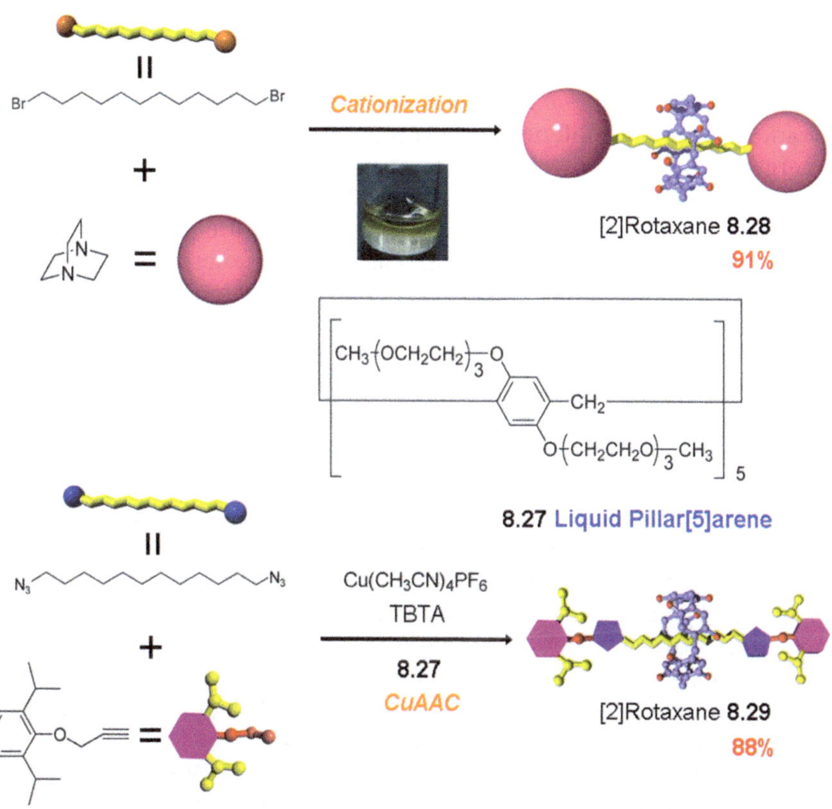

Scheme 8.5 Pillar[*n*]arene-based [2]rotaxane using liquid pillar[5]arene **8.27**. Reproduced with permission from ref. 13. Copyright 2012 American Chemical Society.

liquid pillar[5]arene (**8.27**) as a solvent for complexation. In the liquid pillar[5]arene, host–guest complexation was extremely efficient even in the unfavorable combinations of host–guest complexes with low association constants in common solvents. This is because the liquid pillar[5]arene system is able to maximize the concentration of the host and guest species. Because the host–guest complexation was highly maintained in the liquid pillar[5]arene system, we successfully synthesized [2]rotaxanes in high yield in the liquid pillar[5]arene. [2]Rotaxanes (**8.28** and **8.29**) were obtained in high yields (>88%) by end-capping through cationization and the CuAAC reaction in the liquid pillar[5]arene solvent. In contrast, their yields were quite low (8%) in a normal solvent system. Therefore, the liquid pillar[5]-arene is a useful medium for the synthesis of MIMs even with unfavorable statistical combinations of host–guest complexes. In particular, the CuAAC reaction in liquid pillar[5]arene is beneficial for preparing various MIMs because this reaction has a high yield and is functional-group tolerant and compatible with a wide range of substrates.

We synthesized a [2]rotaxane showing efficient Förster resonance energy transfer.[19] A CuAAC reaction between axle **8.30** and a perylene stopper with azide moiety (**8.31**) in the presence of A1/A2 di-pyrene-appended pillar[5]arene wheel **8.32** gave a [2]rotaxane (**8.33**) in high yield (80%). The [2]rotaxane **8.33** showed strong emission from perylene by excitation of pyrene moieties at 345 nm, indicating that Förster resonance energy transfer from pyrene to perylene took place (Figure 8.2). The emission intensity of [2]rotaxane **8.33** was approximately 26 times and 20 times that of the dumbbell **8.34** and a mixture of wheel **8.32** and dumbbell **8.34**, respectively. The efficiency of the energy transfer was approximately 75% and the calculated Förster radius in the [2]rotaxane **8.33** was 1.8 nm.

Huang and co-workers reported a solvent- and thermo-responsive [2]rotaxane containing an imidazolium cation on the axle (**8.35**) (Figure 8.3).[28] In $CDCl_3$ (nonpolar solvent), the pillar[5]arene was located on the imidazolium station. In contrast, as the polarity increased, the pillar[5]arene ring moved from the imidazolium part to the alkyl moiety. The pillar[5]arene ring on the axle also moved from the imidazolium part to the alkyl moiety upon heating. This is because the relative host–guest complex stability between the imidazolium/pillar[5]arene and alkyl/pillar[5]arene complexes is different and is affected by the solvent and temperature.

Huang and co-workers also reported a solvent-driven molecular spring (**8.36**; Figure 8.4) formed by end-capping a doubly threaded pseudo-rotaxane dimer and urea bond formation.[29] In $CDCl_3$, the pillar[5]arene ring in molecular spring **8.36** was located on the urea moiety, owing to the stronger HN/π interactions when compared with CH/π interactions. When the spring was in a contracted state (a), it had a length of 31 Å. However, as the content of DMSO-d_6 increased (b), the pillar[5]arene ring moved from the urea to the alkyl moiety because NH/π interactions weaken in the polar solvent. When the spring was in an extended state (c), it had a length of 37 Å.

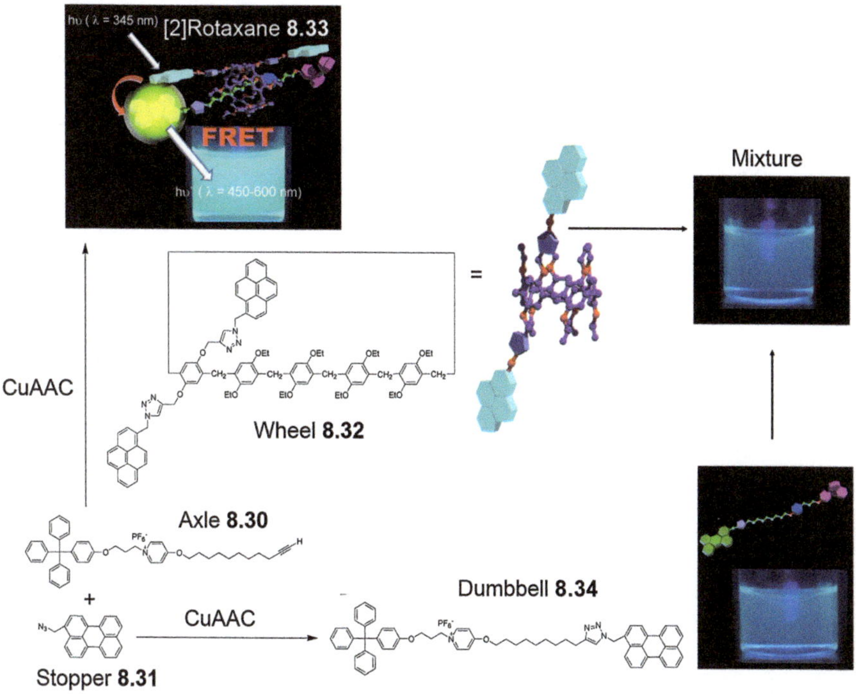

Figure 8.2 [2]Rotaxane **8.33** showing Förster resonance energy transfer from the pyrene moiety in the wheel to the perylene stopper in the axle. Reproduced from ref. 19 with permission from The Royal Society of Chemistry.

Xia and Xue synthesized a pillar[5]arene-based [1]rotaxane using CH/π and cation/anion electrostatic interactions.[30] First, a highly pre-organized pseudorotaxane structure was prepared by mixing an amine guest molecule with a bulky stopper at one end (**8.37**) and a mono-carboxylic acid pillar[5]arene (**8.38**). Amidation between the primary amine group on the guest and the carboxylic acid group on the pillar[5]arene then afforded [1]rotaxane **8.39** in high yield (73%) (Figure 8.5).

8.2.2 Pillar[5]arene-based Polyrotaxanes and Topological Gels

Polyrotaxanes, in which many cyclic molecules are mechanically interlocked onto a polymeric chain, have been synthesized using macrocyclic compounds such as cyclodextrins, crown ethers and blue boxes. Synthesis of pillar[5]arene-based polyrotaxanes consisting of cationic polymers, such as viologen polymer, is relatively easy because pillar[5]arene forms stable host–guest complexes with cationic molecules (see details in Chapter 9);[31,32] however, it is difficult to synthesize nonionic polymer-based polyrotaxanes. To synthesize nonionic polymer-based polyrotaxanes, we used a liquid

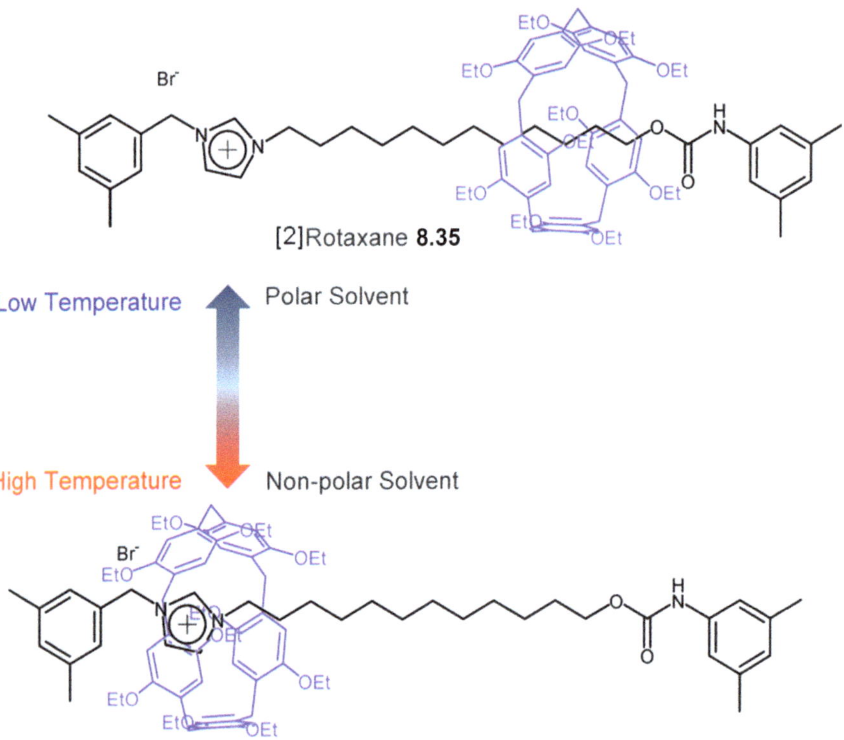

Figure 8.3 Solvent- and thermo-responsive [2]rotaxane **8.35**.

pillar[5]arene system, which enabled an efficient bulk-state synthetic method for rotaxanes even in the weak association constant between the axle and wheels (Scheme 8.5).[33]

A nonionic polymer chain, poly(tetrahydrofuran) (polyTHF), was dissolved in liquid pillar[5]arenes **8.27** and **8.40**. An end-capping reaction was then carried out using the CuAAC reaction (Figure 8.6).[33] Under the optimized conditions, the yield of the nonionic polyrotaxanes **8.41** and **8.42** was moderate (44%). Pillar[5]arene wheels in polyrotaxanes **8.42** have reactive vinyl groups; therefore, through inter-molecular crosslinking of the wheel segments in polyrotaxane **8.42** by olefin metathesis, we obtained a nonionic topological gel (**8.43**). The topological gel (**8.43**) swelled in organic solvents, such as chloroform, dichloromethane and THF, but not in nonpolar hexane or highly polar solvents, such as water, methanol and acetone.

8.3 Pillar[5]arene-based Catenanes

A catenane is one of the MIMs consisting of two or more interlocked macrocycles. We recently reported the first pillar[5]arene-based [2]catenane.[21] A pyridinium salt with alkene moieties at both ends (**8.44**) was prepared. Pillar[5]arene formed a stable host–guest complex with the molecule

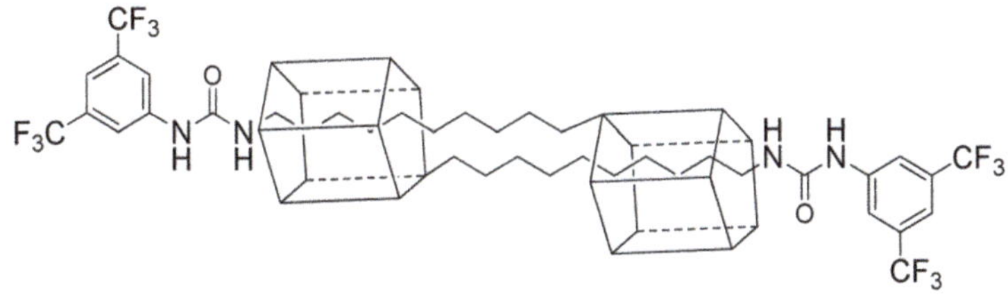

8.36

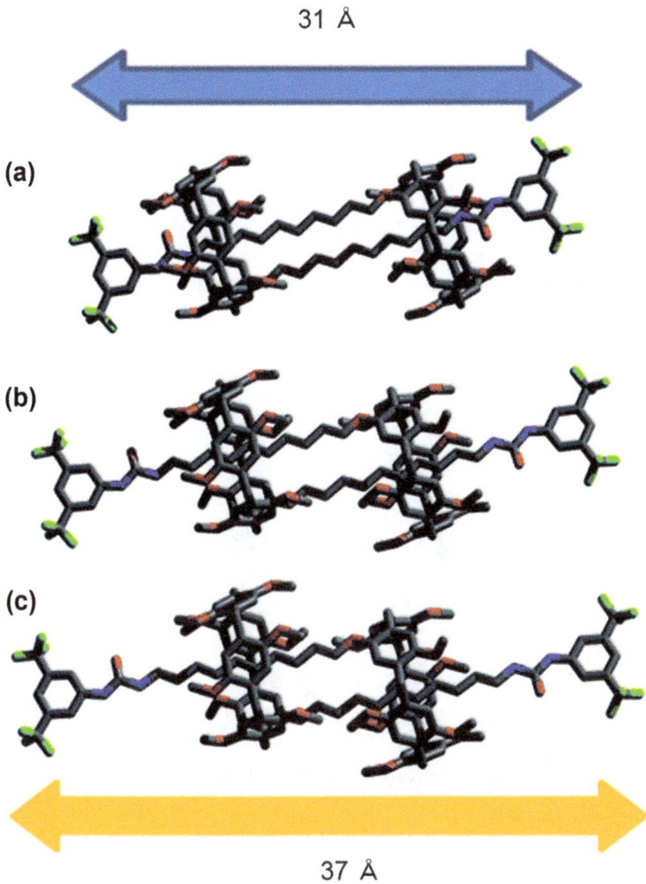

Figure 8.4 Solvent-driven molecular spring **8.36.**
Reproduced from ref. 29 with permission from The Royal Society of Chemistry.

[$K = (1.76 \pm 0.19) \times 10^3$ M^{-1}]. The driving force was the cation/π interaction. The concentrations of the pyridinium salt (**8.44**) and the wheel (**8.6**) are dominant for the synthesis of [2]catenane (**8.45**). A mixture of polymeric products was obtained when a ring-closing metathesis reaction occurred

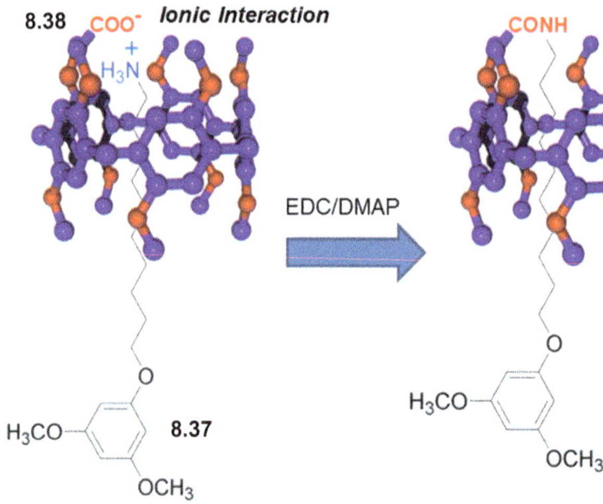

Figure 8.5 High-yield synthesis of a pillar[5]arene-based [1]rotaxane (**8.39**) by pre-organization using ionic interactions between the pillar[5]arene (**8.38**) and the guest (**8.37**).
Reproduced from ref. 22 with permission from The Royal Society of Chemistry.

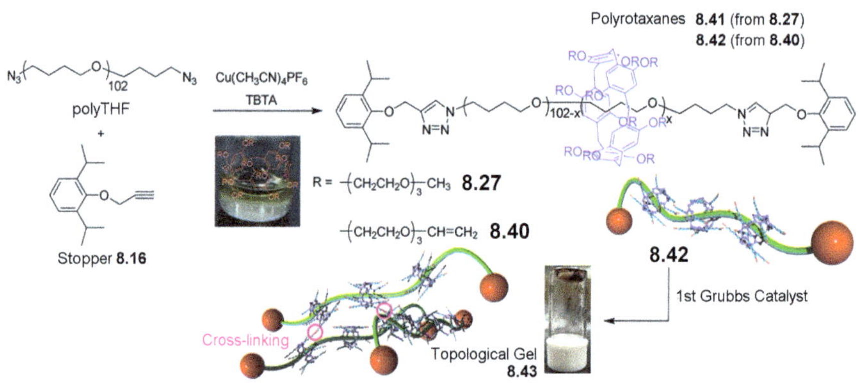

Figure 8.6 Nonionic polyrotaxanes (**8.41** and **8.42**) and a topological gel (**8.43**) prepared from liquid pillar[5]arenes.
Reproduced from ref. 33 with permission from The Royal Society of Chemistry.

between the alkenes at the ends in a high concentration of the pyridinium salt **8.44** (5 mM) in the presence of an excess of pillar[5]arene wheel **8.6** (3 equiv. of **8.44**). In contrast, in a dilute concentration of **8.44** (1 mM) in the presence of an excess of pillar[5]arene wheel **8.6** (100 equiv. of **8.44**), the target [2]catenane **8.45** was obtained in moderate yield (24%; Scheme 8.6). As with the synthesis of [2]rotaxanes, the catenation is diastereoselective. The [2]catenane was a racemic mixture consisting of symmetrical C_5 **1a**/**b** forms of pillar[5]arene wheels.

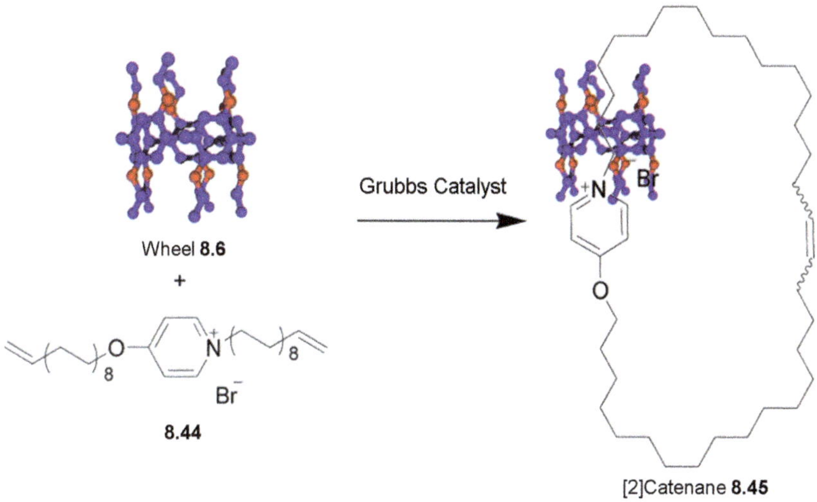

Scheme 8.6 The first pillar[*n*]arene-based [2]catenane **8.45**.

We also synthesized pseudo[1]catenane, which consists of two bicyclic structures.[34] One cycle is a pillar[5]arene ring and the other is a cyclic alkyl chain. The cyclic alkyl chain is connected to one pillar[5]arene unit. The pseudo[1]catenane **8.47** (Figure 8.7) was synthesized by a CuAAC reaction between a pillar[5]arene having two alkyne moieties (**8.46**) and 1,12-diazidodododecane (yield 17%). The pseudo[1]rotaxane has four possible conformers: in-**1a**, in-**1b**, out-**1a** and out-**1b**. We successfully isolated in-**1a** and in-**1b** using chiral column chromatography. The inclusion and de-threading of the alkyl chain is the trigger for the planar-chiral inversion from in-**1a** to out-**1b** or from in-**1b** to out-**1a**. Thus, an achiral guest molecule, 1,4-dicyanobutane, was added to an in-**1a** or in-**1b** solution. The 1,4-dicyanobutane displaced the alkyl chain located in the cavity, and de-threading of the alkyl chain was the trigger to induce planar-chiral inversion. Pillar[5]arenes form complexes with linear solvents, such as acetonitrile and dichloromethane, therefore these linear solvents also induced planar-chiral inversion of the pseudo[1]catenane **8.47**. The inclusion of these linear solvents was the trigger for de-threading of the alkyl chain of **8.47**. However, branch-shaped solvents, such as methanol and chloroform, did not form complexes with **8.47**, and therefore did not induce planar-chiral inversion. Hence, the small size differences between these solvents, such as chloroform and dichloromethane, can be sensed by the planar chirality of pseudo[1]catenane **8.47**.

Huang *et al.* reported a pseudo[1]catenane produced by a combination of pillar[5]arene and crown ether host–guest chemistry.[35] A novel co-pillar[5]arene, bearing two benzo-18-crown-6 units (**8.48**), was synthesized by a CuAAC reaction (Figure 8.8). 1,8-Diaminooctane was added as the guest molecule to thread into the pillar[5]arene cavity to form the

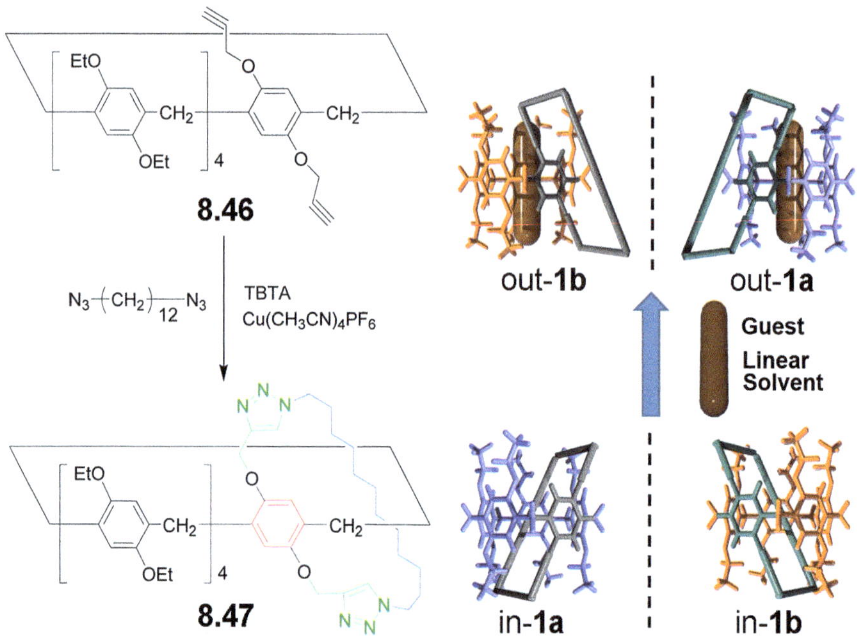

Figure 8.7 Guest and liner solvent-triggered planar chiral inversion of pseudo[1]catenane **8.47**.

[2]pseudorotaxane structure. The addition of 2 equiv. of trifluoroacetic acid to 1,8-diaminooctane resulted in the formation of a 1,8-diammoniumbutane salt. The 1,8-diammoniumbutane salt formed host–guest complexes with the benzo-18-crown-6. The two different host–guest complexes formed between the pillar[5]arene cavity and the alkane chain, and between benzo-18-crown-6 and the 1,8-diammoniumbutane salt, by threading followed by complexation led to the formation of pseudo[1]catenane.

8.4 Pillar[6]arene-based Rotaxanes

Our group reported the first pillar[6]arene-based rotaxane formed by the CuAAC reaction, which is the same method used to synthesize pillar[5]arene-based [2]rotaxanes.[14] A cationic pyridinium derivative with reactive terminal alkyne moiety (**8.49**) was used as an axle, and a bulky trityl moiety with an azide end (**8.50**) was used as a stopper. A CuAAC reaction between the alkyne moiety in the axle (**8.49**) and azido moiety in the bulky trityl stopper (**8.50**) in the presence of the pillar[6]arene wheel **8.51** afforded [2]rotaxane **8.53**. However, owing to the weak association constant between the axle **8.49** and the pillar[6]arene ring **8.51** ($K = 19.1 \pm 2.2\ M^{-1}$), the yield of [2]rotaxane **8.53** was low (14%). In contrast, when we used pillar[5]arene **8.6** as a ring, the yield of the [2]rotaxane **8.52** (Figure 8.9) was high (60%) owing to the stable host–guest complex between pillar[5]arene **8.6** and the pyridinium moiety

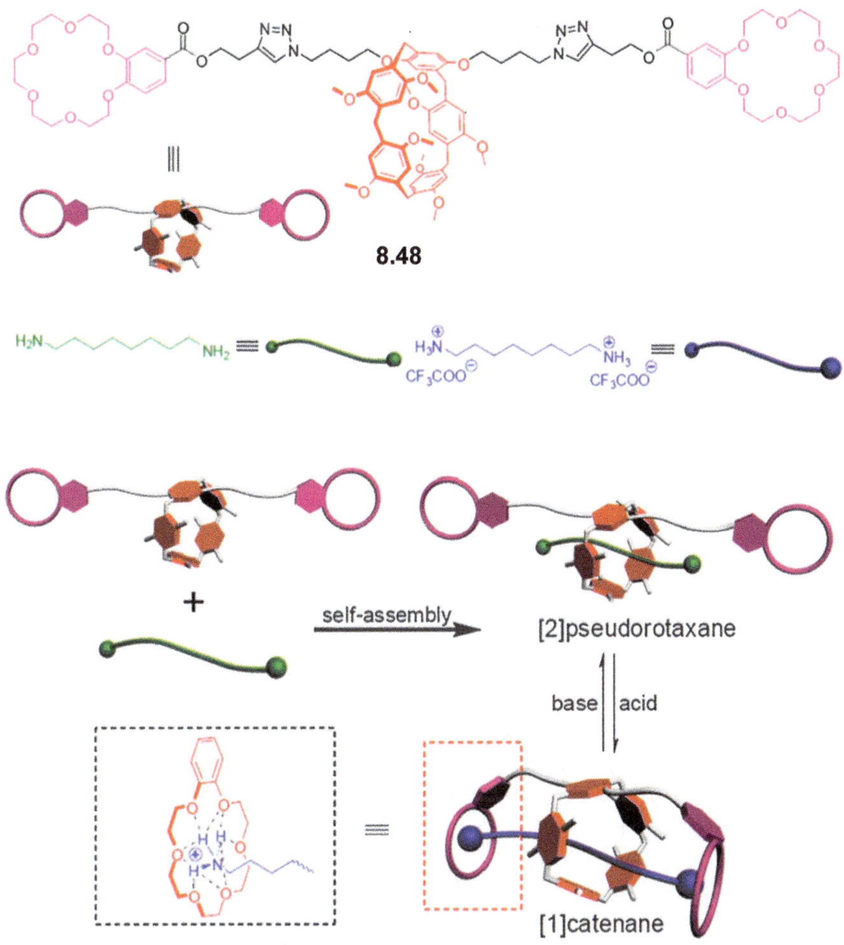

Figure 8.8 Cartoon representation of the pH-responsive dynamic [1]catenane **8.48**. Reproduced from ref. 35 with permission from The Royal Society of Chemistry.

(**8.49**) [$K = (1.01 \pm 0.21) \times 10^4\ M^{-1}$]. The two [2]rotaxanes (**8.52** and **8.53**) have two alkyl linker stations, short and long alkyl linkers between the pyridinium and trityl moieties; these are defined as station I and II, respectively. In the case of the pillar[5]arene-based [2]rotaxane **8.52**, the pillar[5]arene ring was located on station II and did not move to station I even upon heating and cooling because the cavity size of the pillar[5]arene was not sufficient to slip over station I. However, the pillar[6]arene ring moved from station I to station II upon heating, because the cavity size of pillar[6]arene is larger than that of pillar[5]arene. At low temperature, the binding site of the pyridinium moiety was contained in station I, therefore locating at station I was favorable from the entropy point of view. However, in terms of the entropy,

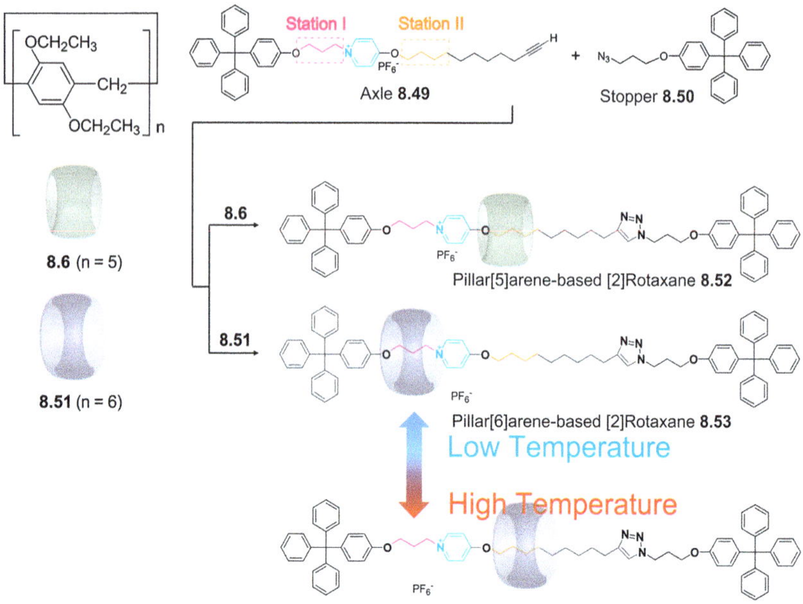

Figure 8.9 Thermo-responsive [2]rotaxane **8.53** constructed from a pillar[6]arene wheel.

location at station II is preferable upon heating. Rotation of the pillar[6]arene units occurred in the [2]rotaxane **8.53** while no rotation of pillar[5]arene units was observed in the [2]rotaxane **8.52**. The large cavity size of the pillar[6]arene in [2]rotaxane **8.53** allowed rotation of the units even with inclusion of the axle segment.

Stoddart *et al.* reported the efficient synthesis of hetero[4]rotaxanes consisting of one axle, two cucurbit[6]uril rings and one pillar[6]arene ring (**8.54**).[17] The methodology for the synthesis of the hetero[4]rotaxanes is the same as for the synthesis of hetero[4,5]rotaxanes consisting of pillar[5]arenes and cucurbit[6]urils.[16] When axles containing a viologen station were used (Figure 8.10, **8.55** and **8.56**), the yields of the hetero[4]-rotaxanes (**8.58** and **8.59**) were 38–39%. Incorporation of a ter-phenyl moiety into the axle (**8.57**) increased the rigidity, which led to increasing the yield of the hetero[4]rotaxane (**8.60**, 68%). The units of the pillar[6]arene ring in the hetero[4]rotaxanes rotated freely at room temperature on account of the large cavity of pillar[6]arene. As described in Chapter 4, pillar[6]arene has 13 conformers. The most stable conformation of the per-hydroxylated pillar[6]arene **8.54** in the hetero-[4]rotaxane is the alternating up-to-down arrangement among the 13 conformers. In the X-ray crystal state (Figure 8.10b), the conformation of the pillar[6]arene ring in the hetero[4]-rotaxanes also formed the up-to-down arrangement, with a structure that was stabilized by an intra-molecular hydrogen bond in the ring and an inter-molecular hydrogen bond between the pillar[6]arene ring and cucurbit[6]urils.

(a)

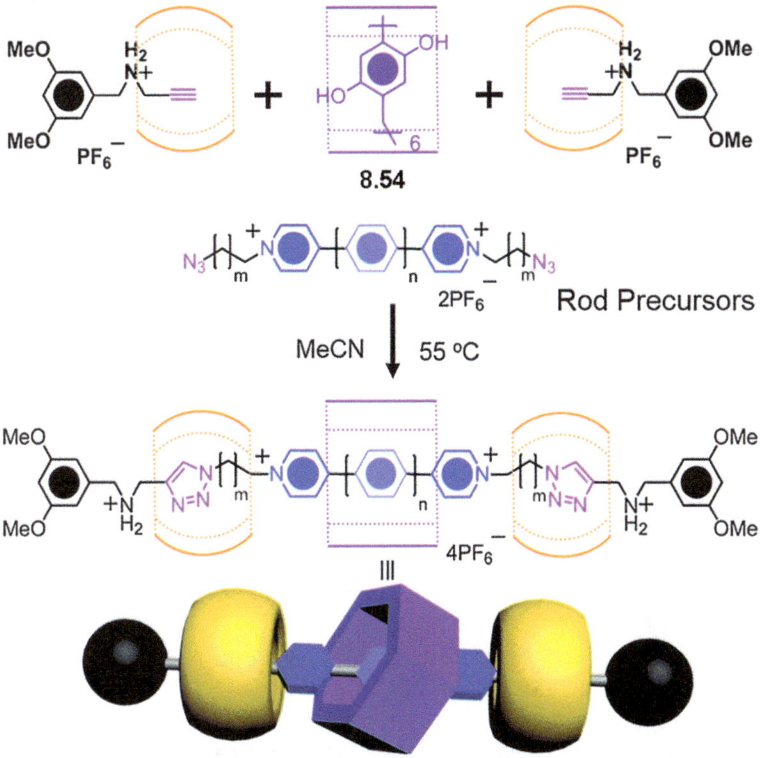

m	n	Rod Precursors	[4]Rotaxanes	Yield (%)
1	0	**8.55**	**8.58**	39
2	0	**8.56**	**8.59**	38
1	1	**8.57**	**8.60**	68

(b)

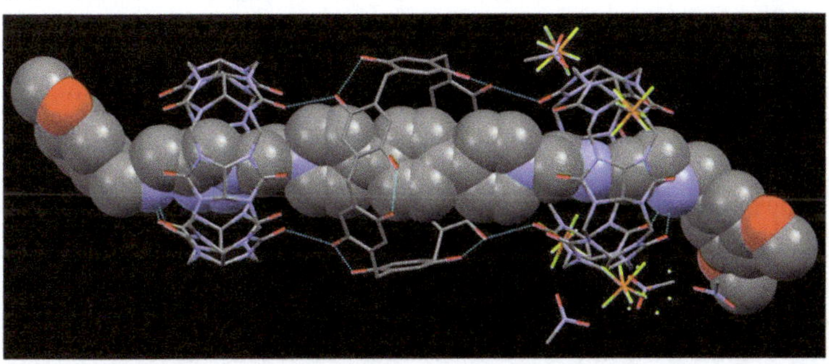

Figure 8.10 (a) Synthesis of the hetero[4] rotaxanes consisting of one pillar[5]arene and two cucurbit[6]uril wheels by copper-free AAC reaction and (b) X-ray crystal structure of **8.60**.
Reproduced from ref. 17 with permission from The Royal Society of Chemistry.

8.5 Conclusion and Outlook

Pillar[*n*]arene-based MIMs, such as rotaxanes, catenanes, polyrotaxanes and topological gels, have been synthesized. The CuAAC reaction is very useful for the synthesis of MIMs because it is a highly efficient reaction, proceeding in many different solvents and under varied conditions, to generate a triazole moiety, which is a good station for pillar[5]arenes. We can design various pillar[*n*]arene MIMs using the CuAAC reaction, therefore one of the challenges for the synthesis of pillar[*n*]arene-based MIMs is the enantioselective synthesis of MIMs. Pillar[*n*]arene-based MIMs should show high thermal stability and useful mechanical properties because pillar[*n*]arenes have many phenolic moieties, such as phenolic resins. Therefore, the application of pillar[*n*]arene-based MIMs for supramolecular materials will be another challenge.

References

1. S. A. Nepogodiev and J. F. Stoddart, *Chem. Rev.*, 1998, **98**, 1959–1976.
2. G. Wenz, B.-H. Han and A. Müller, *Chem. Rev.*, 2006, **106**, 782–817.
3. M. Xue, Y. Yang, X. Chi, X. Yan and F. Huang, *Chem. Rev.*, 2015, **115**, 7398–7501.
4. M. Alvarez-Pérez, S. M. Goldup, D. A. Leigh and A. M. Z. Slawin, *J. Am. Chem. Soc.*, 2008, **130**, 1836–1838.
5. M. J. Langton and P. D. Beer, *Acc. Chem. Res.*, 2014, **47**, 1935–1949.
6. J. P. Sauvage, *Acc. Chem. Res.*, 1998, **31**, 611–619.
7. F. Vögtle, T. Dünnwald and T. Schmidt, *Acc. Chem. Res.*, 1996, **29**, 451–460.
8. M. Fujita, *Acc. Chem. Res.*, 1999, **32**, 53–61.
9. A. Harada, A. Hashidzume, H. Yamaguchi and Y. Takashima, *Chem. Rev.*, 2009, **109**, 5974–6023.
10. Z. Niu and H. W. Gibson, *Chem. Rev.*, 2009, **109**, 6024–6046.
11. L. Chen, Q. Chen, M. Wu, F. Jiang and M. Hong, *Acc. Chem. Res.*, 2015, **48**, 201–210.
12. N. L. Strutt, R. S. Forgan, J. M. Spruell, Y. Y. Botros and J. F. Stoddart, *J. Am. Chem. Soc.*, 2011, **133**, 5668–5671.
13. T. Ogoshi, T. Aoki, R. Shiga, R. Iizuka, S. Ueda, K. Demachi, D. Yamafuji, H. Kayama and T. Yamagishi, *J. Am. Chem. Soc.*, 2012, **134**, 20322–20325.
14. T. Ogoshi, D. Yamafuji, T. Aoki and T. Yamagishi, *Chem. Commun.*, 2012, **48**, 6842–6844.
15. T. Ogoshi, D. Yamafuji, T. Aoki, K. Kitajima, T. Yamagishi, Y. Hayashi and S. Kawauchi, *Chem. – Eur. J.*, 2012, **18**, 7493–7500.
16. C. F. Ke, N. L. Strutt, H. Li, X. S. Hou, K. J. Hartlieb, P. R. McGonigal, Z. D. Ma, J. Iehl, C. L. Stern, C. Y. Cheng, Z. X. Zhu, N. A. Vermeulen, T. J. Meade, Y. Y. Botros and J. F. Stoddart, *J. Am. Chem. Soc.*, 2013, **135**, 17019–17030.

17. X. Hou, C. Ke, C. Cheng, N. Song, A. K. Blackburn, A. A. Sarjeant, Y. Y. Botros, Y.-W. Yang and J. F. Stoddart, *Chem. Commun.*, 2014, **50**, 6196–6199.
18. T. Ogoshi, R. Iizuka, D. Kotera and T. Yamagishi, *Org. Lett.*, 2015, **17**, 350–353.
19. T. Ogoshi, D. Yamafuji, T. Yamagishi and A. M. Brouwer, *Chem. Commun.*, 2013, **49**, 5468–5470.
20. T. M. N. Trinh, I. Nierengarten, M. Holler, J. L. Gallani and J. F. Nierengarten, *Chem. – Eur. J.*, 2015, **21**, 8019–8022.
21. K. Kitajima, T. Ogoshi and T. Yamagishi, *Chem. Commun.*, 2014, **50**, 2925–2927.
22. T. Ogoshi and T. Yamagishi, *Chem. Commun.*, 2014, **50**, 4776–4787.
23. K. Han, Y. Zhang, J. Li, Y. Yu, X. Jia and C. Li, *Eur. J. Org. Chem.*, 2013, **2013**, 2057–2060.
24. P. F. Wei, X. Z. Yan, J. Y. Li, Y. J. Ma, Y. Yao and F. H. Huang, *Tetrahedron*, 2012, **68**, 9179–9185.
25. S. Y. Dong, C. Y. Han, B. Zheng, M. M. Zhang and F. H. Huang, *Tetrahedron Lett.*, 2012, **53**, 3668–3671.
26. P. G. Young, K. Hirose and Y. Tobe, *J. Am. Chem. Soc.*, 2014, **136**, 7899–7906.
27. T. Ogoshi, T. Aoki, K. Kitajima, S. Fujinami, T. Yamagishi and Y. Nakamoto, *J. Org. Chem.*, 2011, **76**, 328–331.
28. S. Y. Dong, J. Y. Yuan and F. H. Huang, *Chem. Sci.*, 2014, **5**, 247–252.
29. Z. B. Zhang, C. Y. Han, G. C. Yu and F. H. Huang, *Chem. Sci.*, 2012, **3**, 3026–3031.
30. B. Y. Xia and M. Xue, *Chem. Commun.*, 2014, **50**, 1021–1023.
31. T. Ogoshi, Y. Nishida, T. Yamagishi and Y. Nakamoto, *Macromolecules*, 2010, **43**, 3145–3147.
32. T. Ogoshi, Y. Nishida, T. A. Yamagishi and Y. Nakamoto, *Macromolecules*, 2010, **43**, 7068–7072.
33. T. Ogoshi, T. Aoki, S. Ueda, Y. Tamura and T. A. Yamagishi, *Chem. Commun.*, 2014, **50**, 6607–6609.
34. T. Ogoshi, T. Akutsu, D. Yamafuji, T. Aoki and T. A. Yamagishi, *Angew. Chem., Int. Ed.*, 2013, **52**, 8111–8115.
35. X. Z. Yan, P. F. Wei, Z. T. Li, B. Zheng, S. Y. Dong, F. H. Huang and Q. Z. Zhou, *Chem. Commun.*, 2013, **49**, 2512–2514.

CHAPTER 9

Supramolecular Assemblies from Pillararenes (Micellar, Vesicular and Tubular Formations)

HUACHENG ZHANG[a] AND YANLI ZHAO*[a,b]

[a] Division of Chemistry and Biological Chemistry, School of Physical and Mathematical Sciences, Nanyang Technological University, 21 Nanyang Link, 637371, Singapore; [b] School of Materials Science and Engineering, Nanyang Technological University, 639798, Singapore
*Email: zhaoyanli@ntu.edu.sg

9.1 Introduction

Pillararenes, a novel type of host molecule discovered in 2008 by T. Ogoshi and T.-A. Yamagishi,[1] are a rising star in supramolecular chemistry. Pillararenes integrate the advantages of other existing host molecules, *e.g.*, similar cavity size to cyclodextrins, similar chemical composition to calixarenes, and similar molecular recognition ability to cucurbiturils. Furthermore, pillararenes possess their own unique characteristics, such as relatively easy preparation and functionalization as well as having highly symmetrical structures, which can recognize specific guests without further modification.[2] Up to now, the synthesis,[3] modifications,[4] and molecular recognition[5] of pillararenes have been explored extensively. In particular, pillararene-based assemblies have become more and more attractive on

Monographs in Supramolecular Chemistry No. 18
Pillararenes
Edited by Tomoki Ogoshi
© The Royal Society of Chemistry 2016
Published by the Royal Society of Chemistry, www.rsc.org

account of their great application potential in the fabrication of functional materials.[6] Some novel pillararene-based assemblies with complicated topologies, interesting functions, and promising applications have been reported recently.[6] In particular, micellar, vesicular and tubular morphologies and the sizes of these assemblies are significant when exploring their practical applications, such as in materials science, nanoreactors, and toxicology. Therefore, it is necessary to summarize and highlight the new achievements in pillararene-based micellar, vesicular and tubular assemblies in order to inspire more fascinating research in this area.

In this chapter, we highlight recent advancements in pillararene-based micellar, vesicular, and tubular assemblies. First, we briefly discuss the driving forces and factors that control and influence the formation of these multidimensional assemblies, which include solvent effects, geometric packing models and host–guest interactions.[1,6] Second, several pillararene-based micellar, vesicular and tubular assemblies are classified according to different preparation methods and various fabrication strategies. In addition, potential applications of these assemblies are discussed along with the related assembly architectures. Finally, current challenges and prospects for future research directions and applications in this area are predicted.

9.2 Factors that Influence the Formation of Micellar, Vesicular and Tubular Assemblies

Many factors can influence the formation of pillararene-based micellar, vesicular and tubular assemblies. Among them, the solvent effect[6] plays quite an important role. Up to now, all reported pillararene-based micellar, vesicular and tubular assemblies have been formed in solution. Therefore, the formations are related to the solubility of the building units of these assemblies, which depends on the nature of their structures and conformation. In particular, the solvophobic effect usually determines whether the fabrication of macromolecular assemblies is successful or not. Modifying the structures of pillararenes by introducing functional groups with different solvophobic effects in the same solvent, and choosing an appropriate solvent mixture to make the fixed pillararene structure have solvophobic effects can tune the solvophobic effects of the targeted pillararenes to form interesting assemblies.[6] Furthermore, the solvent effect can significantly influence the formation of the building units of these assemblies, especially for those building units formed by supramolecular interactions because the association strength during the course of host–guest complexation is quite different in various solvents. Either tuning solvent polarity or influencing solvophobic effects can change the association affinity of pillararenes with guests.[6]

According to geometric packing models,[6] the geometries of conventional amphiphiles are important factors dominating the morphologies of microaggregates in solution. Thus, the building units in pillararene-based assemblies, mainly some micellar and vesicular morphologies, are required to form the cylinder-like shape. Several methods can be used to prepare the cylinder-shaped building units, such as forming supramolecular dimers by employing supramolecular interactions, and extending the highly symmetrical, rigid and electron-donating structures of pillararene cavities by modification with appropriate functional groups (Scheme 9.1).

The host–guest interactions that can be influenced by counter ion effects, pH effects and solvent effects, as well as the structural nature of the hosts and guests, are also significant in the formation of pillararene-based assemblies. The host–guest interactions not only provide the construction strategy – for example, the host-guest inclusion complexes can serve as the building blocks for assemblies – but also stabilize and tune the assemblies obtained – for example, the recognition abilities can be changed or enhanced by introducing different host–guest pairs.[6]

9.3 Micellar and Vesicular Assemblies (Zero Dimensional Morphology)

9.3.1 Building Units Constructed by Cylindrical Supramolecular Dimers

As mentioned above, the building blocks for micellar and vesicular assemblies are normally required to have the cylinder-shaped structures. The pillararene cavity is highly symmetrical, rigid and cylinder-shaped, with electron-donating properties. However, the dimensions of this type of native macrocycle, including the length and diameter, are not large enough to aggregate spherically to form micellar and vesicular morphologies in solution. As reported,[7,8] DMpillar[5]arene (1,4-dimethoxypillar[5]arene) cannot form regular assemblies in either a THF/H_2O mixture (v/v, 2 : 1),[8] or a Me_2CO solution.[7] Interestingly, after introducing a biviologen guest with two recognition sites (**G9.1** in Scheme 9.2),[7] DMpillar[5]arene could form a complex with **G9.1** in a 2 : 1 stoichiometry, and the resulting complex aggregated into vesicular architectures in Me_2CO (Figure 9.1). It was proposed that, after forming the cylinder-like inclusion complex, it could further aggregate in an orderly way into vesicular lamellar structures according to the geometric packing model.

Thus, the formation of a cylinder-like supramolecular dimer is the key step in the construction of some micellar and vesicular assemblies. Several methods can be employed to make building blocks having cylinder-shaped structures. For example, cylinder-like supramolecular dimers can be initially formed from mono-functionalized pillararenes through supramolecular interactions such as π–π stacking. Functional groups can be used to modify pillararenes, and the pillararene derivatives obtained may form daisy-chain type dimers through the complexation. However, such dimers prefer to form further supramolecular polymers rather than spherical assemblies at certain concentrations.[6] Therefore, it should be noted that the functional groups

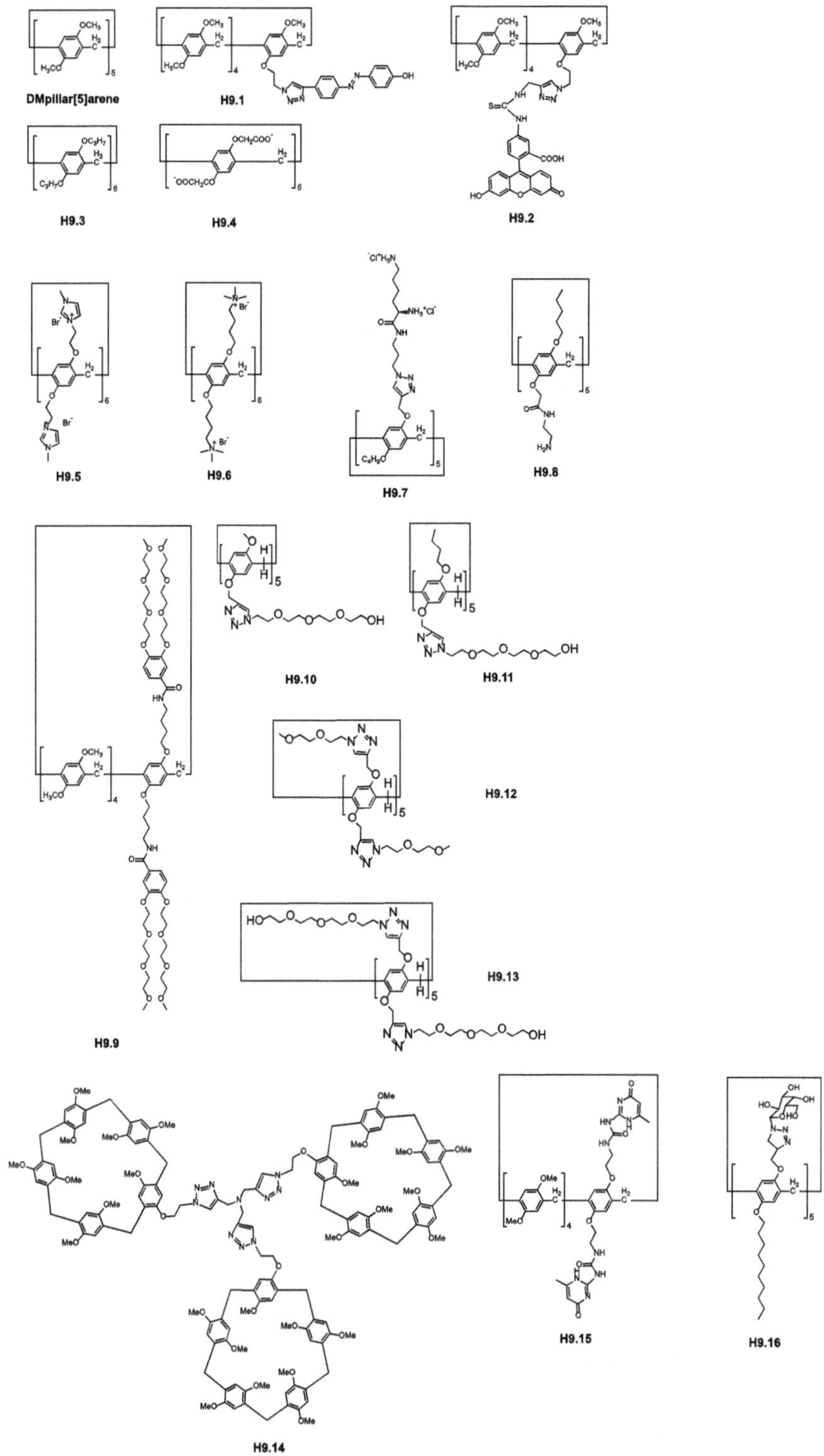

Scheme 9.1 Structural representations of pillararene derivatives **H9.1–H9.19.**

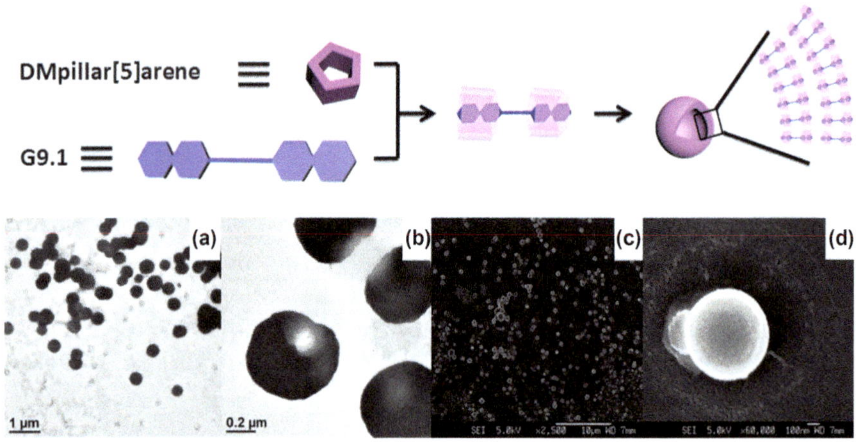

Figure 9.1 (Top) Structural representations of DMpillar[5]arene and guest **G9.1**, as well as the proposed mechanism for the formation of vesicular assemblies. (Bottom) Microscopic morphology of the vesicular assemblies formed from the complex between DMpillar[5]arene and **G9.1** in Me_2CO (1 mM). Negative-stained transmission electron microscopy (TEM) images with scale bars of (a) 1 μm and (b) 0.2 μm. Gold sputtering scanning electron microscopy (SEM) images with scale bars of (c) 10 μm and (d) 100 nm.
Adapted from Ref. 7 with permission from The Royal Society of Chemistry.

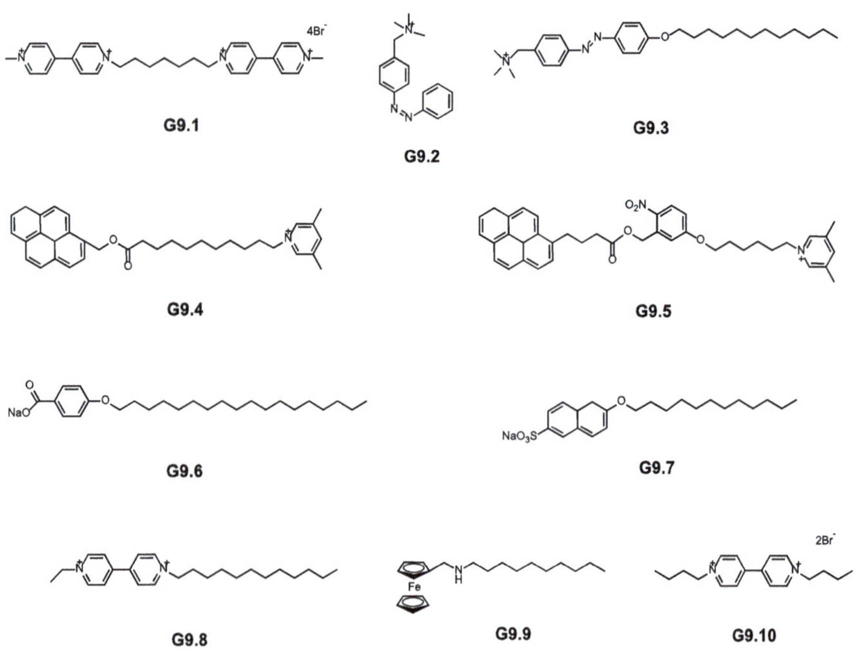

Scheme 9.2 Structural representations of guests **G9.1–G9.10**.

introduced to modify pillararenes to form building blocks are normally required to have larger molecular size than the pillararene cavity.

Given that there is no obvious interaction between the azobenzene unit and the pillar[5]arene cavity, and that the molecular size of the azobenzene unit is larger than the cavity, the azobenzene functionalized pillar[5]arene (**H9.1** in Scheme 9.1) cannot form a self-inclusion complex or supramolecular polymer in a THF/H_2O mixture (v/v, 2 : 1).[8] According to microscopic observations, hollow spherical vesicles were formed from **H9.1** in the mixture solution. In contrast, azobenzene-modified cyclodextrins formed dimeric inclusion complexes rather than spherical aggregates in solution. Thus, the azobenzene unit that is not included in the cavity of pillar[5]arene plays an important role in forming spherical assemblies. Given that **H9.1** has different solubility in THF and in H_2O, the driving force for the formation of the assembly in the mixture solution was proposed to be the solvophobic effect. In addition, as a result of its cylinder-like shape, **H9.1** was assumed to self-assemble into orderly bilayered membranes, where the azobenzene units were inside the bilayers and the pillar[5]arene units faced both the inner and outer solvents (Figure 9.2), similar to conventional amphiphiles. Using the same strategy, after introducing another large functional group, fluorescein isothiocyanate (FITC), onto pillar[5]arene through a

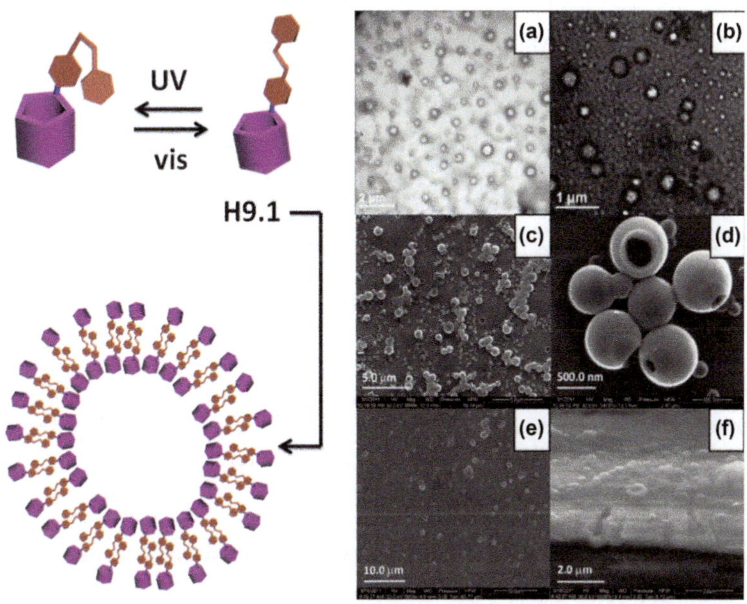

Figure 9.2 Schematic representation of azobenzene mono-functionalized pillararene **H9.1** and its UV-vis responsive assembly behavior. Micro-morphology of **H9.1** in a mixture solution under visible light: negative stained TEM image (a), cryo-TEM image (b), gold sputtering SEM images with (c) scale bar of 5 mm and (d) scale bar of 500 nm, and ESEM (environmental scanning electron microscope) images with (e) pressure of 5 Torr and (f) pressure of 3 Torr.
Adapted from Ref. 8 with permission from The Royal Society of Chemistry.

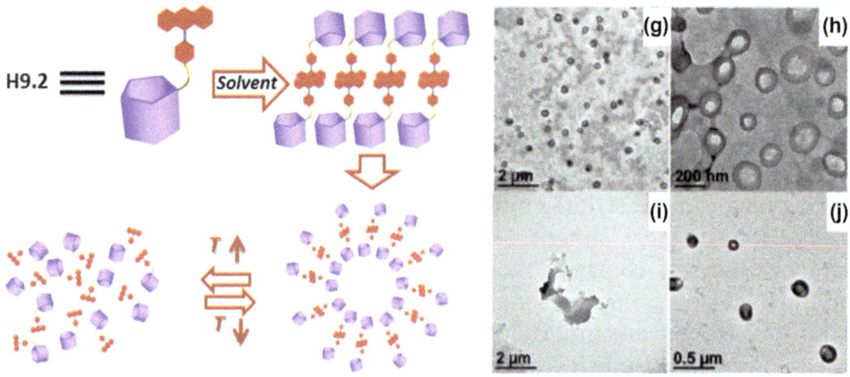

Figure 9.3 Schematic representation of **H9.2** and its thermo-responsive assembly behavior. Micro-morphology of **H9.2** in Me_2CO (0.020 mM) determined by negative-stained TEM at room temperature (g, h), after heating the sample to 54 °C (i), and cooling it to room temperature (j).
Adapted from Ref. 9 with permission from The Royal Society of Chemistry.

copper(I) catalyzed Huisgen-type azide–alkyne cycloaddition, the FITC mono-functionalized pillar[5]arene obtained (**H9.2**) could also aggregate into vesicular spheres in Me_2CO solution.[9] Owing to the cylindrical shape of the pillararene unit, **H9.2** was also proposed to self-assemble into ordered bilayer membranes, where the FITC units were inside the bilayers associated with intermolecular $\pi\cdots\pi$ stacking interactions and the pillararene units were on the surfaces facing the inner and outer Me_2CO solvent molecules (Figure 9.3).

Interestingly, these spherical vesicular aggregates could be used as smart materials. Because the azobenzene unit shows a reversible *trans* and *cis* isomerization property in response to alternating irradiation with visible and UV light,[8] the azobenzene mono-functionalized pillararene **H9.1** formed hollow spherical vesicles under visible light, which could be transferred into solid particles by exposure to UV light. This photo-responsive pillararene-based assembly presents an application potential as a UV sensor. In another example,[9] the fluorescent vesicles formed by **H9.2** were sensitive to variable temperatures owing to the formation and dissociation of the $\pi\cdots\pi$ stacking interactions, showing thermo-responsive emission at different temperatures.

9.3.2 Building Units Constructed by Supramolecular Amphiphiles

Conventional amphiphiles often assemble into micellar and vesicular structures by supramolecular interactions such as hydrophilic and hydrophobic interactions in H_2O. In some cases, amphiphiles themselves are also prepared from various precursors through supramolecular interactions, *e.g.*, host–guest interaction, and the supramolecular amphiphiles obtained then form more complicated architectures.[10] As a matter of fact, the supramolecular dimer mentioned above is also a kind of supramolecular amphiphile.

In comparison with the azobenzene modified pillararene **H9.1**,[8] UV/visible-responsive assemblies can also be prepared from supramolecular amphiphiles in a complementary way. For example, pillar[6]arene, having a bigger cavity size than pillar[5]arene, forms a complex with *trans*/*cis*-azobenzene under different geometries. A spontaneous vesicular assembly was constructed in a mixture solution of $CHCl_3$ and MeCN (v/v, 10 : 1) just by employing a host–guest complex between short alkyl chain perfunctionalized pillar[6]arene **H9.3** (Scheme 9.1) and ionic azobenzene *cis*-**G9.2** · PF_6^- (Scheme 9.2) under 365 nm UV light.[11] This assembly could be reversibly photo-controlled to disassemble and re-assemble through alternating irradiation with visible and UV light, respectively, on account of the *trans*–*cis* isomerization of the azobenzene guest (Figure 9.4), giving a morphological switch between irregular and vesicular aggregates. Similarly, *trans*-**G9.3** · Br^- could also be included in water soluble **H9.4** · $10NH_4^+$ (Scheme 9.1) to form a host–guest complex in H_2O.[12] It was found that the addition of

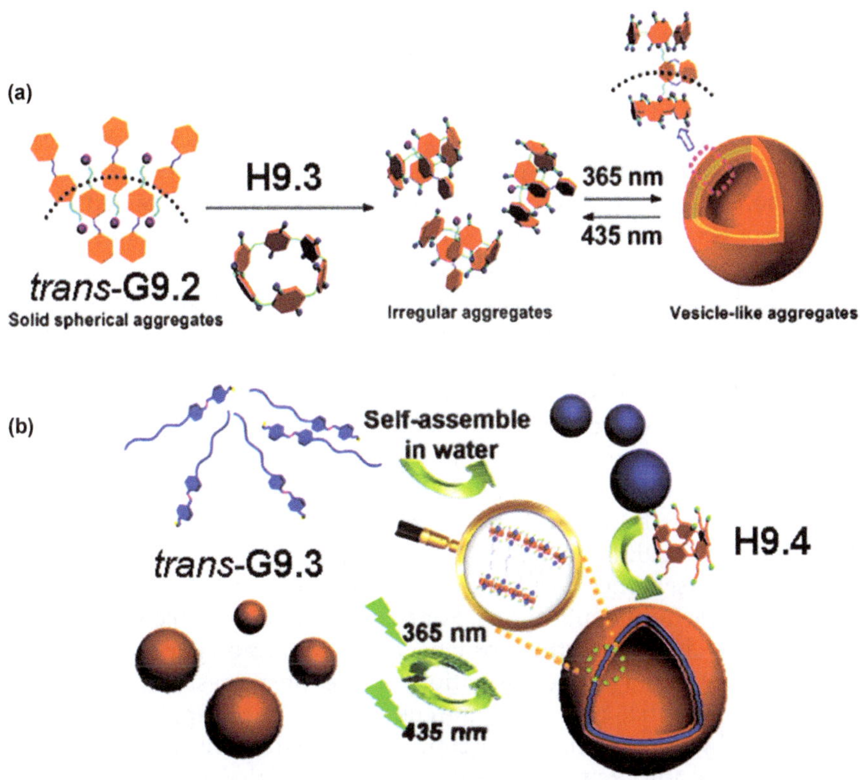

Figure 9.4 (a) Proposed mechanism for the formation of the irregular and vesicular aggregates from **H9.3** and **G9.2** controlled by UV and visible light irradiation. Reproduced with permission from ref. 11. Copyright 2012 American Chemical Society. (b) Structural representation of the photo-responsive assemblies constructed from **H9.4** and **G9.3** in H_2O. Adapted from Ref. 12 with permission from The Royal Society of Chemistry.

H9.4 · $10NH_4^+$ could change the morphology of the micellar aggregate formed from the amphiphile *trans*-**G9.3** · Br^- into a vesicular one under visible light (435 nm, Figure 9.4). Furthermore, the morphologies of these vesicular assemblies could be switched between vesicular and solid spheres under UV and visible light irradiation, respectively. It should be noted that the conditions, *e.g.*, UV and visible light, for the formation of vesicular morphologies are totally different between the two cases of **H9.3** with *cis*-**G9.2** · PF_6^- and **H9.4** · $10NH_4^+$ with *trans*-**G9.3** · Br^-. The difference is caused by different host–guest interactions between the azobenzene derivatives and the pillar[6]arene cavity under different external light sources.[12]

In addition, **H9.4** · $10Na^+$ could be used to tune the tubular morphologies assembled from a cationic pyrene compound **G9.4** · Br^- (Scheme 9.2) in H_2O.[13] After adding **H9.4** · $10Na^+$, the tubular assemblies were changed into vesicular architectures. The mechanism for forming vesicles by mixing **G9.4** · Br^- and **H9.4** · $10Na^+$ is similar to that for the mixture of **H9.4** · $10NH_4^+$ and *trans*-**G9.2** · Br^-. Moreover, the mixture of **G9.4** · Br^- and **H9.4** · $10Na^+$ could be employed for the dispersion of multi-walled carbon nanotubes in H_2O. In another example,[14] the addition of **H9.4** · $10Na^+$ changed the assembly morphologies formed from a cationic pyrene compound **G9.5** · Br^- (Scheme 9.2) from nanosheets to nanospheres in H_2O on account of the host–guest interactions between the pillararene cavity and the pyridinium cation in **G9.5** · Br^-. Interestingly, because **G9.5** · Br^- contains a UV-responsive 2-nitrobenzyl ester unit, the nanospheres obtained by the mixture of **H9.4** · $10Na^+$ and **G9.5** · Br^- changed back into nanosheets under the irradiation with UV light (365 nm).[14]

Using a similar design strategy, another water-soluble host (**H9.5**; Scheme 9.1) complexed with a sodium *p*-hydroxybenzoate derivative **G9.6** (Scheme 9.2) to form a supramolecular amphiphile, which further self-assembled into vesicular architectures.[15] In another example, because water-soluble **H9.6** (Scheme 9.1) showed a high affinity for sodium 2-naphthalenesulfonate, it could complex with an amphiphilic guest **G9.7** (Scheme 9.2) containing a 2-naphthalenesulfonate unit. In this case, **H9.6** performed as the hydrophilic portion to change the aggregation of **G9.7** in H_2O, forming regular and uniform micellar superstructures.[16]

Vesicular assemblies formed by supramolecular amphiphiles may be used for biomedical applications. For example, water-soluble **H9.4** · $10Na^+$ significantly inhibited the toxicity of a viologen guest **G9.8** · $2Br^-$ (Scheme 9.2) through the formation of the host–guest complex, indicating the potential use of pillararenes in biological applications.[17] As confirmed by fluorescent microscopy, upon gradual introduction of the viologen guest into Raw 264.7 cells, the shapes of the cells were changed obviously, and the cell membrane was seriously damaged, indicating that the viologen guest was toxic to these cells. In contrast, following the addition of the host–guest complex formed from **H9.4** · $10Na^+$ and **G9.8** · $2Br^-$ at the same concentration, the toxicity of the guest for the cells was significantly reduced upon the formation of the stable inclusion complex. The

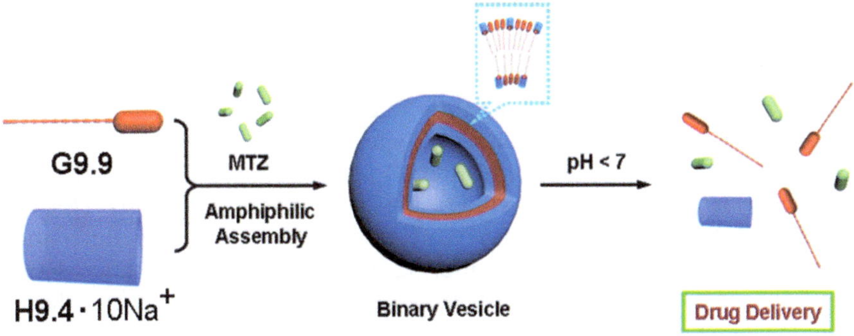

Figure 9.5 Schematic illustration of the mechanism for the formation of supramolecular vesicles from **H9.4**·10Na$^+$ and ferrocene guest **G9.9**, as well as the pH-responsive drug release.
Reproduced with permission from ref. 20. Copyright 2013 American Chemical Society.

significant reduction in toxicity was confirmed by analysis of the relative cell viability using the 3-(4,5-dimethylthiazol-2-yl)-2,5-diphenyl tetrazolium bromide (MTT) assay.[17]

In addition, a potential drug delivery system based on supramolecular vesicles[10] was prepared from the host–guest inclusion complex between **H9.4**·10Na$^+$ and ferrocene derivative **G9.9** (Scheme 9.2)[18,19] in H_2O (Figure 9.5),[20] which displayed significant pH-responsive behavior. Drug loading and *in vitro* drug release studies confirmed that these vesicular assemblies were able to encapsulate mitoxantrone (MTZ) to achieve a MTZ-loaded vesicular system that exhibited a rapid release of MTZ in a low pH environment ($pH < 7$). The cellular uptake of these pH-responsive MTZ-loaded vesicles by cancer cells was observed by living cell imaging techniques. Interestingly, the cytotoxicity assay showed that unloaded vesicles had low toxicity for normal cells, which could dramatically reduce the toxicity of MTZ upon loading, similar to the case of **H9.4**·10Na$^+$ and **G9.8**·2Br$^-$ discussed above.[17] Meanwhile, MTZ-loaded vesicles exhibited comparable anticancer activity *in vitro*, freeing MTZ under the experimental conditions.

9.3.3 Building Units Constructed from Covalently Extended Pillararenes

The highly symmetrical, rigid and electron-donating cavity of pillararenes can be extended to enlarge the cylinder-like architecture through covalent modifications with appropriate functional groups, *e.g.*, hydrophilic and hydrophobic groups. Similar to the building blocks based on the supramolecular dimers and supramolecular amphiphiles, the pillararenes with extended cavities also behave as amphiphiles to form orderly aggregates in the appropriate solvents, owing to the solvophobic effects.

For example, an amphiphilic pillararene derivative, **H9.7** (Scheme 9.1), was designed and synthesized by employing polar lysine head group as the hydrophilic part to extend the pillararene cavity.[21] Based on hydrophobic and hydrophilic interactions, amphiphilic **H9.7** could spontaneously self-assemble into stable micellar morphologies in H_2O above a critical micelle concentration (CMC) of around 0.1 mM, with alkyl tails facing each other and the hydrophobic parts covered by the long hydrophilic groups. This was confirmed by small angle X-ray scattering (SAXS), field flow fluctuation coupled with multi-angle light scattering (FFF-MALS) and other microscopic methods. In another case, a water-soluble amphiphilic pillararene derivative, **H9.8** (Scheme 9.1), was synthesized by introducing five primary amine groups as the hydrophilic head and five alkyl chains as the hydrophobic tail.[22] Pillararene **H9.8** could self-assemble to form vesicles initially and, after 4 months, it formed tubular morphologies in water. The vesicular assemblies could further encapsulate calcein within their interior cavities under neutral conditions and release it in response to a decrease in pH. The tubular assemblies with primary amine groups on their surfaces were found to have the capacity to adsorb trinitrotoluene (TNT) through donor–acceptor interactions.

In addition to the per-functionalization of all hydroquinone reactive sites on pillararenes to extend the cavity as mentioned above, selective functionalization of two hydroquinone reactive sites by amphiphilic groups can produce bola-amphiphilic pillararenes, *e.g.*, **H9.9** (Scheme 9.1).[23] As confirmed by dynamic light scattering (DLS) and microscopic methods (Figure 9.6), pillararene **H9.9** formed responsive multi-lamellar giant vesicles in nonpolar solvents such as $CHCl_3$ at around 1.70 mg mL^{-1}; KPF_6 and benzo-18-crown-6 were used to tune the assembly and disassembly of these vesicles. Interestingly, a gel could be formed by **H9.9** in a mixture of polar solvents such as H_2O–THF (5 : 1, v/v) at around 0.8 mg mL^{-1}, which showed reversible gel–sol phase transition behavior by either heating and cooling, or adding and removing potassium cation.

Covalently functionalized amphiphilic pillararenes may also be applied in biomedicine.[24,25] For example, biocompatible pillararene-based assemblies have been used to deliver mixed dyes for dual bioimaging.[26] A series of tadpole-like and bola amphiphilic pillararenes **H9.10–H9.13** (Scheme 9.1) were synthesized by selectively employing water-soluble ethylene glycols and hydrophobic alkyl units as the starting materials. In comparison with their monomers, these amphiphilic pillararenes not only showed improved biocompatibility for cells, but were also able to form homogeneous supramolecular self-assemblies. Interestingly, different types of amphiphilic pillararene-based assemblies exhibited different performance in the delivery of dyes with different water solubility. All the assemblies could deliver water-soluble rhodamine B (RhB) to cells, while tadpole-like amphiphilic pillararene-based assemblies performed better when delivering hydrophobic FITC for imaging (Figure 9.7). In addition, the pillararene derivatives **H9.10, H9.12** and **H9.13** could complex with a viologen guest (**G9.10** in Scheme 9.2),

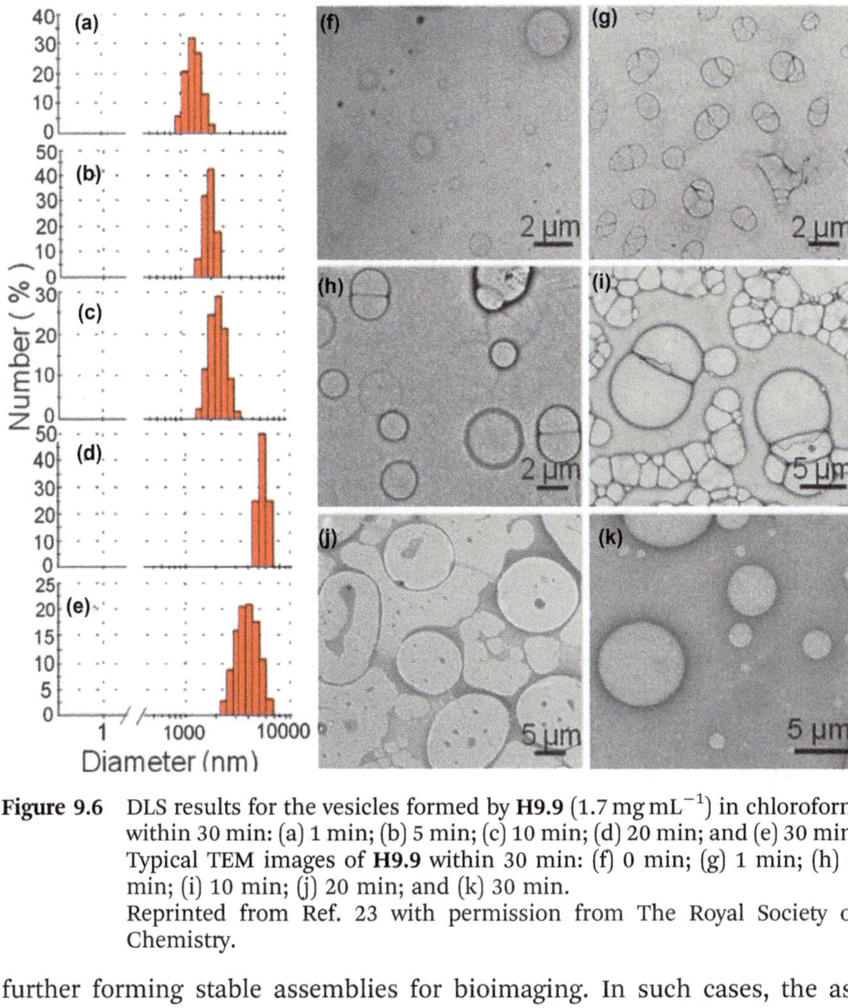

Figure 9.6 DLS results for the vesicles formed by **H9.9** ($1.7\ mg\ mL^{-1}$) in chloroform within 30 min: (a) 1 min; (b) 5 min; (c) 10 min; (d) 20 min; and (e) 30 min. Typical TEM images of **H9.9** within 30 min: (f) 0 min; (g) 1 min; (h) 5 min; (i) 10 min; (j) 20 min; and (k) 30 min.
Reprinted from Ref. 23 with permission from The Royal Society of Chemistry.

further forming stable assemblies for bioimaging. In such cases, the assembly formed from the complex of tadpole-like amphiphile pillararene **H9.10** with the viologen guest performed better in delivering mixed dyes. Finally, the anticancer drug doxorubicin (DOX) was successfully delivered to cells by using pillararene-based assemblies (Figure 9.8). This work has demonstrated the capacity of pillararene-based assemblies to deliver different dyes for bioimaging *in vitro* and paves the way for using these biocompatible carriers in combined cancer therapy.

9.3.4 Other Cases

In addition to the three cases discussed above, in which the building blocks based on functionalized pillararenes and pillararene complexes performed as the amphiphiles, pillararene-based supramolecular networks can also be wrapped into hollow vesicular spheres. For example, pillararene trimer

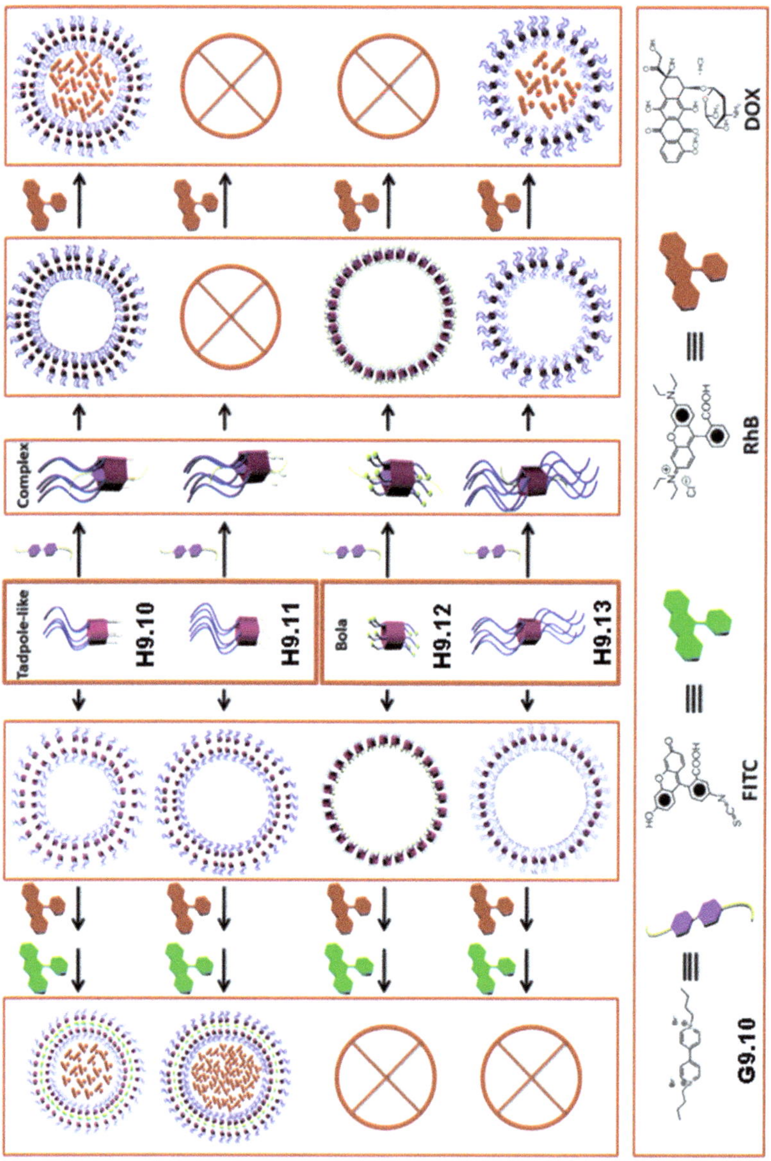

Figure 9.7 Illustrative mechanism for dual bioimaging by delivering **G9.10**, FITC, RhB, and DOX from biocompatible pillararene-based assemblies formed by **H9.10–H9.13**.
Reproduced with permission from ref. 26. Copyright 2013 American Chemical Society.

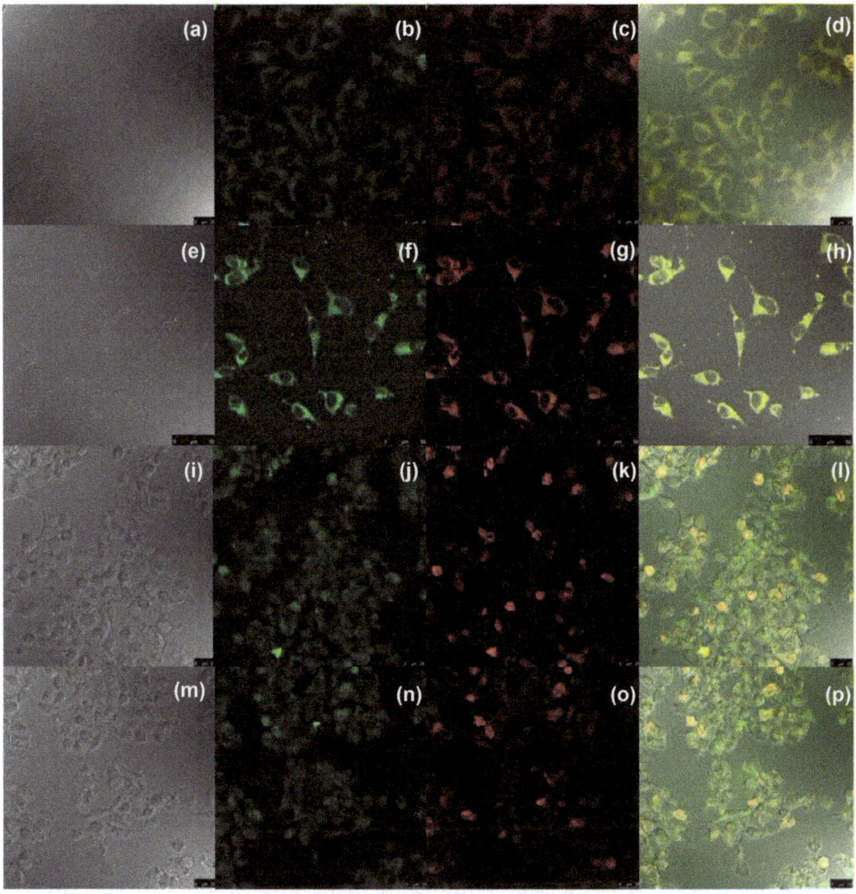

Figure 9.8 Confocal laser scanning microscopy images of HeLa cells, after treatment with biocompatible pillararene-based assembly **H9.10** (40 μM) containing FITC (4 μM) and RhB (4 μM) for 24 h: (a) bright field, (b) FITC channel, (c) RhB channel, and (d) merged image of (a–c), scale bar 25 μm; and after treatment with biocompatible pillararene-based assembly **H9.11** (40 μM) containing FITC (4 μM) and RhB (4 μM) for 24 h: (e) bright field, (f) FITC channel, (g) RhB channel, and (h) merged image of (e–g), scale bar 50 μm. Confocal laser scanning microscopy images of HeLa cells, after treatment with biocompatible pillararene-based assembly **H9.10** (40 μM) containing guest **G9.10** (40 μM) and RhB (4 μM) for 24 h: (i) bright field, (j) green field (excited at 488 nm), (k) RhB channel, and (l) merged image of (i–k), scale bar 25 μm; after treatment with biocompatible pillararene-based assembly **H9.13** (40 μM) containing guest **G9.10** (40 μM) and RhB (4 μM) for 24 h: (m) bright field, (n) green field (excited at 488 nm), (o) RhB channel, and (p) merged image of (m–o), scale bar 25 μm.
Reproduced with permission from ref. 26. Copyright 2013 American Chemical Society.

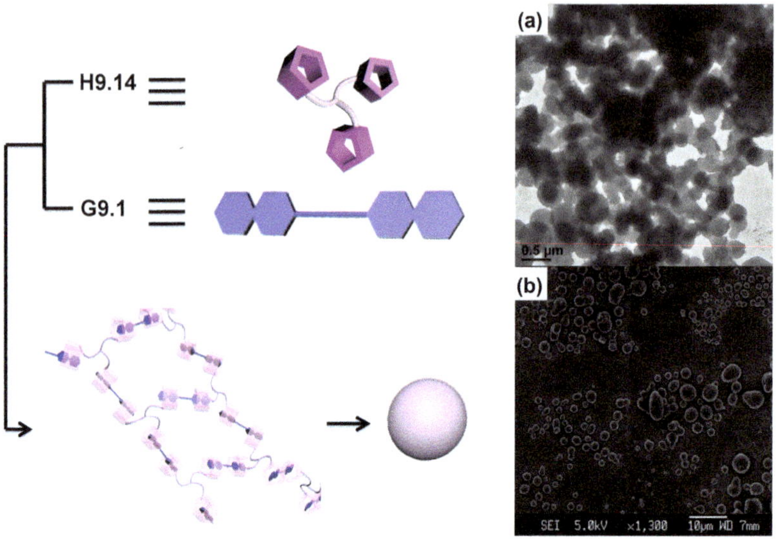

Figure 9.9 (Left) Structural representations of pillararene trimer **H9.14**, guest **G9.1**, and the proposed mechanism for the formation of vesicular assemblies under low concentration. (Right) Morphologies of the complex between pillararene trimer **H9.14** and guest **G9.1** at low concentration. (a) Negative-stained TEM image (scale bar 0.5 µm), and (b) gold sputtering SEM image (scale bar 10 µm) at a sample concentration of 1 mM.
Adapted from Ref. 7 with permission from The Royal Society of Chemistry.

H9.14 (Scheme 9.1) integrated with three recognition sites has been designed and synthesized; it can include the linear guest **G9.1** (Scheme 9.2) consisting of two bipyridinium units and form small pieces of supramolecular network at low concentration (Figure 9.9).[7] After reaching the critical assembly concentration (CAC), it was observed to form vesicular spherical assemblies. According to microscopic observations and DLS, it was supposed that small building blocks with sizes ranging from 10 to 40 nm were formed when the sample concentration was around 0.5 mM. Upon increasing the concentration to 1 mM, flat pieces were formed, which rolled into hollow spherical structures with sizes ranging from 150 to 500 nm.[7]

9.4 Tubular Assemblies (One-dimensional Morphology)

9.4.1 Pillararene Cavity Performing as a Tubular Channel

Tubular assemblies can be prepared using suitable pillararenes connected through supramolecular interactions, where the pillararene cavity performs as the tubular channel in this case.[27] The formation of pillararene-based supramolecular polymers is a good strategy for preparing tubular morphologies. For example, because ureidopyrimidinone (UPy) possesses a highly directional quadruple hydrogen-bonding unit, UPy bifunctionalized pillar[5]arene derivative **H9.15** (Scheme 9.1) is able to form a linear tubular supramolecular polymer at high concentrations (>21 mM) in $CHCl_3$ through intermolecular hydrogen-bonding interactions provided by the UPy unit (Figure 9.10).[28]

Not only can the channel of tubular assemblies be used for the delivery of small neutral and ionized molecules as mentioned above, but their functionalized surface can also be employed in biomedicine, *e.g.*, in antibacterial applications. For example, a sugar-functionalized amphiphilic pillar[5]arene derivative **H9.16** (Scheme 9.1) with galactose groups as the hydrophilic part and alkyl chains as the hydrophobic part can self-assemble in H_2O owing to the existence of intermolecular hydrogen-bonding interactions between the galactose molecules and the van der Waals interactions between the alkyl chains, producing vesicles immediately and tubular superstructures after 1 week of transformation (Figure 9.11).[29,30] After coating with biocompatible galactose on the surface, the nanotubes obtained showed interesting biofunctions, with low toxicity for both cancerous and normal cells, and could

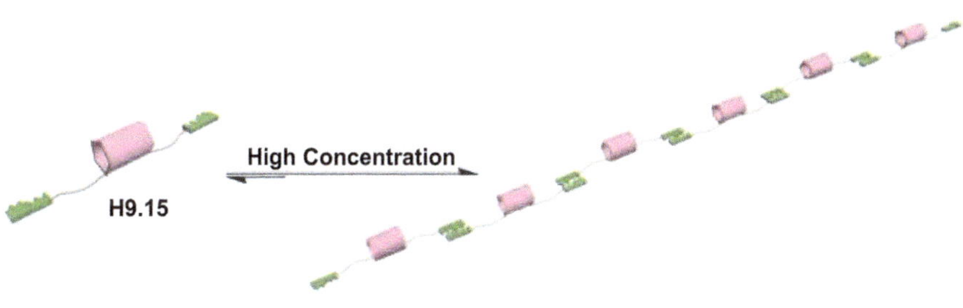

Figure 9.10 Linear supramolecular polymers formed by **H9.15** through hydrogen-bonding interactions at high concentration.[28]

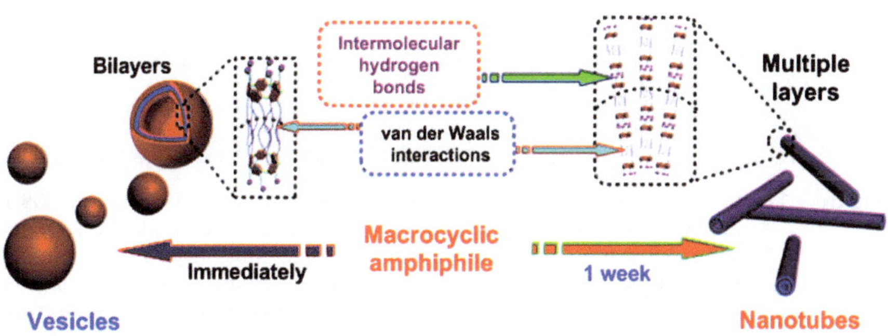

Figure 9.11 Schematic representation of the mechanisms for the self-assembly and transformation process based on sugar-functionalized amphiphilic pillar[5]arene derivative **H9.19**.
Reproduced with permission from ref. 29. Copyright 2013 American Chemical Society.

be utilized as an excellent cell glue to agglutinate the bacterium *Escherichia coli*. It was supposed that the presence of galactose on these tubular superstructures provided multivalent binding sites that have high affinity for carbohydrate receptors on *E. coli*. These results indicated that supramolecular self-assemblies composed of simple binding sites driven by noncovalent interactions are distinctive chemical tools for capturing living bacteria in solution.

9.4.2 Pillararenes Performing as Building Blocks for Constructing Large Tubular Channels

As mentioned above,[7] the pillararene trimer **H9.14** (Scheme 9.1) can include the linear guest **G9.1** (Scheme 9.2) to form vesicular spheres at relatively low concentration (see Figure 9.9). However, the formation of vesicular spheres was only the beginning of the morphology development. By continuously increasing the concentrations, one-dimensional (1D) tubular structures, two-dimensional (2D) layers and three-dimensional (3D) multiple layers were observed one after another (Figure 9.12). For example, when increasing the concentrations from 1 to 2 mM, the morphology of the assembly changed from the previous spherical superstructures (Figure 9.9) to tubular ones

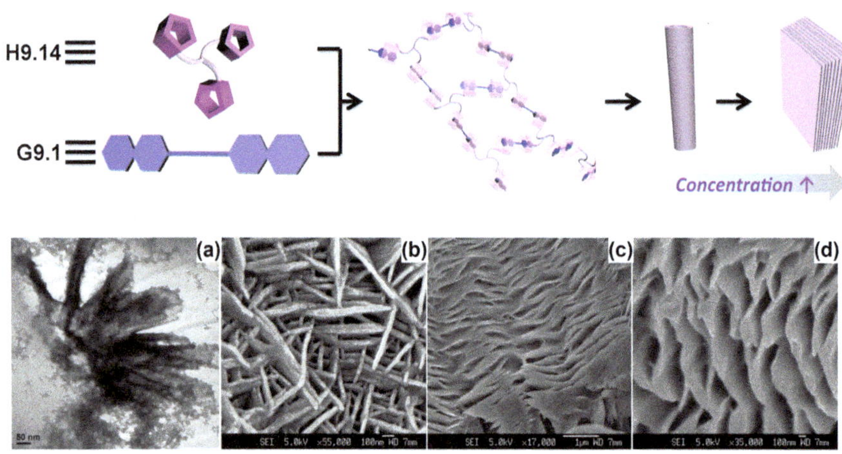

Figure 9.12 (Top) Structural representations of pillararene trimer **H9.14**, guest **G9.1**, and the proposed mechanism for the formation of multidimensional assemblies under different concentrations. (Bottom) Morphologies of the complex between **H9.14** and **G9.1** at variable concentrations. (a) Negative-stained TEM image (scale bar 50 nm) and (b) gold sputtering SEM image (scale bar 100 nm) at the sample concentration of 2 mM. Gold sputtering SEM images at a sample concentration of 5 mM with scale bars of (c) 1 μm and (d) 100 nm.
Adapted from Ref. 7 with permission from The Royal Society of Chemistry.

(Figure 9.12), as observed by microscopic methods. Both variable concentration (VC) UV-vis spectra and VC fluorescent emission spectra confirmed that the intensity slopes obviously increased upon increasing the concentrations over 1.5 mM, demonstrating that the multidimensional morphological changes occurred at this concentration. Large spherical superstructures fused further into tubes at this concentration (1.5 mM), as observed by microscopic studies (Figure 9.12). Thus, the functional pillararene in this case performed as building blocks for constructing large tubular channels.

By tuning the concentrations, further morphological transformation occurred, and high dimensional morphology appeared in this case.[7] The ^{1}H NMR spectra of the complex began to broaden above a concentration of 2 mM, suggesting that the morphological changes did not stop at this stage. Owing to the star-shaped structure of **H9.14**, with three recognition sites, linear fibers formed flat layers, and then the layers stacked together regularly at a high concentration of 5 mM (Figure 9.12). This was the first example of dynamic supramolecular self-assembly with switchable morphologies, attributable to the trimeric nature of **H9.14**, which was quite different from the assemblies formed by DMpillar[31] and pillararene dimer.[32] The work also provided a successful example of the achievement of multi-dimensional morphology changes using the pillararene complex, and presented a novel strategy for developing functional materials with a switchable morphology control.[7]

9.5 Conclusions and Outlook

Recent research advances in pillararene-based micellar, vesicular and tubular assemblies have been described in this chapter. As discussed at the beginning of the chapter, many factors can influence the formation of these assemblies. For instance, the solvent effect not only affects the solubility of pillararene derivatives and their inclusion complexes that can be used as the building blocks for further assembly, but also provides the key driving force – the solvophobic effect – for the formation of supramolecular assemblies in solution. In addition, whether or not the building blocks possess cylinder-like shapes has been deemed to be a significant factor determining the formation of micellar, vesicular and tubular assemblies according to geometric packing models. In many cases of the assemblies shown in this chapter, the host–guest interactions, influenced by counter ion effects, pH effects, and solvent effects, as well as the structural nature of the hosts and guests, provide not only the construction strategy but also the building blocks. All these factors are important driving forces in the formation of these assemblies.

Subsequently, pillararene-based micellar, vesicular and tubular assemblies have been fabricated, based on these driving forces, with a view towards potential applications. Several preparation strategies have been discussed in detail for fabricating multidimensional supramolecular architectures. For example, in order to obtain a cylinder-like shape for forming stable vesicular assemblies, various methods have been explored, including the construction of pillararene-based supramolecular dimers and supramolecular amphiphiles through supramolecular interactions, as well as the extension of the pillararene cavities by covalent modifications with appropriate functional groups. Furthermore, the multidimensional assemblies obtained have been investigated for potential applications in the fields of smart materials, sensors, biomimicking and biomedicine.

Based on these significant developments, several challenges and prospects for pillararene-based micellar, vesicular and tubular assemblies should also be noted. First, theoretical tools such as computational calculations[6] can be employed to investigate the dynamic processes, as well as the mechanism of formation of these assemblies. Some critical issues regarding pillararene-based inclusion complexes and multidimensional supramolecular assemblies need to be further addressed.

Second, more interesting and advanced superstructures based on pillararenes and their inclusion complexes can be designed and synthesized for various functions. For example, new pillararenes with larger cavity sizes have been reported recently,[6] and the host–guest complexes based on these pillararenes should be explored in detail. Novel assembled architectures may be achieved by using these new host–guest complexes.[20]

Third, in addition to paying continuous attention to the preparation of controllable micellar, vesicular and tubular assemblies, novel types of assembly can be explored for fabricating various supramolecular materials such as stimuli-responsive smart materials.[33] The amphiphilic pillararene derivatives reported provide a useful basis for further studies. For example, hydrophilic oligoethylene oxide-modified pillararenes present lower critical solution temperature (LCST) in H_2O, and form aggregates with different average hydrodynamic diameters below and above the clouding point (T_{cloud}).[34] The diameters can also be tuned by the addition of suitable guests or other competitive hosts such as cucurbit[7]uril. This tunable thermo-responsive supramolecular material with LCST behavior has been considered as a good supplement to the existing thermo-sensitive materials such as poly(nisopropylacrylamide). In addition, by integrating various kinds of organic materials with inorganic species, the hybrid materials with combined advantages[35] may lead to novel functional materials with unique application potential in many fields[36] such as biomedicine.[37]

Furthermore, using the unique structures of amphiphilic pillararenes and particular properties of their assemblies, more investigation should be carried out to explore their potential applications in biomedicine,[38] liquid crystals[39,40] and catalysis.[41] It is expected that further research based on pillararenes and their micellar, vesicular and tubular assemblies will boost this emerging area to a new level.[42]

Acknowledgements

This work is supported by the National Research Foundation (NRF), Prime Minister's Office, Singapore under its NRF Fellowship (NRF2009NRF-RF001-015) and Campus for Research Excellence and Technological Enterprise (CREATE) Programme – Singapore Peking University Research Centre for a Sustainable Low-Carbon Future, as well as the NTU-A*Star Centre of Excellence for Silicon Technologies (A*Star SERC No.: 112351500003).

References

1. T. Ogoshi and T.-A. Yamagishi, *Eur. J. Org. Chem.*, 2013, **2013**, 2961–2975.
2. P. J. Cragg and K. Sharma, *Chem. Soc. Rev.*, 2012, **41**, 597–607.
3. H. Zhang, Z. Liu, F. Xin and A. Hao, *Chin. J. Org. Chem.*, 2012, **32**, 219–229.
4. T. Ogoshi, *J. Inclusion Phenom. Macrocyclic Chem.*, 2011, **72**, 247–262.
5. M. Xue, Y. Yang, X. Chi, Z. Zhang and F. Huang, *Acc. Chem. Res.*, 2012, **45**, 1294–1308.
6. H. Zhang and Y. Zhao, *Chem. – Eur. J.*, 2013, **19**, 16862–16879.
7. H. Zhang, K. T. Nguyen, X. Ma, H. Yan, J. Guo, L. Zhu and Y. Zhao, *Org. Biomol. Chem.*, 2013, **11**, 2070–2074.
8. H. Zhang, N. L. Strutt, R. S. Stoll, H. Li, Z. Zhu and J. F. Stoddart, *Chem. Commun.*, 2011, **47**, 11420–11422.
9. H. Zhang, X. Ma, J. Guo, K. T. Nguyen, Q. Zhang, X.-J. Wang, H. Yan, L. Zhu and Y. Zhao, *RSC Adv.*, 2013, **3**, 368–371.
10. H. Zhang, F. Xin, Y. Li, A. Hao, W. An and T. Sun, *Prog. Chem.*, 2010, **22**, 2276–2281.
11. G. Yu, C. Han, Z. Zhang, J. Chen, X. Yan, B. Zheng, S. Liu and F. Huang, *J. Am. Chem. Soc.*, 2012, **134**, 8711–8717.
12. D. Xia, G. Yu, J. Li and F. Huang, *Chem. Commun.*, 2014, **50**, 3606–3608.
13. G. Yu, M. Xue, Z. Zhang, J. Li, C. Han and F. Huang, *J. Am. Chem. Soc.*, 2012, **134**, 13248–13251.
14. J. Yang, G. Yu, D. Xia and F. Huang, *Chem. Commun.*, 2014, **50**, 3993–3995.
15. Y. Yao, J. Li, J. Dai, X. Chi and M. Xue, *RSC Adv.*, 2014, **4**, 9039–9043.
16. Y. Ma, J. Yang, J. Li, X. Chi and M. Xue, *RSC Adv.*, 2013, **3**, 23953–23956.
17. G. Yu, X. Zhou, Z. Zhang, C. Han, Z. Mao, C. Gao and F. Huang, *J. Am. Chem. Soc.*, 2012, **134**, 19489–19497.
18. H. Zhang, W. An, Z. Liu, A. Hao, J. Hao, J. Shen, X. Zhao, H. Sun and L. Sun, *Carbohydr. Res.*, 2010, **345**, 87–96.
19. H. Zhang, Z. Liu, F. Xin, W. An, A. Hao, J. Li, Y. Li, L. Sun, T. Sun, W. Zhao, Y. Li and L. Kong, *Carbohydr. Res.*, 2011, **346**, 294–304.
20. Q. Duan, Y. Cao, Y. Li, X. Hu, T. Xiao, C. Lin, Y. Pan and L. Wang, *J. Am. Chem. Soc.*, 2013, **135**, 10542–10549.
21. T. Nishimura, Y. Sanada, T. Matsuo, T. Okobira, E. Mylonas, N. Yagi and K. Sakurai, *Chem. Commun.*, 2013, **49**, 3052–3054.

22. Y. Yao, M. Xue, J. Chen, M. Zhang and F. Huang, *J. Am. Chem. Soc.*, 2012, **134**, 15712–15715.
23. L. Gao, B. Zheng, Y. Yao and F. Huang, *Soft Matter*, 2013, **9**, 7314–7319.
24. I. Nierengarten, M. Nothisen, D. Sigwalt, T. Biellmann, M. Holler, J. S. Remy and J. F. Nierengarten, *Chem. – Eur. J.*, 2013, **19**, 17552–17558.
25. I. Nierengarten, K. Buffet, M. Holler, S. P. Vincent and J.-F. Nierengarten, *Tetrahedron Lett.*, 2013, **54**, 2398–2402.
26. H. Zhang, X. Ma, K. T. Nguyen and Y. Zhao, *ACS Nano*, 2013, **7**, 7853–7863.
27. M. Barboiu, *Angew. Chem., Int. Ed.*, 2012, **51**, 11674–11676.
28. X.-Y. Hu, X. Wu, S. Wang, D. Chen, W. Xia, C. Lin, Y. Pan and L. Wang, *Polym. Chem.*, 2013, **4**, 4292–4297.
29. G. Yu, Y. Ma, C. Han, Y. Yao, G. Tang, Z. Mao, C. Gao and F. Huang, *J. Am. Chem. Soc.*, 2013, **135**, 10310–10313.
30. G. Yu, Y. Ma, C. Han, Y. Yao, G. Tang, Z. Mao, C. Gao and F. Huang, *J. Am. Chem. Soc.*, 2013, **135**, 14459.
31. W. Si, X.-B. Hu, X.-H. Liu, R. Fan, Z. Chen, L. Weng and J.-L. Hou, *Tetrahedron Lett.*, 2011, **52**, 2484–2487.
32. W. Si, L. Chen, X. B. Hu, G. Tang, Z. Chen, J. L. Hou and Z. T. Li, *Angew. Chem., Int. Ed.*, 2011, **50**, 12564–12568.
33. S. Dong, B. Zheng, Y. Yao, C. Han, J. Yuan, M. Antonietti and F. Huang, *Adv. Mater.*, 2013, **25**, 6864–6867.
34. T. Ogoshi, R. Shiga and T. A. Yamagishi, *J. Am. Chem. Soc.*, 2012, **134**, 4577–4580.
35. H. Zhang, G. Gruner and Y. Zhao, *J. Mater. Chem. B*, 2013, **1**, 2542–2567.
36. J. Zhou, M. Chen, J. Xie and G. Diao, *ACS Appl. Mater. Interfaces*, 2013, **5**, 11218–11224.
37. H. Zhang, X. Ma, K. T. Nguyen, Y. Zeng, S. Tai and Y. Zhao, *ChemPlusChem*, 2014, **79**, 462–469.
38. D. D. Zheng, D. Y. Fu, Y. Wu, Y. L. Sun, L. L. Tan, T. Zhou, S. Q. Ma, X. Zha and Y. W. Yang, *Chem. Commun.*, 2014, **50**, 3201–3203.
39. I. Nierengarten, S. Guerra, M. Holler, J. F. Nierengarten and R. Deschenaux, *Chem. Commun.*, 2012, **48**, 8072–8074.
40. I. Nierengarten, S. Guerra, M. Holler, L. Karmazin-Brelot, J. Barberá, R. Deschenaux and J.-F. Nierengarten, *Eur. J. Org. Chem.*, 2013, **2013**, 3675–3684.
41. T. Ogoshi, N. Ueshima and T.-A. Yamagishi, *Org. Lett.*, 2013, **15**, 3742–3745.
42. T. Ogoshi and T. Yamagishi, *Chem. Commun.*, 2014, **50**, 4776–4787.

CHAPTER 10

Hybrid Materials Based on Pillararenes

NAN SONG AND YING-WEI YANG*

International Joint Research Laboratory of Nano-Micro Architecture Chemistry (NMAC), College of Chemistry, Jilin University, Changchun 130012, P.R. China
*Email: ywyang@jlu.edu.cn

10.1 Introduction

With the development of modern science and technology, materials fabricated from a single component cannot satisfy the demands of industrial applications. Therefore, hybrid materials composited from two or more components have been widely explored and show a broad application prospect. Hybrid materials usually contain two or more nano- or molecularly different components, including organic, inorganic and metallic constituents, and exhibit composite functions as well as complementary and optimized properties.[1] Hybrid materials also overcome the limitations of single-component materials and present abundant new features in terms of optical, mechanical, electrical and electrochemical properties. It can be predicted that hybrid materials will be the dominant materials in the 21st century.[2]

As an emerging class of synthetic macrocycles, pillararenes and their derivatives have been one of the most important research directions of supramolecular chemistry during the past several years.[3] Their unique symmetrical rigid structures and typical host–guest properties inspire researchers to construct functionalized materials based on pillararenes and their derivatives.[4] Thus, pillararenes have been hybridized with several

Monographs in Supramolecular Chemistry No. 18
Pillararenes
Edited by Tomoki Ogoshi
© The Royal Society of Chemistry 2016
Published by the Royal Society of Chemistry, www.rsc.org

inorganic and metallic materials to fabricate some new composites including metal–organic frameworks, pillararene-stabilized metal nanoparticles, carbon materials dispersed by pillararene derivatives, and so on. There is no doubt that pillararene-hybridized materials have exhibited very desirable properties and features, and have been applied to recognition, stabilization, dispersion, controlled drug release and absorption.

10.2 Pillararene-based Porous Frameworks

10.2.1 Introduction to Metal–Organic Frameworks (MOFs)

Metal–organic framework (MOF) materials, constructed from metal ion clusters and organic linkers, are an important class of porous solid materials with a regular crystalline lattice and specific pore structures.[5] The area of MOFs has become one of the most attractive fields in modern chemistry owing to their interesting structures and functional tunability. MOFs have been the focus of much research interest in the past few years because of their various industrial applications. The ever-escalating number of MOF structures and functions has been studied and reported.[6] The developments in MOFs have been centered on advances in structure determination, preparation of pertinent ligands, maturation of functional modifications, potential applications and interdisciplinary connections of MOF research with different fields such as chemical engineering, supramolecular chemistry, and materials science. Diverse organic bridged linkers and metal ions can be used to construct MOF materials with exceptionally high porosity and well-defined absorption sites. Researchers have put great effort into the syntheses of superior ligands. As a new class of synthetic macrocyclic hosts, pillar[*n*]arenes have also been considered as a building strut to prepare MOF materials. The pillar[5]arene-based MOF materials, *e.g.*, enantiomeric MOFs and homochiral MOFs, have been successfully prepared by Stoddart and coworkers, and have active domains and exhibit selective recognition toward several electron-poor guests.[7,8] Yang *et al.* first prepared pillar[5]arene-based supramolecular organic frameworks for highly selective capture of carbon dioxide under ambient conditions. In this section, we will introduce the above-mentioned research.[9]

10.2.2 A1/A2 Difunctionalized Pillar[5]arene-based MOFs

The A1/A2-difunctionalized pillar[5]arene strut **1** (Scheme 10.1), with a rigid structure, has been successfully prepared and incorporated into a MOF, *i.e.*, P5A-MOF-1 (Figure 10.1a),[7] with Zn_4O secondary building units (SBUs) in a conventional manner, *via* heating a DMF solution of a mixture of strut **1** and Zn-$(NO_3)_2 \cdot 6H_2O$ at 100 °C for 24 hours. The cubic and transparent crystals of P5A-MOF-1 can be observed by optical microscopy and confirmed by powder X-ray diffraction (PXRD) (Figure 10.1b). Moreover, their thermal stability has

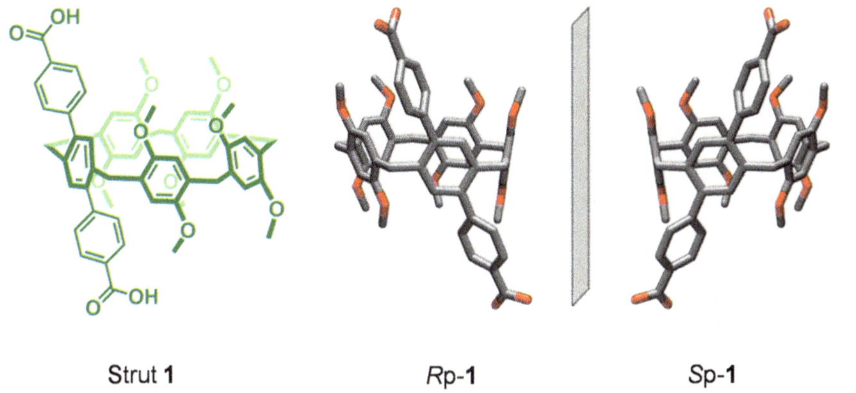

Scheme 10.1 (Left) Structure of strut **1**. (Right) Solid-state structure of strut **1** that displays planar chirality and does not racemize between the *R*p-**1** (left) and *S*p-**1** (right) enantiomers.
Adapted with permission from ref. 7. Copyright 2012 American Chemical Society.

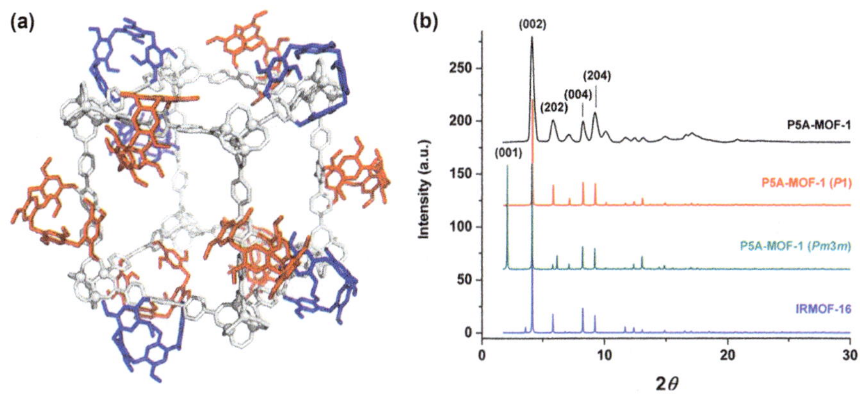

Figure 10.1 (a) Model of P5A-MOF-1. (b) Experimental PXRD pattern for P5A-MOF-1 and IRMOF-16, calculated PXRD patterns for P5A-MOF-1 in a space group of *P*1 and a space group of *Pm*3*m*.
Adapted with permission from ref. 7. Copyright 2012 American Chemical Society.

been determined by thermogravimetric analysis (TGA) and the TGA profile indicates that P5A-MOF-1 remains stable to 450 °C.

Pillar[5]arenes can rotate freely around every terphenylene linker, which results in the disorder within P5A-MOF-1 itself. On the other hand, the random distribution of "enantiomeric" pillar[5]arenes corresponding to their planar chirality also contributes to the disorder. Therefore, single-crystal X-ray data is unable to discern the solid structure of P5A-MOF-1 as the extended framework. Non-interpenetrated IRMOF-16, incorporated with pillar[5]arenes with randomly distributed chirality and orientations around the terphenylene linkers, is used as the model structure of P5A-MOF-1. The cubic unit cell with dimensions of $a = b = c = 42.980$ Å and *P*1 space group can be provided by the optimized geometry of the predicted structure. The close match between P5A-MOF-1 and the model structure can be observed by the simulated PXRD patterns. Another model structure with a space group of *Pm*3*m* has also been introduced in order to determine whether the model with higher symmetry could also match the experimental MOF data. In the second model, there are four pillar[5]arenes in the shape of both enantiomers in two orientations around each organic strut. The cubic cell dimensions are coincident with those of the first model structure, but there are extra peaks presented by the simulated PXRD pattern. Therefore, according to the experimental data, the lower symmetry model structure provides a better match.

Further experiments have been carried out to investigate the ability of P5A-MOF-1 to take up guests. First of all, the association constants (K_a) of the strut **1** with three guests, *e.g.*, *N*-hexylpyridinium cation ($\mathbf{G1}^+$), *N,N′*-dihexyl-4,4-bipyridinium dication ($\mathbf{G2}^{2+}$), and the neutral 1,4-dinitrobenzene (**G3**), were determined by ^{1}H NMR titrations in CD_3COCD_3. They are 43.2 ± 2.9 ($\mathbf{G1}^+$), 170 ± 50 ($\mathbf{G2}^{2+}$), and 66.2 ± 1.9 M^{-1} (**G3**), respectively. Second, the excess of DMF within the frameworks has been removed by suspending P5A-MOF-1 in Me_2CO. The guest uptake experiments are conducted by introducing the samples of P5A-MOF-1 into saturated solutions of $\mathbf{G1}^+ \cdot PF_6$, $\mathbf{G2}^{2+} \cdot 2PF_6$, and **G3** in Me_2CO. The samples of crystals with $\mathbf{G2}^{2+} \cdot 2PF_6$ and **G3** undergo an immediate color change from faint yellow to deep orange because the charge transfers through P5A-MOF-1 to the guests (Figure 10.2). In the control experiment IRMOF-16-OPX, without an active domain, is used and the guest uptake experiments are performed under the same conditions. Compared with IRMOF-16-OPX, P5AMOF-1 takes up much larger amounts of each guest, which can be observed by the mole ratios of guest to organic strut in P5A-MOF-1 and IRMOF-16-OPX obtained from ^{1}H NMR spectra. The reason may be that the active domain of P5A-MOF-1 can form host–guest complexation through efficient noncovalent bonding interactions.

The final experiment is performed by suspending P5A-MOF-1 in acetone with equimolar concentrations (40.0 mM) of two different guests so as to investigate the competitive uptake that can be predicted according to the different association constants. P5A-MOF-1 takes up almost twice the amount of $\mathbf{G2}^{2+} \cdot 2PF_6$ when compared with $\mathbf{G1}^+ \cdot PF_6$; $\mathbf{G2}^{2+} \cdot 2PF_6$ with the active domain of P5A-MOF-1 exhibits a larger K_a than $\mathbf{G1}^+ \cdot PF_6$. In addition, under the same conditions, the uptake of these two guests by IRMOF-16-OPX is very similar, as expected.

10.2.3 Enantiopure Pillar[5]arene-based MOFs

Many examples of homochiral MOFs, with either stereogenic centers or axes of chirality, have been reported. In this field, the incorporation of planar chirality into homochiral MOFs is a relatively unexplored area. Stoddart *et al.*

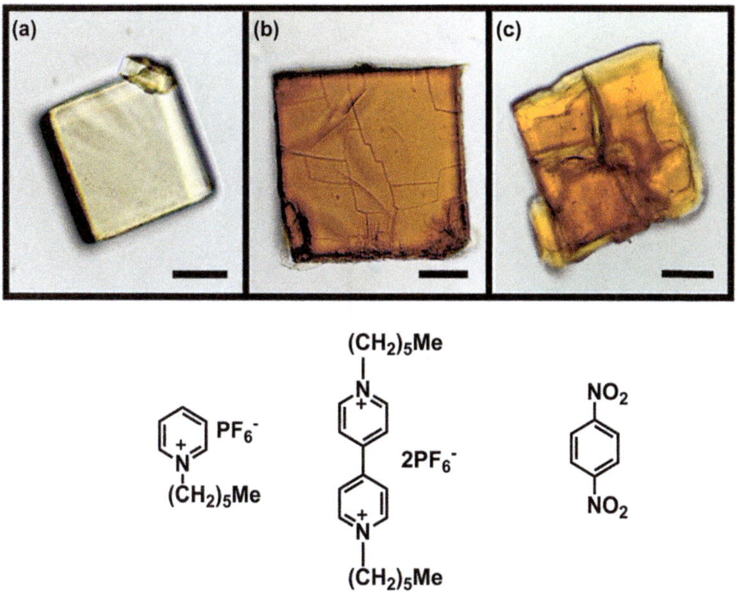

$G1^+ PF_6$ $G2^{2+} 2PF_6$ G3

Figure 10.2 (Top) Optical microscopy images of P5A-MOF-1 (a) with no guest (scale bar, 200 μm), (b) after uptake of $\mathbf{G2}^{2+} \cdot 2PF_6$ (scale bar, 100 μm), and (c) after uptake of **G3** (scale bar, 100 μm). (Bottom) Structures of $\mathbf{G1}^+ \cdot PF_6$, $\mathbf{G2}^{2+} \cdot 2PF_6$, and **G3**.
Reprinted with permission from ref. 7. Copyright 2012 American Chemical Society.

reported the resolution of *rac*-**1** into two enantiopure struts, (*R*p)-**1** and (*S*p)-**1**, and incorporated them into homochiral porous structures of (*R*p)-P5A-MOF-1 and (*S*p)-P5A-MOF-1 (Scheme 10.2).[8] In order to resolve *rac*-**1** on the gram scale without the use of chiral chromatography, the enantiomers (*R*p)-**1** and (*S*p)-**1** are functionalized with an enantiomerically pure chiral auxiliary. Following this, the enantiomers are converted into two diastereoisomers that will be separated by normal phase chromatography. Finally the pure enantiomers (*R*p)-**1** and (*S*p)-**1** can be obtained by removing the chiral auxiliaries.

The enantiomerical structures of (*R*p)-**1** and (*S*p)-**1** have been confirmed by circular dichroism (CD) spectroscopy, where the positive cotton effect for (*R*p)-**1** and the negative cotton effect for (*S*p)-**1** were observed (Figure 10.3).

In addition, (*R*p)-**1** and (*S*p)-**1** have been used to synthesize homochiral frameworks, (*S*p)-P5A-MOF-1 and (*R*p)-P5A-MOF-1, respectively, by heating a mixture of (*R*p)-**1** or (*S*p)-**1** and $Zn(NO_3)_2 \cdot 6H_2O$ in DMF at 100 °C. In common with *rac*-P5A-MOF-1, single-crystal X-ray diffraction (SXRD) analysis of the structures of (*S*p)-P5A-MOF-1 and (*R*p)-P5A-MOF-1 could not be resolved to discern individual atoms in the solid-state structure well enough, which was ascribed to the disorder within the frameworks that resulted from the rotational freedom of pillar[5]arenes with respect to the terphenylene linker.

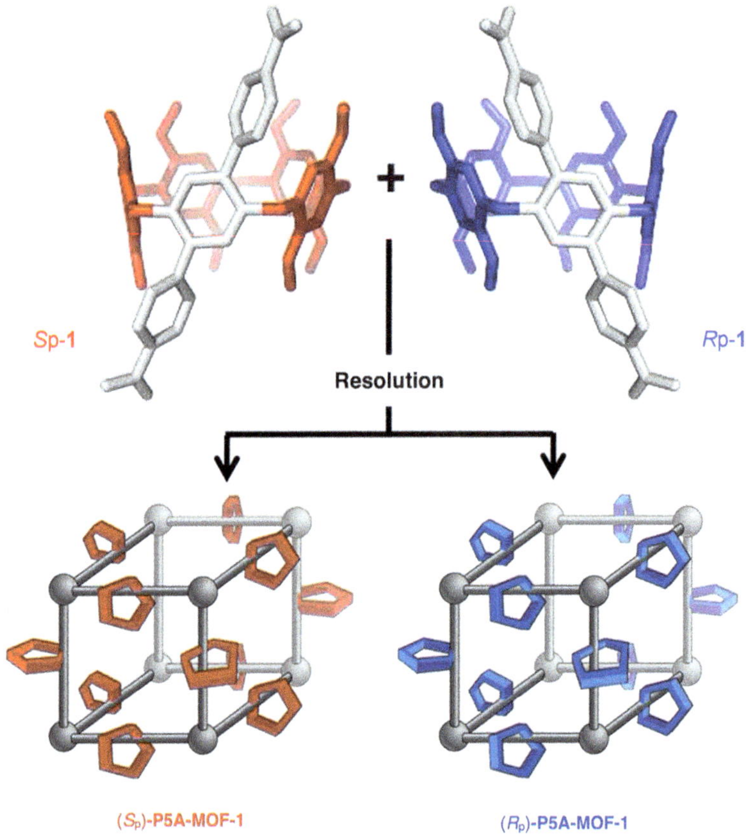

Scheme 10.2 Enantiomers (*R*p)-**1** and (*S*p)-**1** can be used in separate reactions to form homochiral (*R*p)-P5A-MOF-1 and (*S*p)-P5A-MOF-1. Reproduced from ref. 8 with permission from The Royal Society of Chemistry.

Thus, homochiral MOFs containing enantiopure pillar[5]arene active domains have been obtained, which are early instances of porous frameworks incorporating active domains and possessing planar chiralities. Meanwhile, an efficient route to large-scale resolution of racemic pillar[5]-arene derivatives has been developed in the preparation of (*S*p)-P5A-MOF-1 and (*R*p)-P5A-MOF-1.

10.2.4 Pillar[*n*]arene-based Supramolecular Organic Frameworks

Yang and co-workers reported a new perhydroxyl-pillar[5]arene-based supramolecular organic framework (P5-SOF) material which can absorb CO_2 selectively and reversibly (Scheme 10.3).[9] The honeycomb-like structure of P5-SOF resulted from triclinic arrangement of P5 rings consisting of

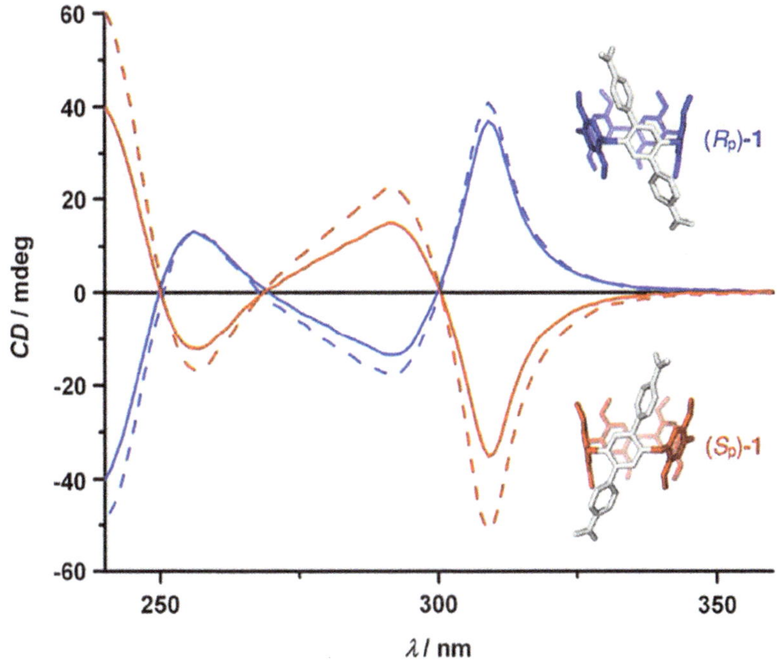

Figure 10.3 Circular dichroism (CD) spectroscopy of *S*p-**1** and *R*p-**1**. Reproduced from ref. 8 with permission from The Royal Society of Chemistry.

three-dimensional (3D) channels such as 1D round channels (diameter 7.10 Å), slit channels (5.14×9.14 Å), and isolated cages (diameter 6.76 Å), as observed by SXRD analysis of pillar[5]arene. On the other hand, the single crystal structure indicates that the porous 3D cage material exhibits exceptional stability due to the O–H···O hydrogen bonds between hydroxyl groups on the rim of neighboring pillar[5]arene. Its high thermal stability is also revealed by TGA.

All of above results suggest that such pillararene-based SOF materials with permanent porosity can be expected to have potential application in gas absorption. As displayed by the gas sorption isotherms measured up to 1 atm at 298 K, P5-SOF exhibits remarkably higher absorption capacity for CO_2 through the dipole–dipole interactions between CO_2 and hydroxyl groups, compared with CH_4 and N_2 under the same condition. The selectivity for CO_2 is estimated to be 339 : 1 over N_2, and 375 : 1 over CH_4. It is also worth mentioning that this is the first report of a pillararene-based SOF material with a permanent porous structure and with extraordinarily selective and reversible absorption capacity for CO_2 compared with that of most of the reported SOF materials with intrinsic cavities. Furthermore, it also provides a glimpse into the novel potential of P5-SOF materials in gas capture and further environmental engineering.

Compared with pillar[5]arenes, pillar[6]arenes possess a larger cavity size of *ca.* 6.7 Å. Ogoshi and co-workers reported the use of highly ordered

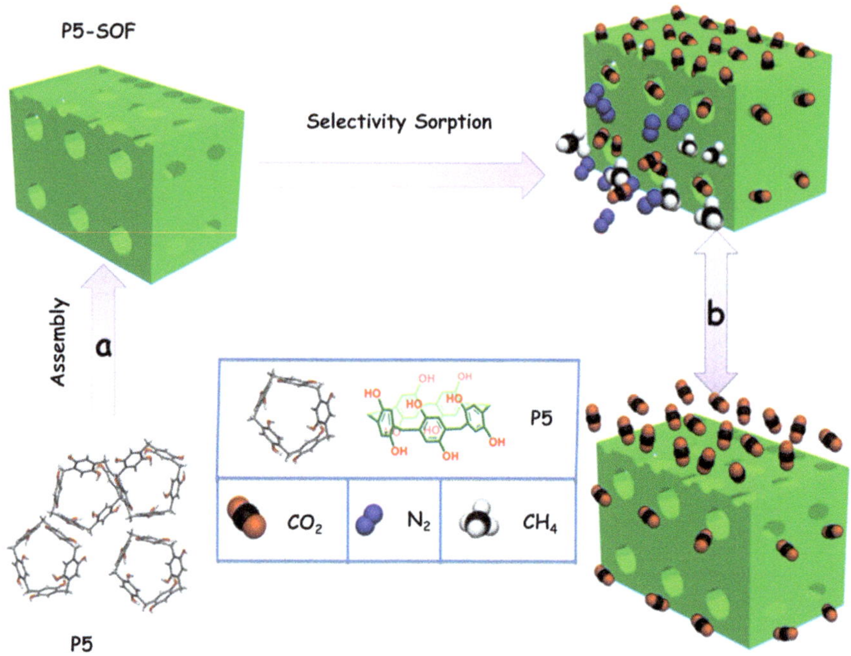

Scheme 10.3 Schematic diagrams of P5-SOF and its desired mixed gas sorption processes. In the actual experiments, we ran the sorption for each gas, *i.e.*, CO_2, N_2, CH_4, independently. (a) P5-SOF assembled from P5. (b) CO_2 can be released by heating or reducing pressure, then P5-SOF can be reused for gas-selective sorption.
Used with permission from ref. 9. Copyright 2014 Wiley.

one-dimensional channels, based on per-hydroxylated pillar[6]arene molecules, for the capture of various gases such as CO_2, N_2 and *n*-butane.[10] According to their investigation, the absorption capacity of the 1D channels for cyclohexane was higher than that for *n*-hexane owing to the fitness of the cavity sizes.

10.2.5 MOFs Gated by Pillar[*n*]arene-based Supramolecular Switches

Stimulated by the developments of materials science and drug delivery systems, MOF-based drug delivery systems are also under investigation.[11] Our group reported the first stimuli-responsive theranostic hybrid platform, consisting of mechanized nanoMOFs immobilized with carboxylatopillar[5]arene (CP[5]A)-based supramolecular switches.[12] This novel CP[5]A-gated nanoMOF can act as a smart cargo delivery system in response to pH and competitive binding agents. MOFs have large pore sizes for cargo loading but premature release is an important challenge for their practical utilizations. However, the premature release was negligible in this

CP[5]A-gated nanovalve system. Furthermore, MTT cytotoxicity assay of 293 cells has been performed; it suggested that the MOF-based platform possesses negligible cytotoxicity before and after capping with CP[5]A and indicated that MOF-based nanovalves have great potential in controlled release of therapeutic agents and the targeted delivery of drugs.

10.3 Pillararene–Metal Nanoparticle Composites

10.3.1 Introduction to Metal Nanoparticles

Metallic nanoparticles are a special class of chemical object that are ubiquitous in many fields of science, such as physics, chemistry, materials science, biochemistry and nanotechnology. Metallic nanoparticles exhibit many special diverse characteristics such as optical, electronic, magnetic and related properties.[13] Most research attention has been focused on the control of the size and shape of nanoparticles, aiming at their utilization in well-defined chemistry.

Among them, gold nanoparticles have attracted a large amount of attention owing to their significance in sensors, catalysis and surface-enhanced Raman scattering (SERS) spectroscopy.[14] Like other metallic nanoparticles, gold nanoparticles are only kinetically stable and easily aggregate into larger particles, which results in high surface energy and low reaction activity.[15] Therefore, the stabilization of gold nanoparticles has been an urgent requirement for their application. Stabilizing ligands with typical functional groups, such as amine ($-NH_2$), carboxyl (–COOH), or sulfhydryl (–SH) groups, are used to synthesize and stabilize gold nanoparticles.[16–20] Silver nanoparticles are very similar to gold nanoparticles.[21]

Quantum dots (QDs), with obvious fluorescence properties, exhibit superior properties to those of traditional dyes, such as strong fluorescence emission intensity, broad excitation, and long fluorescence time.[22–24] Pillar[5]arenes with carboxylate groups can be introduced as a stabilizer of QDs to expand their applications.

Nanoparticles of Fe_3O_4 are magnetic and therefore can be used in absorption and extraction *via* the magnetic response.[25–27] Pillar[5]arenes hybridized with Fe_3O_4 nanoparticles are expected to show special properties and immense application potential.

10.3.2 Pillararene–Gold Nanoparticle Composites

In 2012, Huang and co-workers successfully synthesized a kind of water-soluble pillar[5]arene with 10 imidazolium groups (**IP5A**) substituted on the rims of the macrocyclic ring.[28] The newly synthesized imidazolium-functionalized pillar[5]arene is fully characterized by ^1H NMR, ^{13}C NMR and HIRESMS. Typically, the HIRESMS spectrum reveals a series of peaks at m/z 1170.65, 754.13, 545.37, 420.51, 336.93 and 277.53, corresponding to $[M - 2Br]^{2+}$, $[M - 3Br]^{3+}$, $[M - 4Br]^{4+}$, $[M - 5Br]^{5+}$, $[M - 6Br]^{6+}$ and $[M - 7Br]^{7+}$, respectively, which indicates that the pillar[5]arene derivative can lose different numbers of bromide anions and thus form segments with a series number of positive charges. Depending on the macrocyclic platform and the weak coordinating cations and anions of imidazolium groups, the water-soluble pillar[5]arene can provide a shell of imidazolium cations and bromide anions around the gold nanoparticles so that it can act as the stabilizer for preparation of gold nanoparticles *via* binding on the gold surface.

The gold nanoparticles were synthesized through a typical experiment, *i.e.*, the reduction of $HAuCl_4$ in aqueous dispersions of IP5As. The aqueous sodium borohydride was injected into the aqueous solution of HAuCl4 and water-soluble pillar[5]arenes under strong stirring. The pillar[5]arene-stabilized gold nanoparticles were obtained when the color of the solution changed to wine red (Figure 10.4). The monomer of water-soluble

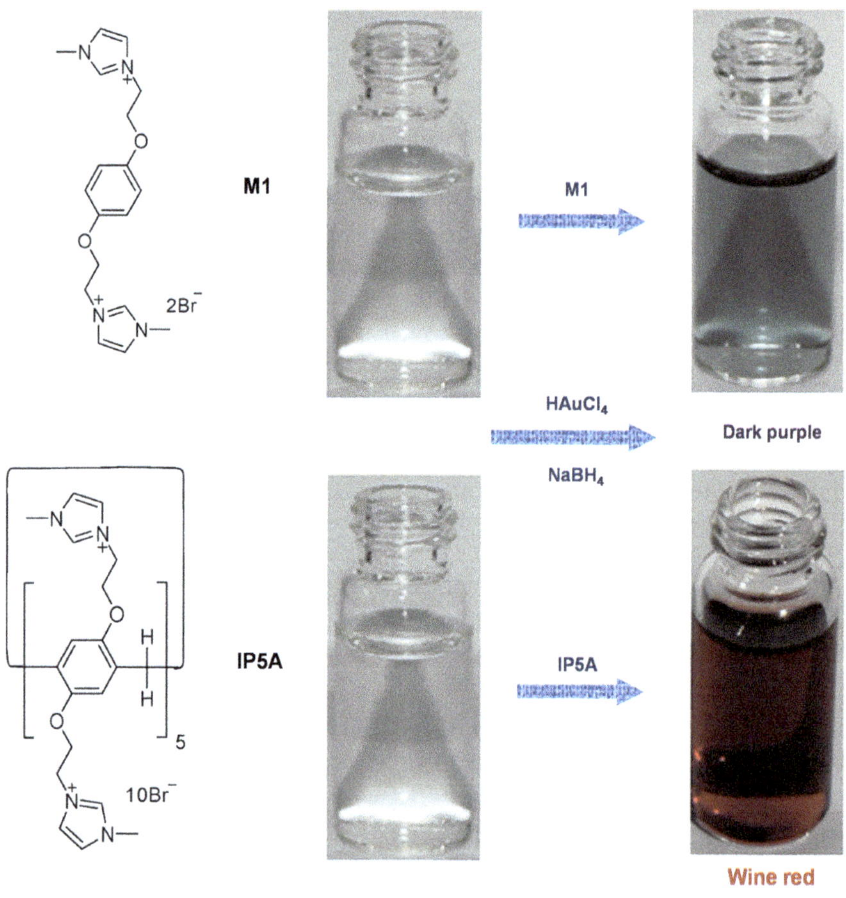

Figure 10.4 (Left) Structures of **IP5A** and **M1**. (Right) Photographs of solutions of gold nanoparticles stabilized by **IP5A** and **M1**.
Adapted from ref. 28 with permission from The Royal Society of Chemistry.

pillar[5]arene sharing the same cationic and anionic groups was also prepared and used to stabilize gold nanoparticles as a comparison. A dark purple solution was obtained by using the monomer, instead of wine red nanoparticles, which demonstrates the crucial significance of the rigid macrocyclic framework of water-soluble pillar[5]arene in the stabilizing of gold nanoparticles.

According to the UV-vis spectra and the transmission electron microscopy (TEM) images, it can be concluded that the size of gold nanoparticles is decreased upon increasing the concentration of water-soluble pillar[5]arenes. As shown in the UV-vis spectra, the surface plasmon absorption band of the gold nanoparticles undergoes a slight blue shift from 560 nm to 512 nm upon increasing the concentration of water-soluble pillar[5]arenes from 2 μM to 200 μM. Furthermore, a decreasing concentration of water-soluble pillar[5]arenes results in a decrease in the intensity of absorption of gold nanoparticles, which indicates that large-size gold nanoparticles can be obtained at a lower concentration of water-soluble pillar[5]arenes. The same conclusion can be drawn intuitively from the TEM images of gold nanoparticles at different concentrations of water-soluble pillar[5]arenes.

The TEM image of gold nanoparticles stabilized by water-soluble pillar[5]arenes at a concentration of 200 μM reveals a spherical shape with an average diameter of 1.88 ± 0.58 nm and the relatively narrow particle size distribution. The well-defined crystalline planes can be observed from the high-resolution TEM image.

The pillar[5]arene-stabilized gold nanoparticles were then used as a catalyst for the borohydride reduction of 1,4-nitroaniline to 1,4-diaminobenzene (Figure 10.5). Even a very small amount of water-soluble pillar[5]-arene-stabilized gold nanoparticles can induce the discoloration corresponding to the occurrence of the reduction reaction, while $NaBH_4$ itself is unable to reduce the 1,4-nitroaniline. The catalytic reduction was conducted on the surface of the nanoparticles and the mechanism can be described as follows: (1) the gold nanoparticles react with the borohydride ions and result in metal hydride; (2) the nitroaniline molecules absorb onto the metal surface; (3) the catalytic reaction of the gold nanoparticles is carried out upon electron transfer from the BH_4^- anion to nitro compounds, mediated by the large Fermi level shift of nanoparticles. Different sizes of gold nanoparticles show different catalytic efficiency. Smaller particles exhibit more efficient activity in catalyzing the reduction reaction than larger ones.

Another type of water-soluble pillar[5]arene, carboxylatopillar[5]arene sodium salts (CP[5]A), has also been used as the stabilizer for *in situ* preparation of gold nanoparticles, according to the work of Yang's group.[29] CP[5]A possesses 10 carboxyl (–COOH) groups, which are essential to the synthesis and stabilization of gold nanoparticles (Scheme 10.4). The water-soluble nature of CP[5]A also confers the ability to prepare gold nanoparticles in aqueous phase. The gold nanoparticles are synthesized by reducing $HAuCl_4$ in aqueous dispersions of water-soluble pillar[5]arenes at selected

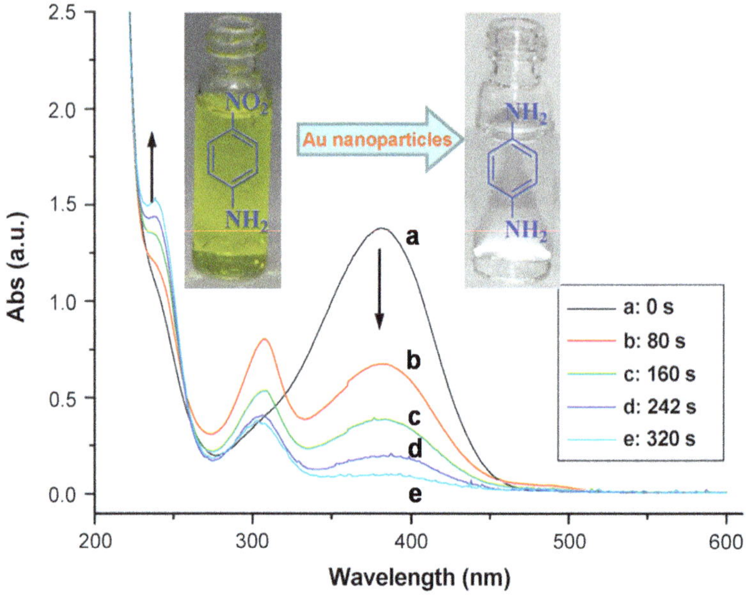

Figure 10.5 Photographs of 4-nitroaniline solution and the solution after reduction, and successive UV-vis absorption of the reduction of 4-nitroaniline by excess $NaBH_4$ using pillar[5]arene-stabilized gold nanoparticles as the catalyst.
Reproduced from ref. 28 with permission from The Royal Society of Chemistry.

concentrations. According to the UV-vis spectra (Figure 10.6), the stable CP[5]A-modified gold nanoparticles possess a typical surface plasmon resonance (SPR) peak at around 500 nm. The blue shift of the SPR peak from 516 nm to 487 nm indicates the smaller average diameter of the gold nanoparticles upon increasing the proportion of CP[5]A to $HAuCl_4$. When [CP[5]A]/[$HAuCl_4$] is varied from 0.075 to 0.3, there is a slight change of wavelength and intensity, indicating that CP[5]A is excess; continuing addition has a negligible influence on the size of the gold nanoparticles, and the surfaces of the gold nanoparticles have been covered completely.

The influence of the macrocyclic framework can be observed in controlled experiments. Upon addition of viologen I to form stable complexation with CP[5]A, the gold nanoparticles prepared without stabilizing ligands and with the noncyclic monomers of CP[5]A aggregate and precipitate from the solution and the samples become dark gray, while the CP[5]A-modified gold nanoparticle sample has neither color change or participation. The experimental results suggest that the CP[5]A can stabilize gold nanoparticles efficiently owing to the rigid framework, and that gold nanoparticles cannot be as stable in the absence of stabilizers as in the presence of noncyclic monomers.

The TEM images also describe the same experimental results (Figure 10.7). Gold nanoparticles prepared in the absence of stabilizing ligands aggregate into large ones with a diameter of 7.7 ± 1.9 nm, while the

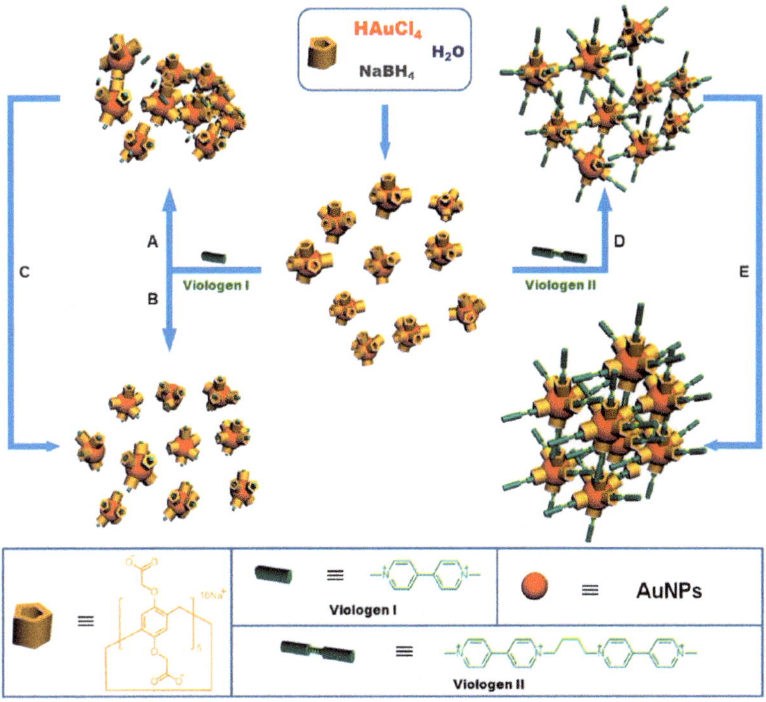

Scheme 10.4 Schematic illustration of CP[5]A-modified gold nanoparticles and their supramolecular self-assembly upon addition of viologen I and viologen II: (A) 5 min and (C) 24 h after the addition of viologen I (10 mM) to a solution of CP[5]A-stabilized gold nanoparticles; (B) 5 min after the addition of viologen I (75 mM) to a solution of CP[5]A-stabilized gold nanoparticles; (D) 5 min and (E) 24 h after the addition of viologen II (3.8 mM) to a solution of CP[5]A-stabilized gold nanoparticles.
Reprinted with permission from ref. 29. Copyright 2013 American Chemical Society.

gold nanoparticles prepared in the presence of CP[5]A possess narrow size distributions and a smaller diameter of 3.1 ± 0.5 nm, which can be observed in the visual field of TEM images. On the one hand, CP[5]A provides carboxylate groups to form strong carboxyl–gold binding interactions for further stabilization of the initial gold nanoparticles. On the other hand, the rigid macrocyclic backbone of CP[5]A inhibits the further growth of gold nanoparticles. The high resolution TEM image shows the well-defined crystalline planar d spacing at 0.235 nm, corresponding to the (111) planes of face-centered cubic gold.

The CP[5]A-stabilized gold nanoparticles can be used as optical probes to detect herbicide. Viologen I can stably complex with CP[5]A at a 1 : 1 molar ratio. Upon addition of different amounts of viologen I, the solution color of CP[5]A-modified AuNPs changes in five minutes. Because the CP[5]A-modified gold nanoparticles have negatively charged surfaces due to the

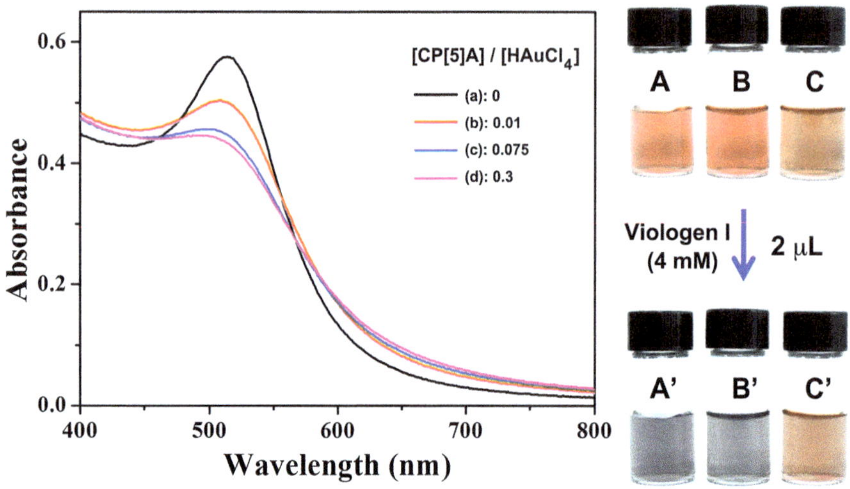

Figure 10.6 (Left) UV-vis spectra of CP[5]A-modified gold nanoparticles synthesized with different [CP[5]A] : [$HAuCl_4$] ratios: (a) 0, (b) 0.01, (c) 0.075, (d) 0.3. (Right) Photographs of solutions of (A and A′) bare gold nanoparticles, (B and B′) noncyclic monomer of CP[5]A-modified gold nanoparticles, (C and C′) CP[5]A-modified gold nanoparticles before and after the addition of viologen I, respectively.
Reprinted with permission from ref. 29. Copyright 2013 American Chemical Society.

carboxylate anions of CP[5]A, cationic viologen I can combine with CP[5]A and reduce the potential of surfaces of the modified gold nanoparticles. Thus, small amounts of viologen I can cause the aggregation of CP[5]A-stabilized gold nanoparticles in a short time, related to the charge variations of the gold nanoparticles. However, with increased time, well mono-dispersed gold nanoparticles can be obtained and the spherical particles become slightly larger, smoother and rounder. Given that the viologen I can induce aggregation of CP[5]A-modified gold nanoparticles in a short time, a potential application of CP[5]A-modified gold nanoparticles is in the detection of paraquat, a herbicide and another name for viologen I. The Δ ratio A : D is introduced for the highly sensitive detection of paraquat. A stands for the integral of the UV-vis spectra ranging from 550 to 700 nm (aggregated area), while D stands for the integral of the UV-vis spectra ranging from 490 to 540 nm (dispersed area). Quantitative detection of paraquat was achieved by calculating the change of the Δ ratio A : D for the area under the plasmon resonance peak after ultralow amounts of paraquat were added. A detection limit of 0.2 μM ($r_2 = 0.996$) can be realized by this method. The ability to detect concentrations below 0.2 μM will be limited, owing to the sensitivity of the UV-vis spectra, and the Δ ratio A : D will show large error. This new technique is a faster and more convenient method for real-time quantitative determination of paraquat according to the change of the Δ ratio A : D, when compared with the traditional testing method.

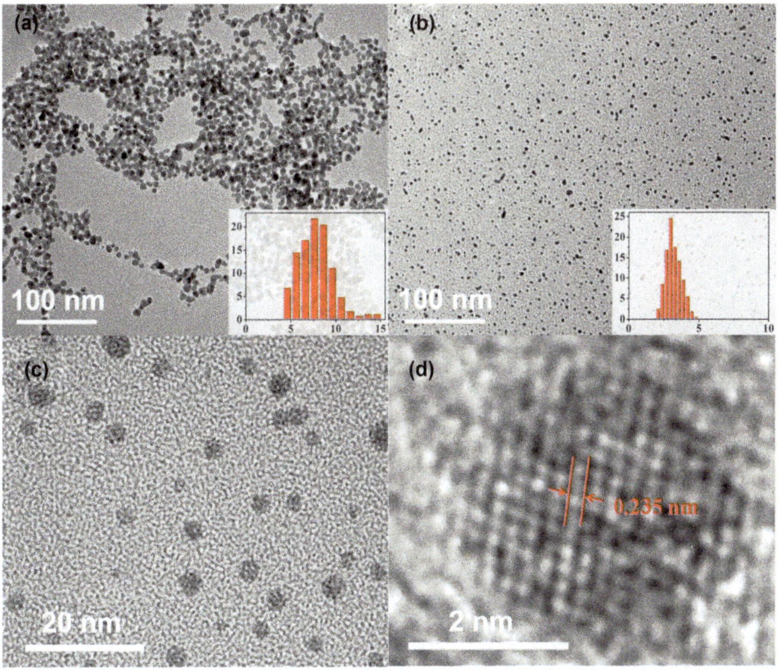

Figure 10.7 Transmission electron microscopy (TEM) images of (a) bare gold nanoparticles ([CP[5]A] : $[HAuCl_4] = 0$) and (b) CP[5]A-modified gold nanoparticles ([CP[5]A] : $[HAuCl_4] = 0.075$). (Insets: corresponding histograms of the gold nanoparticle sizes based on 300 particles.) (c) Higher resolution image of parts of (b), and (d) high resolution TEM images of CP[5]A-modified gold nanoparticles.
Reprinted with permission from ref. 29. Copyright 2013 American Chemical Society.

The ditopic guest viologen II can bridge CP[5]A-modified AuNPs for self-assembly into patterned AuNPs (Figure 10.8). Instead of simple mass aggregation, viologen II can induce one-dimensional chain-like assemblies as well as large three-dimensional assemblies under appropriate conditions.

Subsequently, Huang and co-workers have successfully prepared organic/inorganic supramolecular hybrid nanostructures by the self-assembly of water-soluble pillar[6]arene-stabilized gold nanoparticles with a hydrophobic chain functionalized paraquat derivative.[30] Micelles, onion-like disks and vesicles can be obtained by simply changing the amounts of the hydrophobic paraquat derivative. Thus, they provided a facile method to prepare various organic/inorganic hybrid nanostructures. Meanwhile, they also constructed supramolecular hybrid vesicles based on water-soluble pillar[6]arene-stabilized gold nanorods, and utilized them to encapsulate cargo molecules for delivery and controlled release in response to pH decrease or near-infrared (NIR) irradiation.

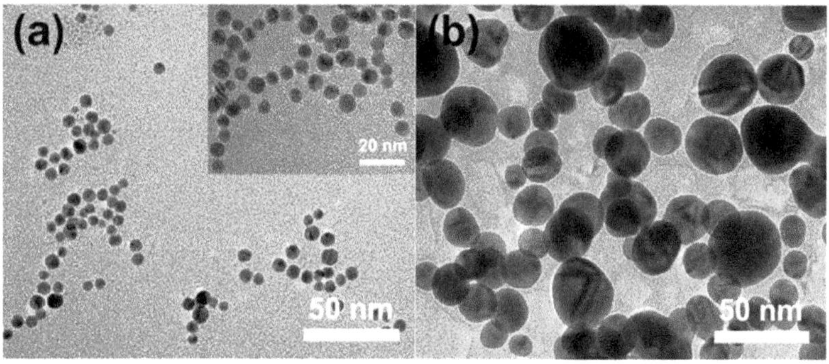

Figure 10.8 TEM images of CP[5]A-stabilized gold nanoparticles (a) 5 min and (b) 24 h after putting CP[5]A-stabilized gold nanoparticles with viologen II (3.8 μM).
Reprinted with permission from ref. 29. Copyright 2013 American Chemical Society.

With regard to carboxylated pillar[*n*]arenes, Park, in collaboration with Kim *et al.*, reported a facile one-pot hydrothermal method for the synthesis of gold nanoparticles stabilized by carboxylated pillar[5]arene that were highly stable against pH variations.[31] The $HAuCl_4$ (0.01%) was refluxed in water (10 mL) for 10 min, followed by the addition of CP[5]A solution (1%, 0.3 mL H_2O). After refluxing the solution mixture for 1 h, spherical gold nanoparticles were obtained.

Spherical gold nanoparticles in a wide range of sizes were also successfully synthesized, mediated by ammonium pillar[5]arene.[32] The immobilization of ammonium pillar[5]arene onto the surface of gold nanoparticles was demonstrated by surface-enhanced Raman scattering (SERS) spectroscopy. Zhou and Jiang *et al.* reported monosulfhydryl functionalized pillar[5]arene-stabilized gold nanoparticles and their reversible aggregation and disassembly resulting from the [4 + 4] cycloaddition of anthracene.[33]

10.3.3 Pillararene–Silver Nanoparticle Composites

According to the research of Yao and co-workers, water-soluble pillar[5]arene (**WP5**)-stabilized silver nanoparticles were synthesized.[34] Similar to the synthesis of gold nanoparticles, **WP5**-modified silver nanoparticles are obtained by the reduction of $AgNO_3$ (10 mM) with $NaBH_4$ (50 mM) in the dispersion of a series of concentrations of **WP5** in aqueous solution (Scheme 10.5). The typical SPR peak at around 400 nm indicates the resistant **WP5**-stabilized silver nanoparticles. The SPR peak also undergoes a blue shift from 430 nm to 400 nm upon gradually increasing the concentration of **WP5**, suggesting that the silver nanoparticles are smaller. When the concentration of **WP5** is increased from 0.400 mM to 400 mM, the maximum of the SPR peak maintains the same wavelength, indicating that the surfaces of the silver nanoparticles have been covered completely and that continuous addition of excess **WP5** has no influence on the diameter of the nanoparticles.

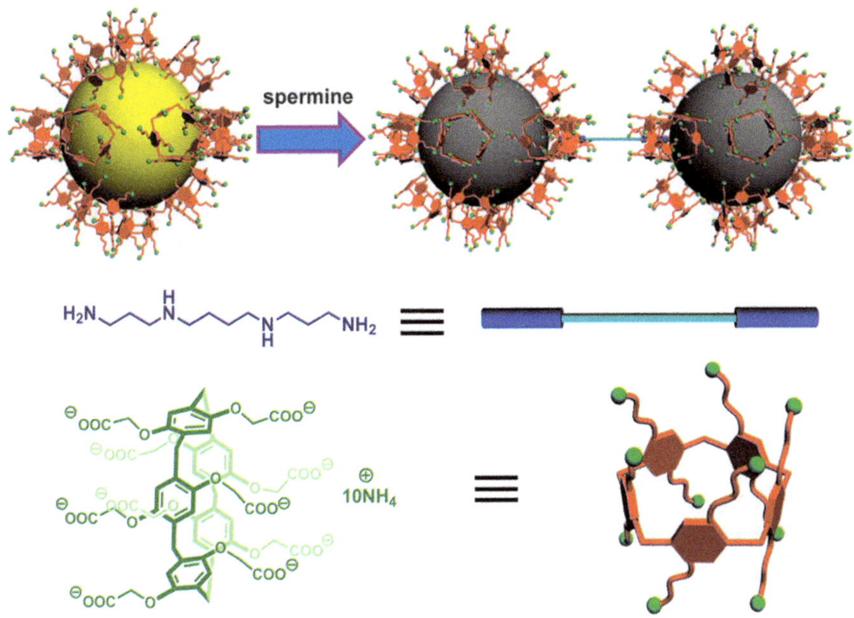

Scheme 10.5 Schematic presentation of the application of the **WP5**-stabilized silver nanoparticles in visual detection of spermine.
Adapted from ref. 34 with permission from The Royal Society of Chemistry.

As shown in the TEM images (Figure 10.9), the average diameter of **WP5**-modified silver nanoparticles is 18.7 ± 2.18 nm, and the spherical shape of the particles with a narrow distribution can also be observed. Moreover, as characterized by the energy-dispersive X-ray spectroscopy (EDX), the characteristic peaks of silver can be observed, which indicates the existence of silver nanoparticles. The **WP5**-stabilized silver nanoparticles exhibit good stability and are resistant to elevated temperature. Furthermore, the silver nanoparticles synthesized remain yellow in color and are monodispersed without aggregation after 2 weeks of storage.

Spermine and its analogs are toxic and are used as tumor biomarkers. They are also widely distributed in nature and industry. Therefore, the detection of spermine is necessary. WP5 can form stable complexation with spermine with a high association constant. When different diamine compounds (10^{-3} M), such as spermine, ursol, tetraethylenepentamine, triethylenetetramine, ethanediamine, 1,12-dodecylamine and 1,6-hexamethylenediamine, are added to an aqueous solution of **WP5**-stabilized silver nanoparticles, respectively, the yellow color of the solution containing spermine and its analogs (spermine, tetraethylenepentamine, and triethylenetetramine) becomes black, accompanied by a great decrease of the absorbance peak at 400 nm. However, other diamine compounds have little influence on the color and absorption peaks. All the experimental facts can be ascribed to the selective response of **WP5**-stabilized silver nanoparticles

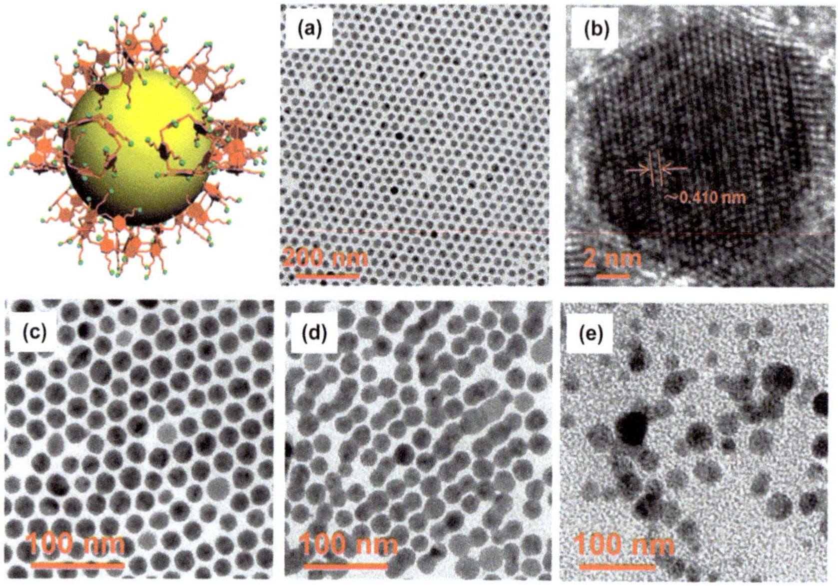

Figure 10.9 (a) TEM image and (b) high-resolution TEM image of silver nanoparticles when the concentration of WP5 is 4.00×10^{-4} M. TEM images of silver nanoparticles after adding (c) ethanediamine, (d) spermine, and (e) spermine for one night.
Adapted from ref. 34 with permission from The Royal Society of Chemistry.

to spermine and its analogs. The aggregation of silver nanoparticles occurs upon addition of spermine and its analogs, and can be observed by the red shift in the wavelength and broadening of the SPR band. The presence of aggregation after addition of spermine and its analogs is also confirmed by TEM experiments (Figure 10.9). One spermine molecule will bind with two **WP5** cavities attached to different silver nanoparticles through electrostatic and hydrogen bonds, and C–H···π interactions. Following this, the aggregation of **WP5**-stabilized silver nanoparticles will occur easily upon addition of spermine. Based on the aggregation properties and color changes of **WP5**-stabilized silver nanoparticles, they can be potentially used as a sensor for spermine and its analogs.

10.3.4 Pillararene–CdTe Quantum Dot Composites

Multi-carboxylated (or multi-thiolated and multi-carbonyl) compounds, as well as synthetic macrocyclic compounds, are always introduced to stabilize metal nanoparticles. For example, quantum dots (QDs) have been hybridized with cyclodextrin and calixarene derivatives to construct hybrid materials used as sensors for ions, organic molecules and chiral compounds.[35,36] Thus, CP[5]A with 10 carboxyl anion groups on both rims of the cavity has

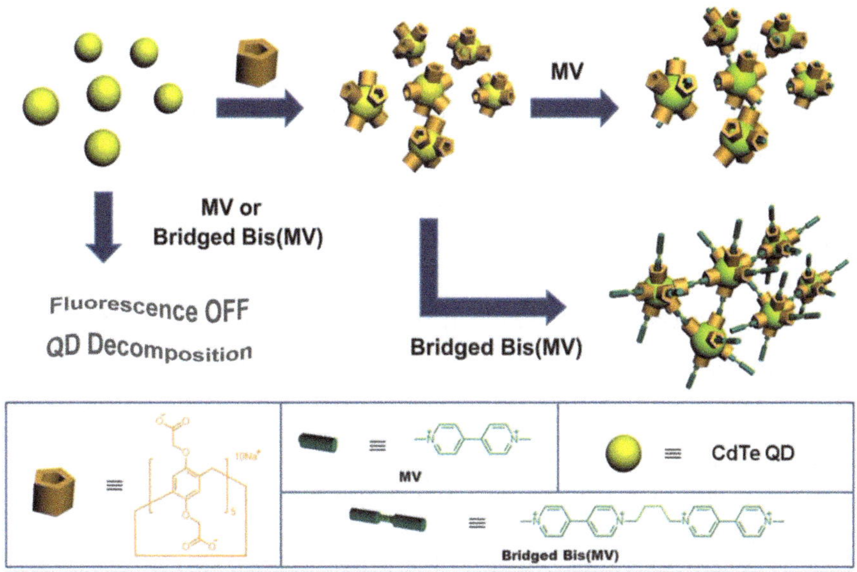

Scheme 10.6 Illustration of the formation of CP[5]A@QDs and its self-assembly mediated by viologens through host–guest interactions.
Reproduced from ref. 37 with permission from The Royal Society of Chemistry.

been functionalized onto the surface of CdTe QDs (Scheme 10.6) *via* a ligand-exchange method.[37] The composite material CP[5]A@QD has been obtained. The TEM images have shown the formation of stable CP[5]A-modified CdTe QDs in aqueous solution. The UV-vis spectroscopy and Fourier transform infrared (FT-IR) spectroscopy can provide further evidence for the successful modification of CP[5]A on the QDs surface.

QDs are electron donors, and the stability of CdTe QDs stabilized with 3-mercaptopropionic acid (MPA@QDs) will be dramatically decreased upon addition of an electron-deficient guest, *e.g.*, methyl viologen (MV) or bridged bis(MV), because the electron transfers from MPA@QDs to the electron-deficient MV or bis(MV). Intriguingly, CP[5]A@QDs exhibit admirable chemical stability, and the addition of MV or bridged bis(MV)s has negligible effects on the particles. That property can be ascribed to the ability of CP[5]A on the surface of QDs to capture the electron-deficient guest molecules and thus inhibit the electron transfer from QDs to electron-deficient molecules. When the bridged bis(MV) is in sufficient excess, and upon continuing addition, the stability of CP[5]A@QDs decreases because of the saturation of host–guest complexation between CP[5]As on the surface and the electron-deficient molecules.

The fluorescence intensity of CP[5]A@QDs was also dramatically influenced by the addition of MV and bridged bis(MV). There is an obvious decrease of fluorescence intensity when MV or bridged bis(MV)s is added to the solution of CP[5]A@QDs. Moreover, if an excess of electron-deficient guest compounds is added, the fluorescence of CP[5]A@QDs will be quenched because the loading capacity of CP[5]A on the surface reaches the maximum.

The aggregation of CP[5]A@QDs to a supramolecular network can be induced by addition of bridged bis(MV)s. Host–guest interactions of CP[5]As and bridged bis(MV)s are the driving force for fabrication of the 3D network structure of this hybrid material, and bridged bis(MV)s will act as a bridge to link CP[5]A-stabilized QD nanoparticles.

10.3.5 Pillararene–Fe_3O_4 Magnetic Nanoparticle Composites

According to a report by Yang, Jia and co-workers, carboxylatopillar[5]arene-coated magnetic Fe_3O_4 nanoparticles have been synthesized.[38] This kind of magnetic hybrid material can be used as a magnetic solid phase extraction (MSPE) absorbent for the detection of pesticides in beverage samples, combined with high performance liquid chromatography (HPLC).

The synthetic route for the magnetic hybrid material is shown in Scheme 10.7: amine-functionalized Fe_3O_4 nanoparticles are synthesized by the solvothermal method with some modifications. Then the CP[5]A macrocycles are modified onto the magnetic cores *via* condensation reactions between the amino groups and carboxylic groups. That is, the neutral form of CP[5]As is attached onto the Fe_3O_4 nanoparticles by covalent bonding.

As shown in the TEM and scanning electron microscopy (SEM) images (Figure 10.10), the magnetic CP[5]A–Fe_3O_4 nanoparticle hybrid materials are spherical in shape with an average diameter of 390 nm. On the other hand, the magnetic Fe_3O_4 cores and the organic shell around the cores can be easily distinguished. Weight loss, possibly representing water or organic

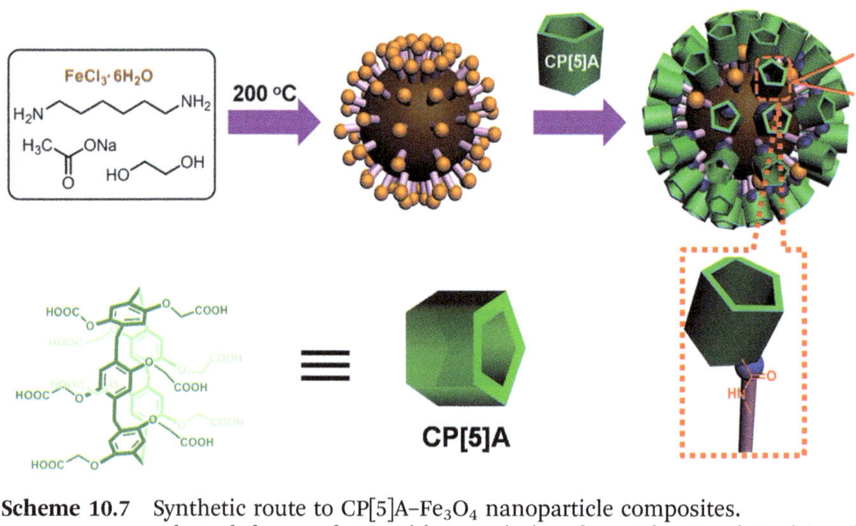

Scheme 10.7 Synthetic route to CP[5]A–Fe_3O_4 nanoparticle composites. Adapted from ref. 38 with permission from The Royal Society of Chemistry.

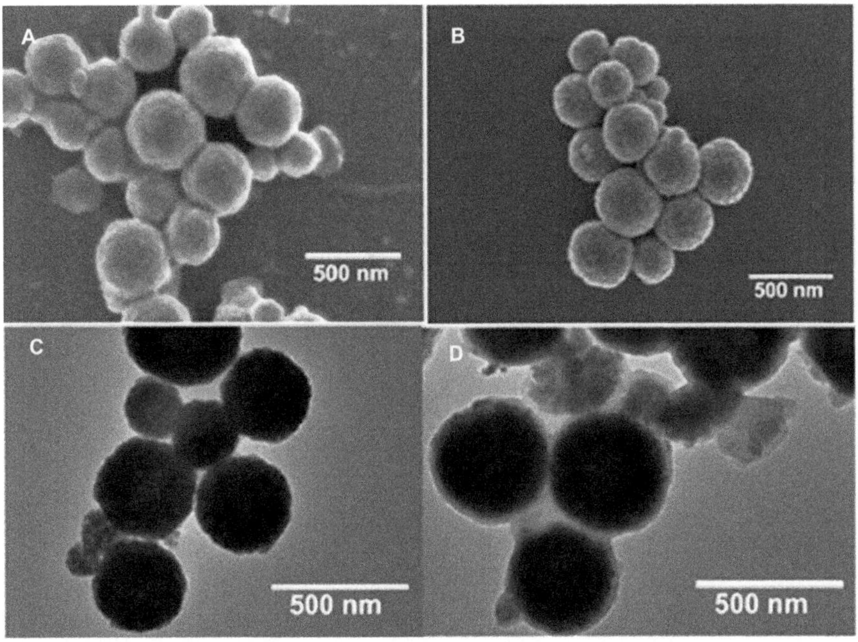

Figure 10.10 SEM images of Fe_3O_4–NH_2 (A) and Fe_3O_4/CP[5]A (B); TEM images of Fe_3O_4–NH_2 (C) and Fe_3O_4/CP[5]A (D) nanoparticles.
Reproduced from ref. 38 with permission from The Royal Society of Chemistry.

components, can also be observed from the TGA curve of the hybrid materials.

The superparamagnetic properties of CP[5]A–Fe_3O_4 nanoparticle composites have been characterized using a vibrating sample magnetometer (VSM), because there is negligible remanence and coercivity, as well as no hysteresis (Figure 10.11). The saturation magnetization of CP[5]A-functionalized Fe_3O_4 nanoparticles is 25.64 emu g^{-1}. The above experimental effects indicate that the hybrid nanomaterial has strong magnetism and is able to be used for magnetic separation. The CP[5]A-functionalized Fe_3O_4 nanoparticles can be isolated from the dispersion with the force of an external NdFeB magnetic field.

The CP[5]A-functionalized Fe_3O_4 nanoparticles possess unique pillar-shaped hollow cavities that can complex efficiently with various analytes through supramolecular binding interactions. The MSPE procedure has been optimized to achieve the best extraction performance.

Seven pesticides, *i.e.*, metalaxyl (MET), dimethomorph (DIM), kresoxim-methyl (KRE), flusilazole (FLU), cyprodinil (CYP), pyrimethanil (PYR), and triflumizole (TRI), have been chosen to perform analyses using Fe_3O_4–NH_2, (monomer of CP[5]A)-functionalized Fe_3O_4 nanoparticles, and CP[5]A-functionalized Fe_3O_4 nanoparticles under the optimized conditions (Figure 10.12). The CP[5]A-functionalized Fe_3O_4 nanoparticles exhibit higher responses to target pesticides than those with other absorbents.

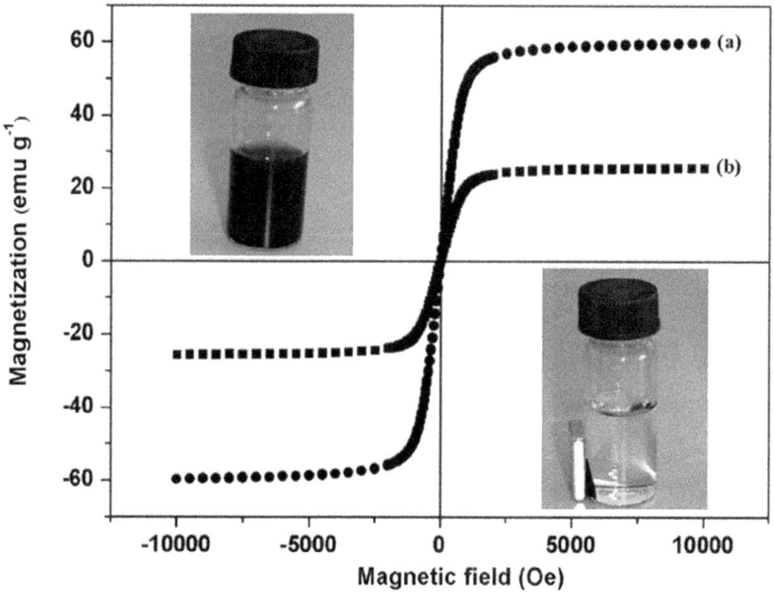

Figure 10.11 Magnetic hysteresis loops of (a) Fe_3O_4–NH_2, and (b) Fe_3O_4/CP[5]A. The insets showed photographs of Fe_3O_4/CP[5]A dispersions before (left inset) and after (right inset) exposure to an external magnetic field. Reproduced from ref. 38 with permission from The Royal Society of Chemistry.

The CP[5]A-functionalized Fe_3O_4 hybrid nanomaterial exhibits more efficient enriching ability for the extraction and enrichment of pesticides than the (monomer of CP[5]A)-functionalized hybrid owing to the specific pillar-shaped cavity and the rich π-electron system of CP[5]A.

Thus, the hybrid magnetic material based on CP[5]As can act as an efficient, rapid and convenient method for detection of trace amounts of MET, DIM, PYR, FLU, CYP, KRE, and TRI in beverage samples, when coupled with HPLC-UV. This newly developed hybrid material shows great potential for the detection of pesticides in order to control food safety.

10.4 Pillararene–Nanoscale Carbon Materials Composites

10.4.1 Introduction to Carbon Materials

Carbon is one of the most important elements in nature and the one most closely related to human beings. Carbon possesses various electronic orbital characteristics such as sp, sp^2 and sp^3 hybridized orbitals. In particular, the anisotropy of sp^2 leads to the anisotropy of crystalline and other diverse arrangements. Therefore, various carbon-based materials have been found and artificially created. In fact, carbon is the only single element that can be

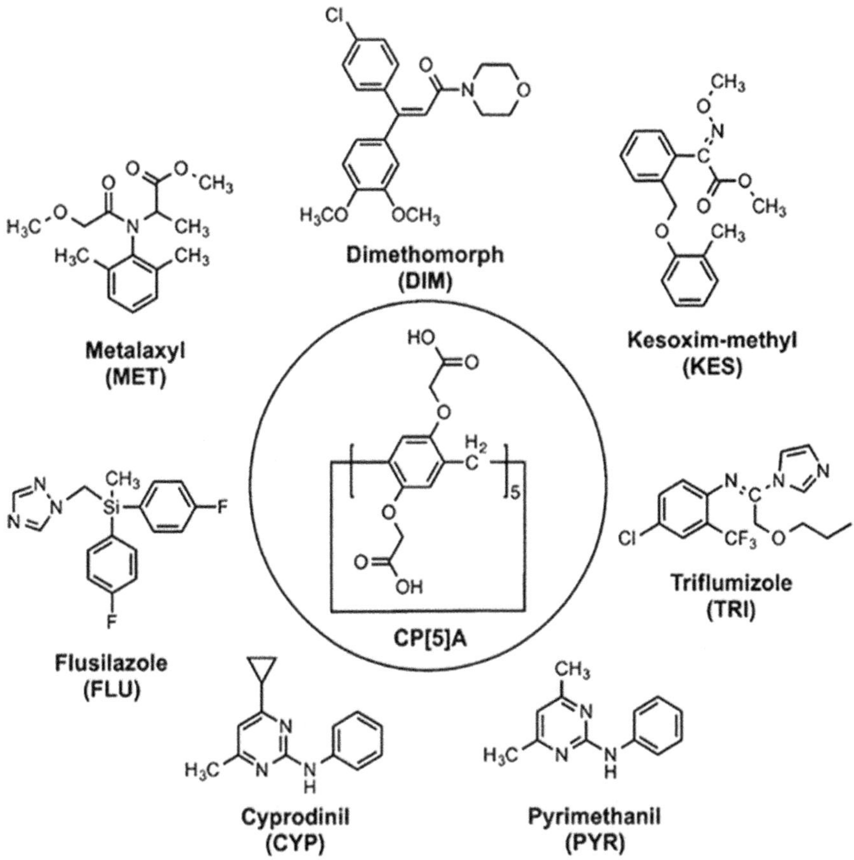

Figure 10.12 Chemical structures of CP[5]A and the target pesticides. Reproduced from ref. 38 with permission from The Royal Society of Chemistry.

used to fabricate abundant attractive materials with very different structures and properties such as zero-dimensional fullerene molecules, one-dimensional carbon nanotubes, and two-dimensional graphene sheets, as well as three-dimensional diamond crystals.[39]

Scientists have gradually found that carbon materials exhibit superior properties to any other materials in many aspects such as hardness, optical properties, temperature resistance, radiation resistance, resistance to chemical medicines, electrical insulation, electrical conductivity, and surface and interface properties. Carbon materials contain nearly all of the features of all materials on the earth, from the hardest to the softest, insulation to good thermal conductivity, and total absorption of light to being totally permeable to light.

Carbon nanotubes are seamless and hollow nanotubes that are curled from a single layer or multi-layer graphite flakes. Carbon nanotubes have many advantages such as high elasticity, low density, good insulation and good infrared absorption. With these special properties, carbon nanotubes have great potential in the preparation of self-healing materials, and the fabrication of bullet-proof warm stealth military equipment, and they are also used as the best hydrogen storage material.

Graphene is a carbon material composed of carbon atoms in a single chip and is the thinnest and hardest nanomaterial on earth. The basic structural units of graphene are the most stable possible, benzene rings, and they form the current ideal of a two-dimensional nanomaterial.[40] Graphene exhibits low resistivity and the most rapid electron transfer speed available. Therefore, it is expected to be utilized for the development of a new generation of thinner, more conductive electronic components or transistors.[39]

However, the largest challenge to the widespread utilization of carbon materials is their poor solubility.[40] Researchers have devoted a great deal of effort to improving their solubility and make them disperse well in typical solvents. Based on the special supramolecular and structural properties of pillar[*n*]arenes, water-soluble pillararenes are considered to disperse carbon nanotubes and graphene in aqueous solutions. The pillararene-based carbon hybrid materials will be introduced in this section.

10.4.2 Dispersion of Multiwalled Carbon Nanotubes (MWNTs) by Pillararenes

The first water-soluble pillar[6]arene (**WP6**) (Schemes 10.8 and 10.9) with controlled solubility was prepared by Huang and co-workers.[41] The solubility of **WP6** can be reversibly adjusted by changing the pH of the solution. When the solution is acidic, all of the carboxylated groups will be insoluble

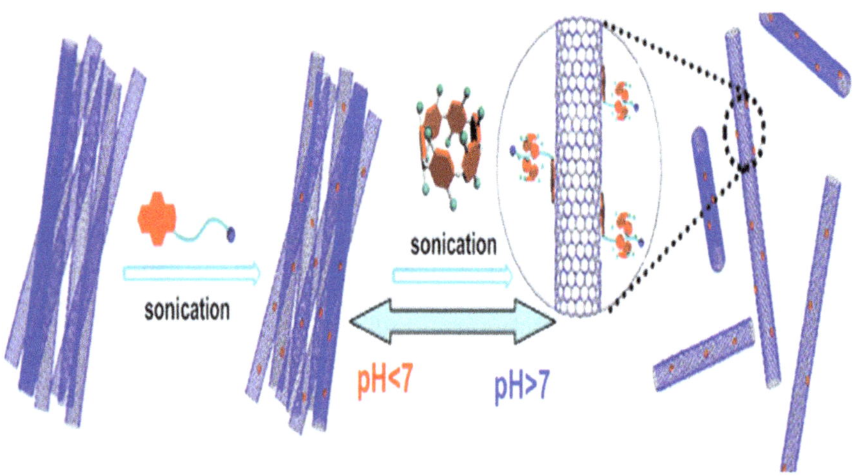

Scheme 10.8 Illustration of the pH-responsive solubility of the MWNTs in the presence of **WP6**$\supset$**G2**.
Reprinted with permission from ref. 41. Copyright 2012 American Chemical Society.

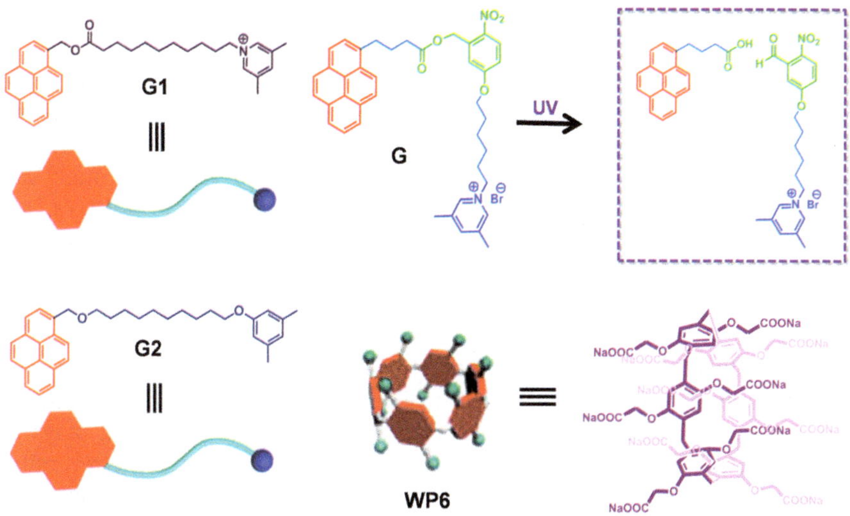

Scheme 10.9 Structures and schematic diagrams of **G1**, **G2**, **G** and **WP6**. Adapted with permission from ref. 41. Copyright 2012 American Chemical Society.

carboxylic acid groups. However, when the pH is increased, insoluble -COOH changes into $-COO^-$, which allows the pillar[6]arene once again to dissolve in water.

The **WP6** can bind strongly with organic pyridinium salt **G1** (Scheme 10.9) in water through hydrophobic and electrostatic interactions. The complexation of **G1**$\subset$**WP6** was investigated using **WP6** and a model compound of **G1**, *i.e.*, 1-octylpyridinium bromide (OPB), and the association constant (K_a) in water was determined to be $(3.26 \pm 0.28) \times 10^5\ M^{-1}$ by fluorescence titration experiments. The fluorescence spectra also indicated that the complexation of **G1**$\subset$**WP6** would be destroyed when the solution is made to be acidic.

The **G1** itself can assemble into regular hollow nanotubes with an average diameter around 300 nm, which can be observed by TEM. Intriguingly, vesicles with an average diameter of 300 nm can be observed upon addition of 1 equiv. of **WP6**. Furthermore, the regular nanotubes reappear when adjusting the solution pH to 6.0. The same experimental results can also be demonstrated by fluorescence experiments. Therefore, the transition between nanotubes and vesicles, driven by complexation and decomplexation of host and guest, will be well controlled by pH variations. The main diameter distribution of **WP6**$\supset$**G1** vesicles is 312 nm, consistent with the TEM experiments. The mechanism of the controlled self-assembly topologies can be described as follows: **G1** alone can assemble into a highly ordered bilayer with high curvature through π–π stacking interactions between pyrenyl groups, forming a tubular structure. Upon complexation, **WP6** is inserted into the membrane of the nanotubes; the straight nanotubular arrays become curved, generating a low curvature, and then form vesicular structures.

A neutral guest **G2** (Scheme 10.9), containing a π-rich pyrenyl ring, has been introduced to form a host–guest complex with **WP6** through hydrophobic interactions: the association constant is $(8.04 \pm 0.68) \times 10^4$. **G2** can attach to the surface of multiwalled nanotubes (MWNTs) through π–π stacking interactions without destroying the structures of the MWNTs. The MWNTs disperse very well in aqueous solution of **WP6⊃G2** and last for a very long time without any changes (Scheme 10.8). However, when the pH of the solution is made acidic, MWNTs precipitate from the solution, and they disperse well again when the pH is raised above 7.0. Furthermore, **G2** alone cannot induce good dispersion of MWNTs. These observations indicate that the dispersion can be reversibly controlled by simply changing the solution pH. On the other hand, these phenomena suggest that **WP6⊃G2** plays an important role in adjusting the solubility of MWNTs owing to the reversible solubility of **WP6** fragments depending on pH. Direct evidence is obtained from SEM and TEM images and photographs of the corresponding mixtures (Figure 10.13).

The results are also confirmed by fluorescence spectroscopy. The solution of **WP6⊃G2** exhibits strong fluorescence emission intensity because of its pyrenyl groups, while the fluorescence quenches after the dispersion of MWNTs. The reason is the π–π stacking between pyrene and the surface of MWNTs.

Following the previous research, another supra-amphiphilic system based on **WP6** has been introduced for dispersion of MWNTs in aqueous solution.[42] The difference is that the dispersion can be controlled by UV irradiation because **G**, bearing the 2-nitrobenzyl ester moiety, is a UV-responsive guest (see Scheme 10.9). TEM images have shown the interaction between **WP6⊃G** and MWNTs. The hydrophobic MWNTs disperse well in water upon addition of the complexation of **WP6⊃G**, because the pyrenyl rings cling to the surface of the MWNTs and the hydrophilic **WP6** dissolves in water. However, when the well-dispersed mixture is irradiated by UV light for about 30 minutes, the MWNTs aggregate and precipitate again as a result of photocleavage of **G** and separation of the hydrophilic segment from the hybrid materials.

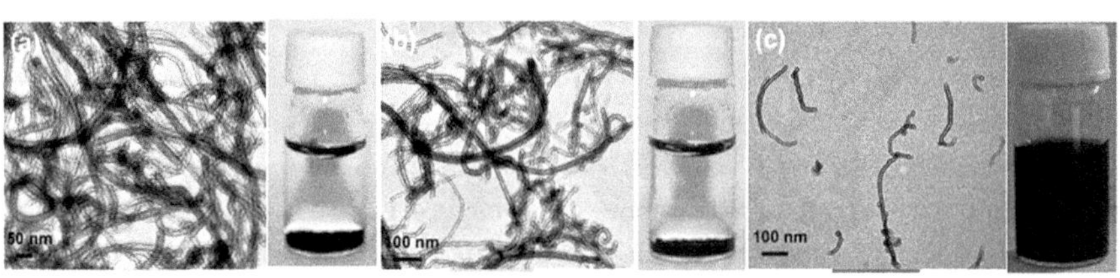

Figure 10.13 TEM images and photographs of (a) MWNTs, (b) **G2** and MWNTs, and (c) **WP6⊃G2**/MWNT complexes.
Adapted with permission from ref. 41. Copyright 2012 American Chemical Society.

10.4.3 Pillararene–Graphene Oxide Composites

Graphene oxide (GO) is another widely studied carbon material, owing to its unique structure and properties. Pillar[5]arene derivatives, *e.g.*, an amphiphilic pillar[5]arene (**AP5**), have been hybridized onto the surface of reduced graphene oxide through covalent amido bonds formed by reaction of amino and carboxyl groups and π–π noncovalent interactions between benzene rings and the flat surface of GO (Scheme 10.10).[43] The **AP5**-modified reduced graphene oxide (RGO) hybrid material has been characterized by FT-IR spectroscopy, UV-vis spectroscopy and TGA. Because of the five amino groups on one rim of **AP5**, the **AP5**–RGO hybrid material is well dispersed in water, very different from RGO alone. Furthermore, gold nanoparticles adhere to the surface of **AP5**–RGO, and therefore RGO–**AP5**–gold nanoparticles, a composite material with three components, has been fabricated and then characterized by high resolution TEM and PXRD experiments.

Considering that the ternary hybrid material RGO–**AP5**–gold nanoparticles combines the superior properties of all three components, such as large surface area, excellent conductivity, host–guest recognition for typical small molecules and the catalytic properties of gold nanoparticles, six electroactive guests, *i.e.*, dopamine (DA), 4-acetamidophenol (APAP), uric acid (UA), methylene blue (MB), tryptophan (Trp), and imidacloprid (IDP), have been chosen to investigate the host–guest recognition abilities of RGO–**AP5**–gold nanoparticles and **AP5**–RGO through their electrochemical behavior, as measured by cyclic voltammetry (CV). Obviously different effects and electrochemical responses are found for the six guests before and after

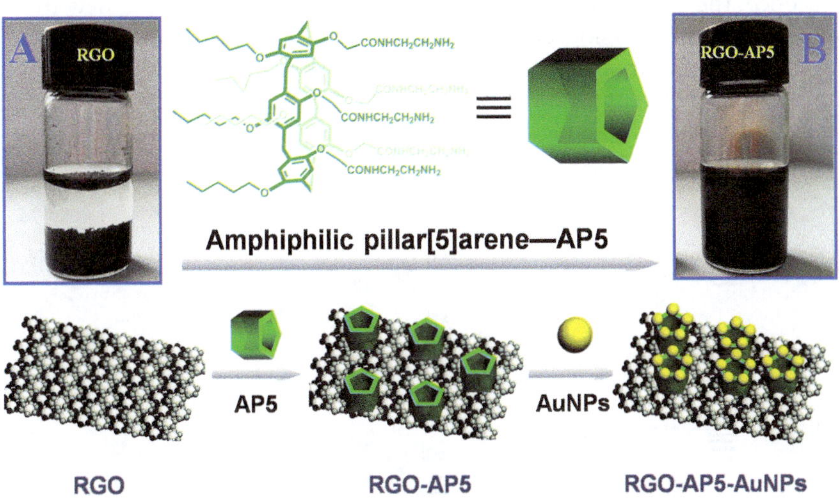

Scheme 10.10 Schematic diagram of RGO, RGO–AP5 and RGO–AP5–AuNPs.[43] (A) Photograph of RGO in aqueous media. (B) Photograph of RGO–AP5 composites in aqueous media.
Adapted with permission from ref. 43. Copyright 2013 American Chemical Society.

modification with **AP5**s. Because DA, APAP, UA, and MB molecules are smaller than the cavity of **AP5** and can be captured by the **AP5**–RGO, these guests will be easily accumulated on the surface of **AP5**–RGO, and therefore the electrochemical response is increased. However, for the guests Trp and IDP, with larger diameter than the cavity of **AP5**, there will be negligible change of the peak current before and after the modification of **AP5**. Interestingly, as for the hybrid materials, RGO–**AP5**–gold nanoparticles, there will be a further increase of the peak currents towards guest molecules that match the cavity size of **AP5** (DA, APAP, UA, and MB). This phenomenon is the result of the catalytic properties of gold nanoparticles that allow them to enhance the electrochemical response. Finally, a novel ternary hybrid material with excellent electrochemical analyzing performance for DA has been constructed. The linear range is broad and the detection limit is very low.

Subsequently, Zhao and co-workers provided a successfully prepared biocompatible hybrid material (Scheme 10.11), pillar[5]arene-functionalized graphene,[44] which was used for dual-mode Raman and fluorescence bioimaging *in vitro* (Figure 10.14). The bola-amphiphilic pillar[5]arene **1** and the tadpole-like amphiphilic pillar[5]arene **2** are attached onto the surface of GO through robust hydrogen bonds, and both the hybrid materials are stable in aqueous solution even after a long time. The good biocompatibility and low cytotoxicity of **1-GO** and **2-GO** can be estimated by MTT essay. **1-GO** exhibits better Raman imaging than **2-GO** when they are endocytosed by HeLa cells for *in vitro* cell imaging.

Moreover, after the addition of a bipyridinium guest (**G3**) with fluorescence, the suspensions of **1-GO** and **2-GO** remain stable. The new hybrid materials with inclusion complexes can also be endocytosed by HeLa cells and fluorescence imaging performed *in vitro*. The **1-GO** complex with **G3** was found to be better than **2-GO** because of the hydrophilic ends of the bola-amphiphilic pillar[5]arenes and the longer stability.

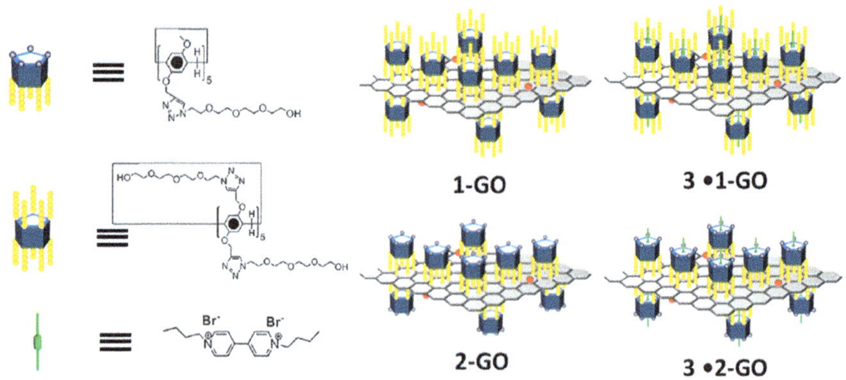

Scheme 10.11 Structural representations of **1-GO**, **2-GO**, **3 · 1-GO** and **3 · 2-GO**. Used with permission from ref. 44. Copyright 2014 Wiley.

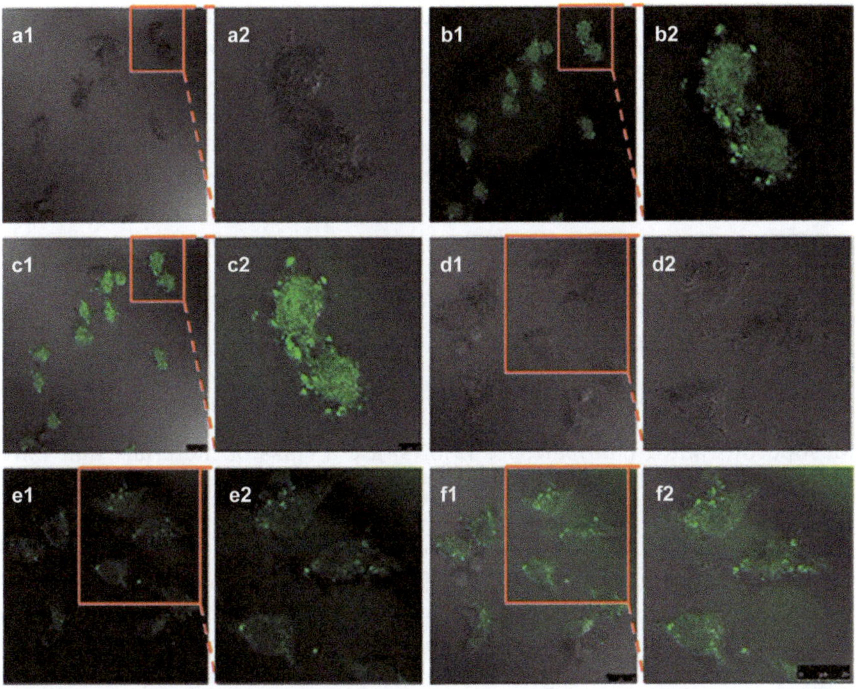

Figure 10.14 Confocal laser scanning microscopy (CLSM) images of HeLa cells after treatment with the **1-GO** · **3** complex (2.5 μg mL^{-1}) for 24 h: (a1) bright field, (b1) green field (λ_{ex} = 488 nm), (c1) merged image of a1 and b1. Scale bar = 25 μm. Enlarged area: (a2) bright field, (b2) green field (λ_{ex} = 488 nm) and (c2) merged image of a2 and b2. Scale bar = 75 μm. CLSM images of HeLa cells after treatment with the **2-GO** · **3** complex (2.5 μg mL^{-1}) for 24 h: (d1) bright field, (e1) green field (λ_{ex} = 488 nm), (f1) merged image of d1 and e1. Enlarged area: (d2) bright field, (e2) green field (λ_{ex} = 488 nm) and (f2) merged image of d2 and e2. Scale bar = 25 μm.
Used with permission from ref. 44. Copyright 2014 Wiley.

Our research group has also successfully modified the surface of RGO with carboxylated pillar[5]arenes through covalent bonds. The pillar[5]arene-functionalized RGO nanosheets, possessing molecular recognition capacity, exhibited much better water-dispersibility.[45] Meanwhile, the pillar[5]arene-modified RGO composites enhanced the fluorescence-quenching property, when compared with bare RGO, which endowed them with the capability to sense and detect analytes.

10.4.4 Pillararene–Fullerene Composites

Fullerene is a condensed ring aromatic compound, possessing extended pi-conjugated systems, which has been applied in various research fields, especially in biology. Ogoshi and co-workers have demonstrated that pillar[10]arene can form a stable complex with C_{60} through host–guest interactions.[46] Han and co-workers have also introduced neutral sugar moieties at the rims of pillar[5]arene through click reactions and obtained a neutral water-soluble pillar[5]arene.[47] The host–guest interactions between this novel pillar[5]arene derivative and the fullerene have also been investigated by the methods of fluorescence spectroscopy, Raman spectroscopy and so on.

10.5 Pillararene–Mesoporous Silica Nanoparticle Composites

10.5.1 Introduction to Mesoporous Silica Nanoparticles

Mesoporous silica nanoparticles (MSNs) are a class of materials that have ordered porosities and can be tunably synthesized. Considering their large pore volumes, high surface areas, controlled pore diameters, good biocompatibility and ability to be easily functionalized, MSNs have been regarded as the most suitable containers available for cargo loading and as platforms for gas absorption.[48] Moreover, much attention has been focused on MSN-based materials coated with nanovalves for controlled capping and release of target molecules depending on external stimuli such as light, pH, enzymes, redox, ultrasound, heating and competitive binding.[49–51] Coupling synthetic macrocycles and supramolecular principles, MSNs with nanovalves have been the most attractive systems for controlled drug delivery and disease therapies.

10.5.2 Drug Delivery Systems Based on Pillararenes and Mesoporous Silica Nanoparticles

Pillar[*n*]arenes, as newly synthesized macrocyclic molecules, have been diversely functionalized and widely reported. Since the water-soluble pillar[5]arenes were successfully prepared, Yang and co-workers have incorporated the pillararene derivative CP[5]A into the nanovalves on the surface of MCM-41 nanoparticles (Scheme 10.12).[52] Positively charged 1-butylpyridinium cations are grafted onto the surfaces of the synthesized MSNs and can also bind with the negatively charged CP[5]As through electrostatic interactions. The association binding constant is determined to be $(2.5 \pm 0.7) \times 10^3$ M^{-1}. Thus, changing the pH and adding competitive binding agents can be used to operate these new mechanized MSNs.

In order to investigate the cargo loading and controlled release of CP[5]A-hybridized MSNs, rhodamine B (RhB) has been encased in MCM-41 nanoparticles, and CP[5]As have been used to encircle the terminals of the stalks. According to the UV-vis absorption spectra, RhB will be trapped and loaded in the pores of the MSNs and will not leak out, owing to the steric hindrance of CP[5]As when the system is under neutral or basic conditions. The flat baseline shows that there is negligible premature release. However, upon

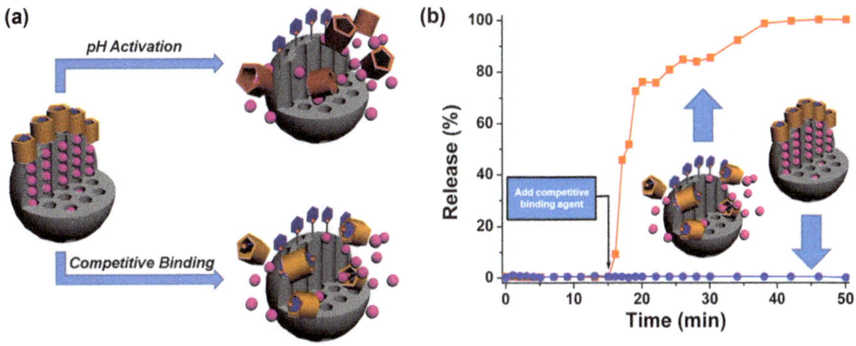

Scheme 10.12 (a) Schematic diagram of carboxylatopillar[5]arene (CP[5]A)-based mechanized MSNs. (b) Release of calcein from CP[5]A-capped mechanized MSNs caused by the addition of methylviologen (the detection wavelength is 492 nm).
Used with permission from ref. 52. Copyright 2013 Wiley.

reducing the pH, CP[5]As will de-thread from the stalks and leave the entrances of the pores; the nanovalves are opened and RhB molecules are released into the external environment. On comparing the release curve under different conditions of pH, we can conclude that the release of cargo molecules is faster when the pH is lower. Based on the above simulative experiments, the release of doxorubicin (DOX) hydrochloride, an anticancer drug, has also been investigated, and a smooth release profile can be observed upon lowering the pH.

The addition of competitive binding agents is another method used to achieve the target of cargo release. RhB or calcein molecules are stored in the nanopores of MSNs. When the competitive molecules, *i.e.*, a methyl viologen salt possessing higher binding affinity for water-soluble CP[5]As ($K_a = 8\times10^4\ M^{-1}$), are added into the drug loading system, the release of RhB or calcein molecules can be observed. The mechanism can be ascribed to the higher binding affinity of water-soluble CP[5]As and methyl viologen salt, which results in the de-threading of CP[5]As from the modified pyridinium groups on the surface of the MSNPs. Subsequently, the cargo molecules can be released without the steric hindrance of CP[5]As. As a consequence, pillar[5]arenes and their derivatives exhibit great potential to be hybridized with inorganic porous materials to construct novel drug delivery systems.

Subsequently, Huang and Du have reported a carboxylate-substituted pillar[6]arene-valved MSN, which was used for stimuli-responsive drug delivery.[53] The release of cargo was triggered by acidic pH or competitive binding agents.

In addition, our group have fabricated another type of pillar[5,6]arene-based material, hybridized with silicon dioxide. The perhydroxyl-pillar[5]-arene and perhydroxyl-pillar[6]arene were attached onto the hydrophilic silica supports through covalent bonds.[54] This pillar[*n*]arene-based composite can act as an absorbent hybrid material for some typical herbicides.

10.6 Conclusions and Outlook

Pillararene-based hybrid materials and their applications have been summarized in this chapter. MOFs with active domains have been successfully prepared, owing to the cyclic structure of pillar[5]arenes, and used for uptake of electron-poor aromatic guests or neutral guests. Given that pillar[5]arenes can be easily substituted by particular groups such as amine ($-NH_2$), carboxyl (–COOH), or sulfhydryl (–SH) groups, they can be used to stabilize metallic nanoparticles such as gold nanoparticles, silver nanoparticles and QDs. The pillararene-stabilized metallic nanoparticles exhibit excellent performance in catalysis, sensing as well as self-assembly because of the rigid structure of pillararenes and their cavities. Moreover, water-soluble pillararenes or amphiphilic pillararenes have been applied to disperse carbon materials such as nanotubes and graphene in aqueous solution. Similar to other synthetic macrocycles, pillararene derivatives have been used to construct nanovalves. By means of hybridization of pillararenes and functionalized MSNs, controlled drug delivery can be achieved by adjusting the pH or adding competitive binding agents.

There are also some other special pillararene-based materials. Huang and co-workers have prepared a hybrid porous material fabricated from carboxylic acid-substituted P[5]As and poly(ionic liquid)s.[55] The hybrid porous material can absorb a series of aliphatic diols from $CDCl_3$ because P[5]As provide it with binding affinity for target molecules through host–guest interactions. Yang, Jia, *et al.* have also successfully packed a neutral form of CP[5]A in a glass microcolumn,[56] and employed it as an absorbent for the separation and preconcentration of trace levels of gold and palladium, coupling with flow injection-flame atomic absorption spectrometry (FAAS).

In conclusion, pillar[*n*]arenes and their derivatives have exhibited great potential to hybridize with many inorganic or metallic materials. One can make full use of their advantages in structural and supramolecular properties to construct hybrid materials with superior functions in recognition, absorption, sensing, detection, and so on.

Acknowledgements

We thank the National Natural Science Foundation of China (51473061, 21272093), the JLU Cultivation Fund for the National Science Fund for Distinguished Young Scholars, and The Fundamental Research Funds for the Central Universities (JCKY-QKJC05), for financial support.

References

1. A. P. Wight and M. E. Davis, *Chem. Rev.*, 2002, **102**, 3589.
2. Y.-W. Yang, Y.-L. Sun and N. Song, *Acc. Chem. Res.*, 2014, **47**, 1950.
3. T. Ogoshi, S. Kanai, S. Fujinami, T. A. Yamagishi and Y. Nakamoto, *J. Am. Chem. Soc.*, 2008, **130**, 5022.

4. T. Ogoshi and T. Yamagishi, *Eur. J. Org. Chem.*, 2013, **2013**, 2961.
5. S. Qiu, M. Xue and G. Zhu, *Chem. Soc. Rev.*, 2014, **43**, 6116.
6. H.-C. Zhou and S. Kitagawa, *Chem. Soc. Rev.*, 2014, **43**, 5415.
7. N. L. Strutt, D. Fairen-Jimenez, J. Iehl, A. M. B. Lalonde, R. Q. Snurr, O. K. Farha, J. T. Hupp and J. F. Stoddart, *J. Am. Chem. Soc.*, 2012, **134**, 17436.
8. N. L. Strutt, H. Zhang and J. F. Stoddart, *Chem. Commun.*, 2014, **50**, 7455.
9. L.-L. Tan, H. Li, Y. Tao, S. X.-A. Zhang, B. Wang and Y.-W. Yang, *Adv. Mater.*, 2014, **26**, 7027.
10. T. Ogoshi, R. Sueto, K. Yoshikoshi and T. Yamagishi, *Chem. Commun.*, 2014, **50**, 15209.
11. L.-L. Tan, H. Li, Y. Zhou, Y. Zhang, X. Feng, B. Wang and Y.-W. Yang, *Small*, 2015, **11**, 3807.
12. L.-L. Tan, H. Li, Y.-C. Qiu, D.-X. Chen, X. Wang, R.-Y. Pan, Y. Wang, S. X.-A. Zhang, B. Wang and Y.-W. Yang, *Chem. Sci.*, 2015, **6**, 1640.
13. F. Vines, J. R. B. Gomes and F. Illas, *Chem. Soc. Rev.*, 2014, **43**, 4922.
14. M. S. Chen and D. W. Goodman, *Chem. Soc. Rev.*, 2008, **37**, 1860.
15. E. Katz and I. Willner, *J. Am. Chem. Soc.*, 2002, **124**, 10290.
16. B. Nikoobakht and M. A. El-Sayed, *Chem. Mater.*, 2003, **15**, 1957.
17. C. H. Munro, W. E. Smith, M. Garner, J. Clarkson and P. C. White, *Langmuir*, 1995, **11**, 3712.
18. X. H. Ji, X. N. Song, J. Li, Y. B. Bai, W. S. Yang and X. G. Peng, *J. Am. Chem. Soc.*, 2007, **129**, 13939.
19. M. P. Rowe, K. E. Plass, K. Kim, Ç. Kurdak, E. T. Zellers and A. J. Matzger, *Chem. Mater.*, 2004, **16**, 3513.
20. Y. B. Zheng, J. L. Payton, T.-B. Song, B. K. Pathem, Y. Zhao, H. Ma, Y. Yang, L. Jensen, A. K.-Y. Jen and P. S. Weiss, *Nano Lett.*, 2012, **12**, 5362.
21. J. J. Storhoff, A. A. Lazarides, R. C. Mucic, C. A. Mirkin, R. L. Letsinger and G. C. Schatz, *J. Am. Chem. Soc.*, 2000, **122**, 4640.
22. U. Resch-Genger, M. Grabolle, S. Cavaliere-Jaricot, R. Nitschke and T. Nann, *Nat. Methods*, 2008, **5**, 763.
23. A. M. Smith and S. Nie, *Acc. Chem. Res.*, 2010, **43**, 190.
24. D. J. Norris, C. B. Murray and M. G. Bawendi, *J. Am. Chem. Soc.*, 1993, **115**, 8706.
25. N. Rastkari and R. Ahmadkhaniha, *J. Chromatogr. A*, 2013, **1286**, 22.
26. Y. Wang, S. Wang, H. Niu, Y. Ma, T. Zeng, Y. Cai and Z. Meng, *J. Chromatogr. A*, 2013, **1283**, 20.
27. Y. B. Luo, Q. W. Yu, B. F. Yuan and Y. Q. Feng, *Talanta*, 2012, **90**, 123.
28. Y. Yao, M. Xue, X. Chi, Y. Ma, J. He, Z. Abliz and F. Huang, *Chem. Commun.*, 2012, **48**, 6505.
29. H. Li, D.-X. Chen, Y.-L. Sun, Y. B. Zheng, L.-L. Tan, P. S. Weiss and Y.-W. Yang, *J. Am. Chem. Soc.*, 2013, **135**, 1570.
30. Y. Yao, Y. Wang and F. Huang, *Chem. Sci.*, 2014, **5**, 4312.
31. C. Park, E. S. Jeong, K. J. Lee, H. R. Moon and K. T. Kim, *Chem. – Asian J.*, 2014, **9**, 2761.

32. V. Montes-García, C. Fernández-López, B. Gómez, I. Pérez-Juste, L. García-Río, L. M. Liz-Marzán, J. Pérez-Juste and I. Pastoriza-Santos, *Chem. – Eur. J.*, 2014, **20**, 8404.
33. Q. Zhou, B. Zhang, D. Han, R. Chen, F. Qiu, J. Wu and H. Jiang, *Chem. Commun.*, 2015, **51**, 3124.
34. Y. Yao, Y. Zhou, J. Dai, S. Yue and M. Xue, *Chem. Commun.*, 2014, **50**, 869.
35. T. Li, Y. Zhou, J. Sun, D. Tang, S. Guo and X. Ding, *Microchim. Acta*, 2011, **175**, 113.
36. Z. B. Shang, S. Hu, Y. Wang and W. J. Jin, *Luminescence*, 2011, **26**, 585.
37. D.-X. Chen, Y.-L. Sun, Y. Zhang, J.-Y. Cui, F.-Z. Shen and Y.-W. Yang, *RSC Adv.*, 2013, **3**, 5765.
38. M.-M. Tian, D.-X. Chen, Y.-L. Sun, Y.-W. Yang and Q. Jia, *RSC Adv.*, 2013, **3**, 22111.
39. J. Coraux, L. Marty, N. Bendiab and V. Bouchiat, *Acc. Chem. Res.*, 2013, **46**, 2193.
40. X. Wang, G. Sun, P. Routh, D.-H. Kim, W. Huang and P. Chen, *Chem. Soc. Rev.*, 2014, **43**, 7067.
41. G. Yu, M. Xue, Z. Zhang, J. Li, C. Han and F. Huang, *J. Am. Chem. Soc.*, 2012, **134**, 13248.
42. J. Yang, G. Yu, D. Xia and F. Huang, *Chem. Commun.*, 2014, **50**, 3993.
43. J. Zhou, M. Chen, J. Xie and G. Diao, *ACS Appl. Mater. Interfaces*, 2013, **5**, 11218.
44. H. Zhang, X. Ma, K. T. Nguyen, Y. Zeng, S. Tai and Y. Zhao, *ChemPlusChem*, 2014, **79**, 462.
45. T. Zhou, H. Yu, M. Liu and Y.-W. Yang, *Chin. J. Chem.*, 2015, **33**, 125.
46. T. Ogoshi, N. Ueshima, F. Sakakibara, T. Yamagishi and T. Haino, *Org. Lett.*, 2014, **16**, 2896.
47. H. Li, Q. Chen, C. Schönbeck and B.-H. Han, *RSC Adv.*, 2015, **5**, 19041.
48. L. T. Gibson, *Chem. Soc. Rev.*, 2014, **43**, 5173.
49. Y.-W. Yang, *MedChemComm*, 2011, **2**, 1033.
50. S. Saha, K. C. F. Leung, T. D. Nguyen, J. F. Stoddart and J. I. Zink, *Adv. Funct. Mater.*, 2007, **17**, 685.
51. K. K. Cotí, M. E. Belowich, M. Liong, M. W. Ambrogio, Y. A. Lau, H. A. Khatib, J. I. Zink, N. M. Khashab and J. F. Stoddart, *Nanoscale*, 2009, **1**, 16.
52. Y.-L. Sun, Y.-W. Yang, D.-X. Chen, G. Wang, Y. Zhou, C.-Y. Wang and J. F. Stoddart, *Small*, 2013, **9**, 3224.
53. X. Huang and X. Du, *ACS Appl. Mater. Interfaces*, 2014, **6**, 20430.
54. T. Zhou, N. Song, H. Yu and Y.-W. Yang, *Langmuir*, 2015, **31**, 1454.
55. Z. Zhang, Q. Zhao, J. Yuan, M. Antonietti and F. Huang, *Chem. Commun.*, 2014, **50**, 2595.
56. S.-Y. Zhou, N. Song, S.-X. Liu, D.-X. Chen, Q. Jia and Y.-W. Yang, *Microchim. Acta*, 2014, **181**, 1551.

CHAPTER 11

Biomedical Applications of Pillararenes

JUN-LI HOU

Department of Chemistry, Fudan University, 220 Handan Road, Shanghai 200433, China
Email: houjl@fudan.edu.cn

11.1 Introduction

Macrocyclic molecules, which are at the heart of supramolecular chemistry, are also the building blocks of a great number of materials.[1] The ability to apply macrocycles to the life sciences, such as designing new molecules to interact selectively with or mimic functions of biomacromolecules, has gained wide interest owing to the vast number of biomedical applications. In the past decades, chemists have employed crown ethers,[2] calixarenes,[3] cyclodextrins,[4] and cucurbiturils[5] as molecular scaffolds to build biosensors and biological models, and for drug delivery, solubilization and stabilization. Given the importance of the macrocycles, it is useful to see the rapid progress in the development of new macrocycles.

Pillararenes (PA), a new kind of paracyclophane, were first synthesized in 2008 as pillar[5]arene (PA[5])[6] and were recognized as a new generation of supramolecular host because of their unique pillar structural feature, nanometer-sized (<1.0 nm) cavity, and multiple functionalizable sites.[7,8] In the last few years, these new types of compound have shown biomedical applications in the construction of artificial channels for transmembrane transport of physiologically active solutes and vesicles for drug delivery, which will be reviewed in this chapter.

Monographs in Supramolecular Chemistry No. 18
Pillararenes
Edited by Tomoki Ogoshi
© The Royal Society of Chemistry 2016
Published by the Royal Society of Chemistry, www.rsc.org

11.2 Transmembrane Transport of Physiologically Active Solutes

The cell membrane forms a barrier to chemical and signal communication between cell and its environment. Therefore, the transport of ions and organic solutes either in or out of a cell is a precisely regulated biological process and relies on specific proteins named channel proteins.[9] The channels that conduct ions and neutral polar molecules across lipid membranes have been of particular interest, because it is challenging to move a charged or polar species through the approximate 35 Å thickness of the hydrophobic layer. Although transmembrane transport of physiologically active solutes has been studied for decades, the structural details of this process are only beginning to emerge. Their importance is obvious from the award of the 2003 Nobel Prize in chemistry for elucidating the detailed structure of the water[10] and K^+ channels.[11] During the past three decades, preparation of synthetic ion channels has been undertaken in an attempt to model protein channels and to develop therapeutic agents.

Numerous examples of artificial channels have been reported since the pioneering work of Tabushi using tetrachained cyclodextrin as a model.[12] Of these examples, the macrocycles, such as crown ethers, calixarenes, cyclodextrins, and cucurbiturils, play an important role.[13,14] The macrocycles provide not only the platforms for construction of whole channels but also the functional sites necessary to achieve transport selectivity and efficiency. From this viewpoint, pillararenes, new macrocycles with unique structural features, may also act as a platform for building such channels, which will give the channels new functions. This hypothesis has been systematically explored by us.

The first generation of pillararene-based artificial channels is the pillar[5]arene decaester **11.1**, which was prepared from the reaction of pillar[5]arene with ethylbromoacetate (Figure 11.1a).[15] In an accidental experiment, **11.1** was found to stack as a solid to form infinite tubular assemblies which are capable of inducing water molecules to form hydrogen-bonded linear water wires inside of tubular cavities (Figure 11.1b).[16] This structure is very interesting, because it is also observed in channel proteins, such as gramicidin A and the influenza M2 channel. These hydrophobic molecules could be incorporated into the lipid bilayer to form transmembrane channels. Considering the length of the tubular cavity of **11.1** (1.6 nm) and the thickness (3.5 nm) of the hydrophobic part of the lipid bilayer, it was proposed that two molecules of **11.1** randomly distributed in the lipid bilayer would stack to form a channel matching the thickness of the bilayer. The stacking of the molecules induced water molecules to form continuous water wires which serve as a track for proton migration by re-assembly of the hydrogen bonding (Figure 11.1c). The water wire-based transport is the structural basis for achievement of highly selective proton transport. Channel switching from

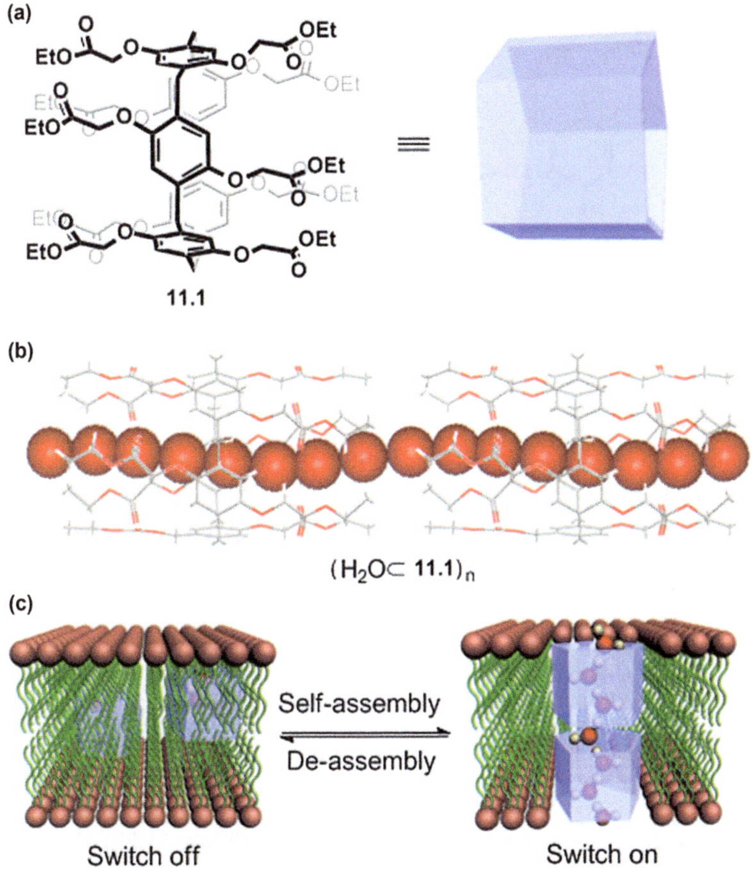

Figure 11.1 (a) Chemical structure of **11.1**. (b) Partial X-ray crystal structure of **11.1**, indicating that water molecules (CPK model) were induced to form a linear water wire in the organic nanotube (stick model). (c) Schematic representation of the switch mechanism for proton transport of **11.1**. Reproduced with permission from ref. 16. Copyright 2011, Wiley-VCH Verlag Gmbh & Co. KGaA, Weinheim.

the open to the closed state is probably due to the disassembly of the water wire or the channel.

The well-arranged oxygen atoms of the pillar[5]arene backbone and ester groups around the cavity provide not only the coordination sites for metal ions but also hydrogen bonding acceptors. It was found that Na^+ and α, ω-alkyldiols can bind to the tubular cavity of **11.1**. The binding of ions and diols to the cavity destroyed the continuous water wire and thus blocked the transport of protons. The stacking of **11.1** in the lipid bilayer is a reversible process, which renders the transmembrane channels unstable. The channel **11.2**, formed by connecting two pillar[5]arene units with a linker of a suitable

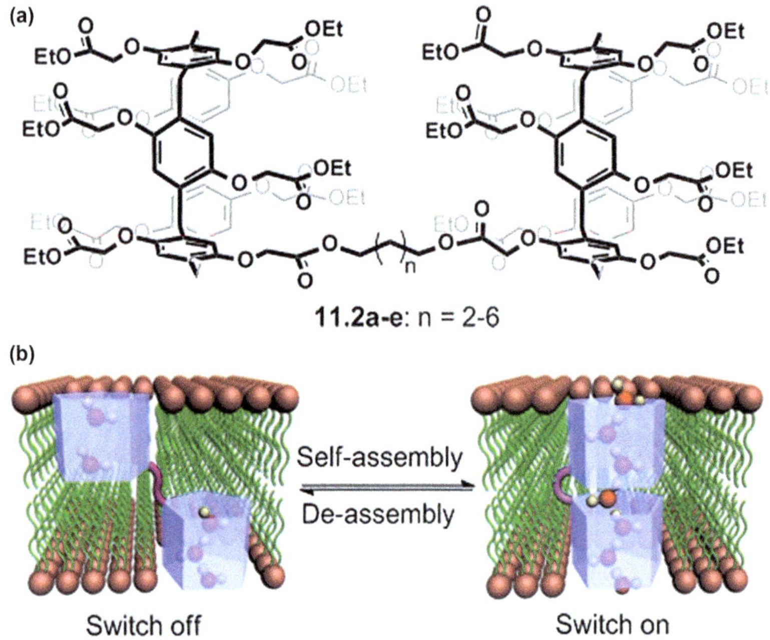

Figure 11.2 (a) Chemical structure of **11.2**. (b) Schematic representation of the switch mechanism for proton transport of **11.2**.
Reproduced with permission from ref. 16. Copyright 2011, Wiley-VCH Verlag Gmbh & Co. KGaA, Weinheim.

length, was able to enhance the channel stability (Figure 11.2a). For example, the open probability of dimer **11.2c** was 64%, more than three times higher than that of monomer **11.1**. When the linker is too long or short, the dimers did not show any activity or were less efficient than **11.1**, implying that their linkers could not enable a good stacking match for connecting the pillar[5]arene units (Figure 11.2b).

Transmembrane water transport is crucially important for regulation of the osmotic pressure of cells.[17] Nature utilizes water channel proteins, called aquaporins, to realize this purpose.[10] The aquaporin channel effectively prevents proton transport while permitting rapid water diffusion.[18,19] This selectivity strongly depends on the channel structure and dimensionality, channel–water affinity, the average pore occupancy and the dynamics of the permeating water.[20] The development of synthetic systems displaying powerful water transport capability not only is fundamentally important but also may lead to the generation of new medicinal and environmental materials.[21] However, there has been less progress in the area of synthetic water channels because of the lack of synthetic architectures that can form narrow and long channels for water. The emergence of pillar[5]arene has offered us the opportunity to construct such channels. First, it provides a platform for building such long tubular molecules by simple elongation of the backbone.

Second, it possesses a cavity of 5.6 Å width,[22] which is suitable for the docking of water molecules, as demonstrated by the crystal structure of **11.1**, and is close to the pore width of aquaporin.

Thus, a second generation of pillararene channels (**11.3** and **11.4**) were prepared by attaching 10 hydrazide-incorporated side-chains to the central pillar[5]arene scaffold (Figure 11.3a,b).[23] The hydrazide units in the side-chains form pentameric cylinders through intramolecular hydrogen bonding to induce the whole molecules to produce tubular structures. In the

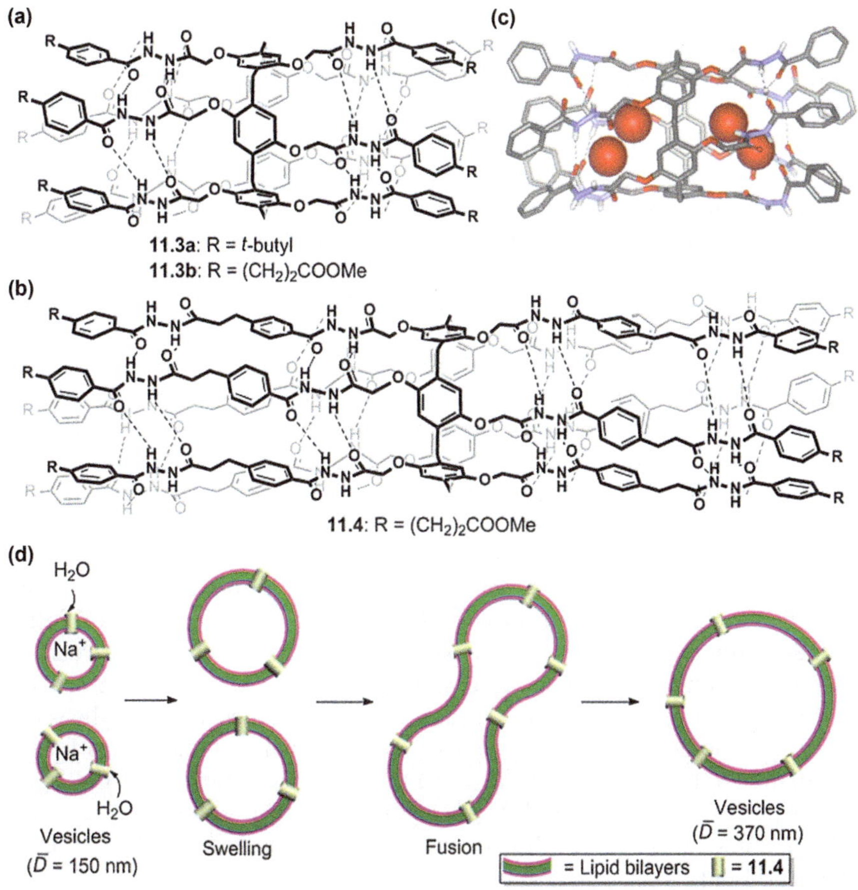

Figure 11.3 (a) and (b) Chemical structure of **11.3** and **11.4**. The hydrazide units of **11.3a–b** and **11.4** form intra-molecular cylinder hydrogen bonds to induce the formation of tubular structures. (c) Crystal structure (stick model) of **11.3a**. The oxygen atoms of four entrapped water molecules are highlighted in the CPK model. The *t*-butyl groups and hydrogen atoms on carbon atoms are omitted for clarity. (d) Schematic representation of vesicle fusion induced by **11.4**-mediated outside-to-inside water transport.
Reproduced with permission from ref. 23. Copyright 2012, American Chemical Society.

crystal structure of **11.3a**, the water is H-bonded and forms only water-dimers near to the hydrophilic regions of the channel (Figure 11.3c). Different from **11.1**, the longer hydrazide-incorporated pillar[5]arene molecules (**11.3** and **11.4**), present alternating hydrophobic/hydrophilic structural domains along their cylindrical structure which disrupt the formation of continuous water-wires within the inner channel core and thus block the proton flux along the discontinuous water wire. The longest channel (**11.4**), of 3.5 nm length, perfectly fits the thickness of the bilayer and shows an excellent transport activity for water and hydroxyl anions *via* a single-file molecular translocation mechanism. The outside-to-inside water transport through the channel can even induce the fusion of the vesicles to form larger vesicles (Figure 11.3d), which supports the notion that channels made from pillararenes are efficient in the transport of water.

Amino acids are the building blocks of proteins and play an important role in living systems. The transmembrane transport of amino acids in biology is realized by specific transporters, such as the adenosine triphosphate (ATP)-binding cassette (ABC) superfamily, the phospholemman (PLM) family, and hypotonically activated amino acid channels (HAAC).[24] During the transport process, the transporters display high substrate or chiral selectivity and show high transport efficiency. The construction of artificial amino acid transporters, especially those that can transport molecules across the lipid membrane, has been a challenge, because amino acids can pass across lipid membranes in a simple diffusion mechanism. The design of any artificial transporters requires an efficiency that is much higher than that of this simple diffusion process. This may be realized by further revising the pillararene-based tubular structure.

In the third generation of pillararene channels, a series of compounds (**11.5**) were synthesized by attaching peptides to the pillararene backbone (Figure 11.4a).[25] It was found that, similar to the second-generation pillararene channels, the molecules also self-assemble to form tubular structures in solution and the lipid bilayer membrane, induced by intra-molecular hydrogen bonding. In **11.5a**, the amino acid residues were in the same L-configurations. This arrangement induced the whole molecule to form a mismatched tubular structure (Figure 11.4b). However, for compound **11.5b**, the peptides were composed of alternating D- and L-amino acid residues, which induced the molecules to form a well organized tubular structure (Figure 11.4c).

Of the tubular structure, the surface is hydrophobic and the both ends are hydrophilic. This structural feature allowed the molecules with tripeptide side-chains to match well with the bilayer and therefore insert into the bilayer to form transmembrane channels. The amide bonds of the peptide chains provide sites for complexing of amino acids, which enable the channels to transport small amino acids, such as glycine and alanine (Figure 11.5a). Remarkably, the channels could achieve chiral selective transport of amino acid enantiomers (Figure 11.5b), which is one of the important functions of natural amino acid channels. This novel type of

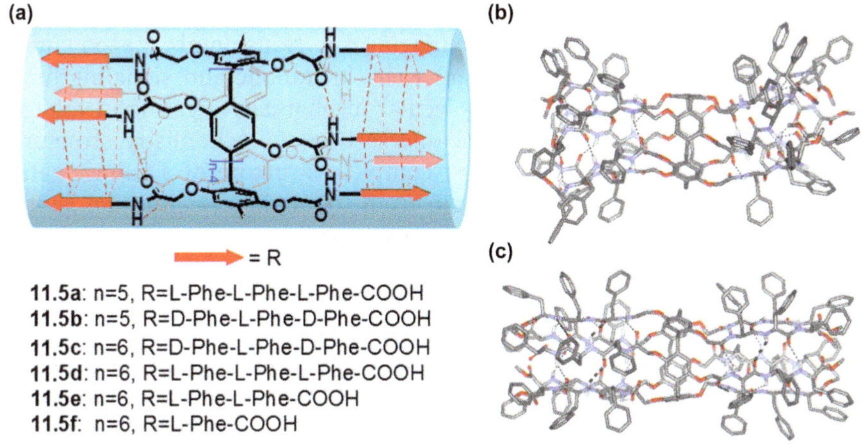

Figure 11.4 (a) Chemical structure of **5**. (b and c) Molecular modeling structure (H atoms on the C atoms have been omitted for clarity) of **11.5a** (b) and **11.5b** (c).
Reproduced with permission from ref. 25. Copyright 2013, American Chemical Society.

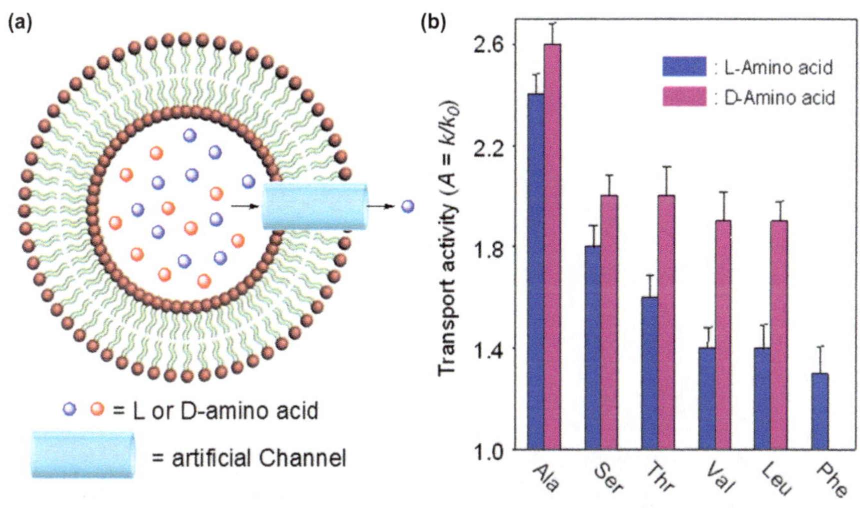

Figure 11.5 (a) Schematic representation of chiral selective transport of **11.5** for amino acid enantiomers. (b) Transport activities of **11.5d** for L- and D-amino acids.
Reproduced with permission from ref. 25. Copyright 2013, American Chemical Society.

artificial channel is very interesting, because the channel width, the length of the cavity, and the hydrophobic–hydrophilic balance can be adjusted by changing the pillararene backbone, the length of the peptide side-chains, and the amino acid sequence, respectively, which might produce functional diversity of the channels.

The transition of the voltage-gated ion channel between the closed and open states can be regulated by changing the transmembrane potential, which is responsible for a variety of crucial metabolic and signaling functions in biological systems. The molecular mechanism of gating in voltage-gated potassium channels lies at the intersection of biochemistry, structural biology, and neurobiology. The first complete structure of a voltage-gated potassium channel, the K_VAP protein from the archeon *Aeropyrum pernix*, was solved in 2003.[26] This channel possesses a unique transmembrane segment, the S4 domain, which contains four to eight positively charged arginine units. Under an electric field, this domain can move across the cell membrane to cause the channel to open and close.[27]

Employing the idea as nature does, we can incorporate positively charged Arg units into the peptide chains of third-generation pillar[5]arene-based transmembrane channels (**11.6**) (Figure 11.6a).[28] These channels undergo voltage-driven movement similar to natural voltage-gated channels. For example, when the channel was added to the *cis* side of the bilayer, a voltage of −50 mV could drive **11.6a** to insert into the lipid bilayer. The interaction between the channel molecules and the lipid is so strong that the channels stayed within the lipid bilayer to enable ion transport (Figure 11.6b).

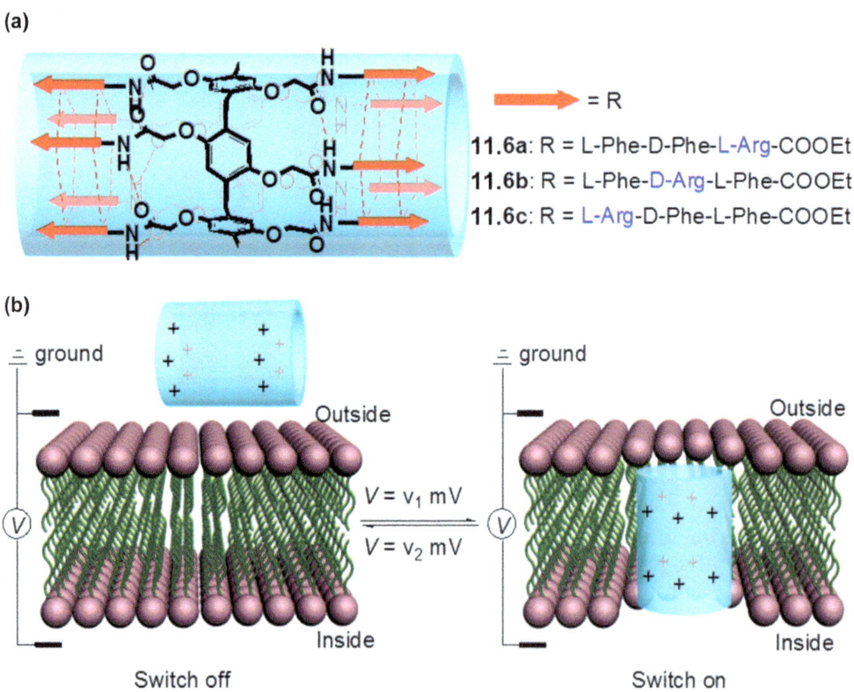

Figure 11.6 (a) Chemical structure of **11.6**. (b) Schematic representation of the voltage-driven channel inserting into and leaving the lipid bilayer. Reproduced with permission from ref. 28. Copyright 2014, Wiley-VCH Verlag Gmbh & Co. KGaA, Weinheim.

A higher reversed potential, such as +80 mV, could drag them out of the bilayer and thus switch off the K^+ transport. Both **11.6b** and **11.6c** could also be driven by the applied voltage to insert into the lipid bilayer. However, in contrast to **11.6a**, once inserted into the bilayer, these two molecules could not be driven out from the bilayer. One possible reason is that their positively charged Arg units aggregate in the hydrophobic inner of the bilayer, which made them more difficult to move out. Under the same conditions, the neutral channel **11.5b** did not show any voltage-responsive behavior.

The voltage-responsive behavior of the pillararene channels provides a promising strategy for the development of antimicrobials, because it has been established that bacteria maintain a negative potential in their cell interior and their cell membranes consist of a large fraction of negatively charged lipids which display a great affinity for cationic compounds.[29] It was found that the channel **11.6b** displayed antimicrobial activity toward *Bacillus subtilis* with a half maximal inhibitory concentration (IC_{50}) of 10 μM, which is comparable to that of natural channel-forming peptide alamethicin. This result raises the possibility of designing new molecules that are even more efficient against bacteria but have low toxicity to mammalian cells by further optimizing the peptide chains.

11.3 Construction of Vesicles for Drug Release

Vesicles are not only important building blocks of all living systems but also potentially useful in the development of drug delivery/release systems, owing to their unique spaces that can efficiently encapsulate drugs.[30] A wide variety of molecular building blocks have been employed for the construction of vesicles, and among them supramolecular amphiphiles with stimuli-responsive properties, used as building blocks to form vesicles, are the most promising. Pillararene has been considered as another useful platform for building such amphiphiles, because it is easy to functionalize both ends of the pillararene backbone to make asymmetrical pillararenes.[31]

In 2012, Huang *et al.* prepared a new type of amphiphilic pillar[5]arene (**11.7**) which can self-assemble to form vesicles in water (Figure 11.7).[32] The vesicles could encapsulate calcein, a model hydrophilic guest, within their interiors under neutral conditions. The vesicular structure appeared to transform into a micellar structure with a decrease in pH. Owing to this transition, the calcein was released into the solution in response to a decrease in pH (Figure 11.7). This is very interesting, because the stimuli-responsive nanocapsule may have potential applications in selective drug delivery in tissues with a lower pH, such as infected tissues and tumor tissues.

The good guest-binding properties of pillararene enable chemists to build supramolecular amphiphiles from simple pillararene backbones. Huang and co-workers also reported that the water-soluble pillar[6]arene **11.8** could bind to paraquat derivative **11.9** in water to form a stable supramolecular amphiphile which is pH-responsive (Figure 11.8a).[33] Reversible

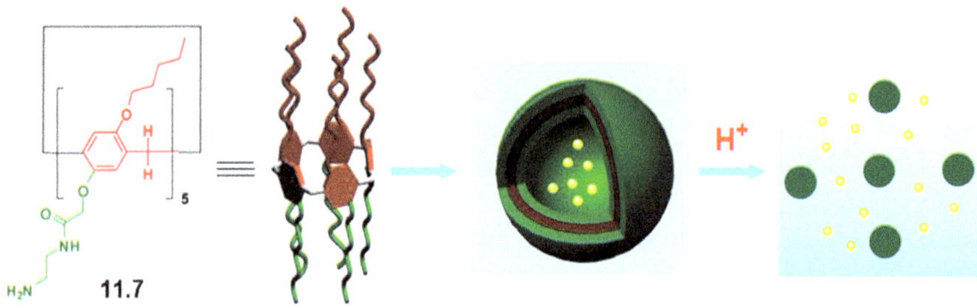

Figure 11.7 Chemical structure of **11.7** and schematic representation of its pH-triggered vesicle-to-micelle transition and the subsequent release of encapsulated calcein.
Reproduced with permission from ref. 32. Copyright 2012, American Chemical Society.

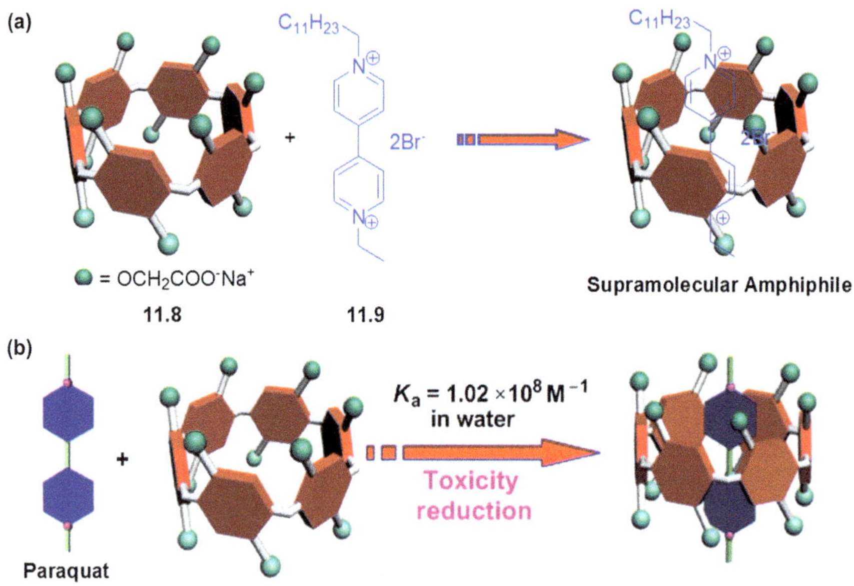

Figure 11.8 Schematic representation of the complexing of **11.8** with (a) **11.9** and (b) paraquat.
Reproduced with permission from ref. 33. Copyright 2012, American Chemical Society.

transformations between micelles based on **11.9** and the vesicles based on the supramolecular amphiphile were realized by adjusting the solution pH. The controlled release of water-soluble dye molecules from the vesicles was achieved by the collapse of the vesicles into the micelles by simply changing the solution pH. It was also found that the high binding affinity between **11.8** and paraquat efficiently reduced the toxicity of paraquat. This is not unexpected, because, after the formation of a stable host–guest complex between **11.8** and paraquat (Figure 11.8b), less opportunity was available for

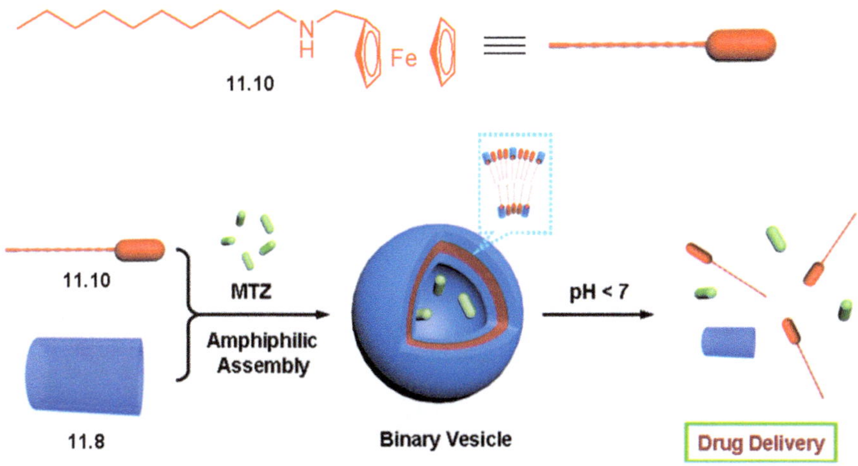

Figure 11.9 Schematic representation of the formation of vesicles and their application in pH-responsive drug release.
Reproduced with permission from ref. 34. Copyright 2013, American Chemical Society.

paraquat to interact with the reducing agents in the cell, which made the generation of its radical cation more difficult, resulting in the efficient reduction of paraquat toxicity.

Wang *et al.* subsequently investigated the drug delivery of such supramolecular vesicles *in vitro*. They constructed the vesicles from an amphihilic host–guest complex between **11.8** and the hydrophobic ferrocene derivative **11.10** in water (Figure 11.9).[34] The studies of *in vitro* drug release demonstrated that these supramolecular vesicles were able to encapsulate the drug mitoxantrone (MTZ) to give MTZ-loaded vesicles which showed rapid MTZ release in a low-pH environment as a result of the pH-responsiveness of the vesicles. These pH-responsive MTZ-loaded vesicles could be taken in by cancer cells, where the microenvironment is acidic. Thus, the rapid release of MTZ from the vesicles can be triggered by the acidic microenvironment, which is extremely significant for specific targeted therapy. Most importantly, the unloaded vesicles showed low toxicity to normal cells. By encapsulation of the MTZ, the toxicity of the drug was dramatically reduced owing to the high stability of the vesicles in normal cells. By further systematic investigation, the same group found that WP6 can also complex with pyridinium amphiphiles to form supramolecular binary vesicles which showed pH-, Ca^{2+}-, and thermal-responsiveness.[35] The anticancer drug doxorubicin hydrochloride (DOX) can be encapsulated by the vesicles, and the resulting DOX-loaded vesicles could efficiently release the encapsulated DOX following a pH or Ca^{2+} stimulus. The supramolecular vesicles exhibit a comparable therapeutic effect on cancer cells to free DOX but with remarkably reduced damage to normal cells. Such multistimuli-responsive supramolecular vesicles may find broad application in the field of controlled drug delivery.

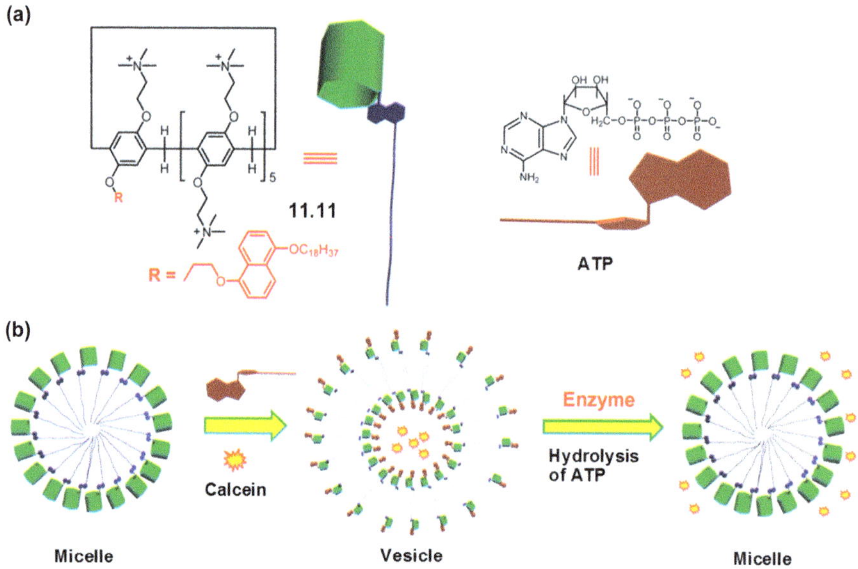

Figure 11.10 (a) Chemical structure of **11.11** and ATP. (b) Schematic representation of the transition between micelle and vesicle induced by complexing with and hydrolysis of ATP.
Reproduced from ref. 36 with permission from The Royal Society of Chemistry.

The sensitivity of the vesicles to bioactive molecules is also important, because it avoids the use of exterior agents and stimuli which might affect the biological system. Diao and co-workers synthesized a new kind of amphiphilic pillar[6]arene (**11.11**) which formed micelles in water and was able to complex with ATP (Figure 11.10a).[36] This host–guest complex further self-assembled to form vesicles (Figure 11.10b). The vesicles were found to become collapsed following the hydrolysis of ATP, catalyzed by phosphatase. This transition can be used to trigger the controlled release of encapsulated calcein, which also provides the possibility of using this pillararene derivative in drug delivery.

Drug resistance in both cancerous and normal cells is currently one of the major problems in chemotherapy. One of the promising strategies to overcome these obstacles is the development of the above mentioned stimuli-responsive vesicles. However, this requires the vesicles to co-carry the drug with small interfering (si)RNA.[37] Therefore, the vesicles should possess the ability to recognize and respond to specific microenvironmental changes. The carrier should be designed to be positively charged, which allows effective loading of polyanionic siRNA by electrostatic interaction, and to respond to intracellular stimuli. Recently, Pei and co-workers synthesized a novel ferrocenium-capped amphiphilic pillar[5]arene (**11.12**) which self-assembled to form cationic vesicles in aqueous solution (Figure 11.11).[38] The cationic vesicles showed significant redox-responsive behavior due to the

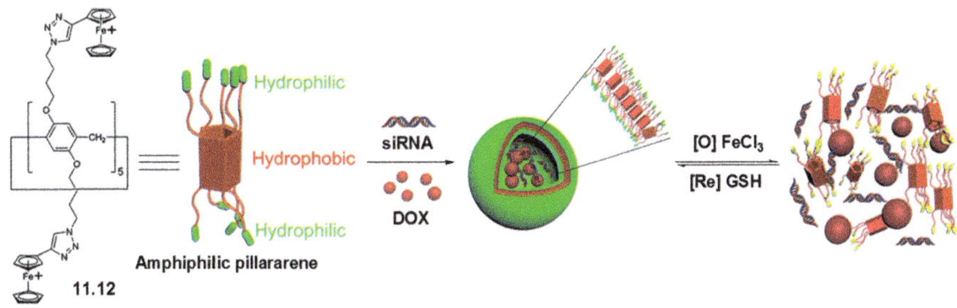

Figure 11.11 Chemical structure of **11.12** and schematic representation of its application in redox-responsive DOX/siRNA release.
Reproduced with permission from ref. 38. Copyright 2014, Wiley-VCH Verlag Gmbh & Co. KGaA, Weinheim.

redox equilibrium between the ferrocenium cations and ferrocenyl groups. The vesicles loaded with DOX and siRNA displayed low cytotoxicity to normal cells. It has been established that the concentration of glutathione (GSH), a reducing agent, in cancer cells is higher than that in normal cells. In cancer cells, the vesicles are rapidly sensitized to the presence of GSH and are disrupted. As a result, the DOX and siRNA are released and the cancer cells are killed efficiently. This new co-delivery system is able to overcome the drug resistance of cancer cells. This is an excellent example of a rational design for an effective stimuli-responsive drug delivery system.

11.4 Conclusions and Outlook

Although pillararenes have been developed for several years, the chemistry of this family is beginning to blossom, as witnessed by the heightened interest in the field for the last several years. In particular, the efficient synthesis of the pillararene backbones and their subsequent direct functionalization have removed one of the major obstacles to pillararene chemistry and henceforth they will be available for exhaustive synthetic manipulations to pursue biomedical applications, including transmembrane transport and drug delivery, as we have demonstrated in this chapter.

The functionalization of pillararene has paved the way to applications of pillararenes. However, developing more efficient functionalization methods, particularly for higher homologs such as pillar[8–10]arenes, is important in expanding their applications.[39] Currently, the field of pillararene-based functional biomaterials is in its infancy, but their prospects are enormous. The unique ability of pillararene in forming exceptionally stable host–guest complexes with high binding constants suggests that the synthetic host–guest systems may be used in the same way as the natural systems in many biochemical applications. With such great promise, the future of pillararene in biomedical applications will be brighter than ever.

References

1. K. Gloe, *Macrocyclic Chemistry: Current Trends and Future Perspectives*, Springer, The Netherlands, 2005.
2. G. W. Gokel, W. M. Leevy and M. E. Weber, *Chem. Rev.*, 2004, **104**, 2723–2750.
3. S. B. Nimsea and T. Kim, *Chem. Soc. Rev.*, 2013, **42**, 366–386.
4. E. M. Martin Del Valle, *Process Biochem.*, 2004, **39**, 1033–1046.
5. K. Kim, N. Selvapalam, Y. H. Ko, K. M. Park, D. Kim and J. Kim, *Chem. Soc. Rev.*, 2007, **36**, 267–279.
6. T. Ogoshi, S. Kanai, S. Fujinami, T.-A. Yamagishi and Y. Nakamoto, *J. Am. Chem. Soc.*, 2008, **130**, 5022–5023.
7. P. J. Cragg and K. Sharma, *Chem. Soc. Rev.*, 2012, **41**, 597–607.
8. M. Xue, Y. Yang, X. Chi, Z. Zhang and F. Huang, *Acc. Chem. Res.*, 2012, **45**, 1294–1308.
9. B. Hille, *Ionic Channels of Excitable Membranes*, Sinauer, Sunderland, 2nd edn, 1992.
10. G. M. Preston, T. P. Carroll, W. B. Guggino and P. Agre, *Science*, 1992, **256**, 385.
11. D. A. Doyle, J. Morais Cabral, R. A. Pfuetzner, A. Kuo, J. M. Gulbis, S. L. Cohen, B. T. Chait and R. MacKinnon, *Science*, 1998, **280**, 69–77.
12. I. Tabushi, Y. Kuroda and K. Yokota, *Tetrahedron Lett.*, 1982, **23**, 4601–4604.
13. J. K. W. Chui and T. M. Fyles, *Chem. Soc. Rev.*, 2012, **41**, 148–175.
14. S. Matle and T. Fyles, *Acc. Chem. Res.*, 2013, **46**, 2741–2742.
15. W. Si, L. Chen, X.-B. Hu, G. Tang, Z. Chen, J.-L. Hou and Z.-T. Li, *Angew. Chem., Int. Ed.*, 2011, **50**, 12564.
16. W. Si, X.-B. Hu, X.-H. Liu, R. Fan, Z. Chen, L. Weng and J.-L. Hou, *Tetrahedron Lett.*, 2011, **52**, 2484.
17. B. L. de Groot, T. Frigato, V. Helms and H. Grubmuller, *J. Mol. Biol.*, 2003, **333**, 279.
18. A. Burykin and A. Warshel, *Biophys. J.*, 2003, **85**, 3696.
19. M. Jensen, U. Rothlisberger and C. Rovira, *Biophys. J.*, 2005, **89**, 1744.
20. M. Barboiu and A. Gilles, *Acc. Chem. Res.*, 2013, **46**, 2814–2823.
21. M. Kumar, M. Grzelakowski, J. Zilles, M. Clark and W. Meier, *Proc. Natl. Acad. Sci. U. S. A.*, 2007, **104**, 20719.
22. C. Han, F. Ma, Z. Zhang, Bi. Xia, Y. Yu and F. Huang, *Org. Lett.*, 2010, **12**, 4360.
23. X.-B. Hu, Z. Chen, G. Tang, J.-L. Hou and Z.-T. Li, *J. Am. Chem. Soc.*, 2012, **134**, 8384.
24. S. Bröer, *Physiol. Rev.*, 2008, **88**, 249.
25. L. Chen, W. Si, L. Zhang, G. Tang, Z.-T. Li and J.-L. Hou, *J. Am. Chem. Soc.*, 2013, **135**, 2152.
26. Y. Jiang, V. Ruta, J. Chen, A. Lee and R. MacKinnon, *Nature*, 2003, **423**, 42.
27. H. C. Lai and L. Y. Jan, *Nat. Rev. Neurosci.*, 2006, **7**, 548.

28. W. Si, Z.-T. Li and J.-L. Hou, *Angew. Chem., Int. Ed.*, 2014, **53**, 4578.
29. F. Wang, L. Qin, C. J. Pace, P. Wong, R. Malonis and J. Gao, *ChemBioChem*, 2012, **13**, 51.
30. B. J. Ravoo, Vesicles in Supramolecular Chemistry, in *Supramolecular Chemistry: From Molecules to Nanomaterials*, ed. J. W. Steed and P. A. Gale, Wiley, Hoboken, 2012.
31. N. L. Strutt, H. Zhang, S. T. Schneebeli and J. F. Stoddart, *Acc. Chem. Res.*, 2014, **47**, 2631.
32. Y. Yao, M. Xue, J. Chen, M. Zhang and F. Huang, *J. Am. Chem. Soc.*, 2012, **134**, 15712.
33. G. Yu, X. Zhou, Z. Zhang, C. Han, Z. Mao, C. Gao and F. Huang, *J. Am. Chem. Soc.*, 2012, **134**, 19489.
34. Q. Duan, Y. Cao, Y. Li, X. Hu, T. Xiao, C. Lin, Y. Pan and L. Wang, *J. Am. Chem. Soc.*, 2013, **135**, 10542.
35. Y. Cao, X.-Y. Hu, Y. Li, X. Zou, S. Xiong, C. Lin, Y.-Z. Shen and L. Wang, *J. Am. Chem. Soc.*, 2014, **136**, 10762.
36. J. Zhou, M. Chen and G. Diao, *Chem. Commun.*, 2014, **50**, 11954.
37. Z. Luo, K. Cai, Y. Hu, L. Zhao, P. Liu, L. Duan and W. Yang, *Angew. Chem., Int. Ed.*, 2011, **50**, 640.
38. Y. Chang, K. Yang, P. Wei, S. Huang, Y. Pei, W. Zhao and Z. Pei, *Angew. Chem., Int. Ed.*, 2014, **53**, 13126.
39. X.-B. Hu, Z. Chen, L. Chen, L. Zhang, J.-L. Hou and Z.-T. Li, *Chem. Commun.*, 2012, **48**, 10999.

CHAPTER 12

Pillararene-related Macrocycles

SEVERIN T. SCHNEEBELI, NATHAN L. STRUTT, CHUYANG CHENG AND J. FRASER STODDART*

Center for the Chemistry of Integrated Systems, Department of Chemistry, Northwestern University, 2145 Sheridan Road, Evanston, IL 60208, USA
*Email: stoddart@northwestern.edu

12.1 Introduction

Pillar[*n*]arenes can be substituted[1,2] readily on the oxygen atoms of their hydroquinone units. More general substitution patterns of the pillararene [1*n*]paracyclophane framework can be achieved either by replacing the methylene bridges of pillar[*n*]arenes with heteroatoms or by attaching additional heteroatoms or carbon substituents directly on to the phenylene units. The possibility of incorporating these more general phenylene substitution patterns into pillararene-related macrocycles opens up opportunities to (1) fine-tune host–guest binding and optoelectronic properties, (2) investigate the electronic coupling and three-dimensional aromaticity in rigid, belt-like structural frameworks, (3) use substitutions of the bridging units themselves to control the nature of the substituents pointing outwards in the radial direction of the macrocycles, and (4) use the additional substituents to direct the growth of metal-directed or covalent tubular frameworks in a growth-from-template approach.

Monographs in Supramolecular Chemistry No. 18
Pillararenes
Edited by Tomoki Ogoshi
© The Royal Society of Chemistry 2016
Published by the Royal Society of Chemistry, www.rsc.org

12.2 [1*n*]Paracyclophanes – The Bare Backbones of Pillar[*n*]arenes

12.2.1 Synthesis and Properties of [1*n*]Cyclophanes

While the later portions of this chapter will focus on enhanced phenylene substitution of pillar[*n*]arene related derivatives, we would like to introduce briefly the chemistry and properties of the bare backbones of the pillar[*n*]arenes **12.2a** and **12.2b**, which are simply comprised (Scheme 12.1) of the [1*n*]paracyclophanes **12.1c** and **12.1d**. These macrocyclic structures combine the rigidity of aromatic rings with the conformational freedom provided by the methylene bridges to allow the creation of molecular rings with diverse sizes and shapes. The parent [1*n*]cyclophanes **12.1a–d** have been synthesized[3–5] with different ring sizes, and *n*, and to date [1*n*]paracyclophanes, with $n = 2$ (**12.1a**),[6,7] 4 (**12.1b**),[8] 5 (**12.1c**),[9] and 6 (**12.1d**)[9] are known. On account of its ring strain, the smallest member of the [1*n*]cyclophane family, [1.1]paracyclophane (**12.1a**), is not stable at room temperature and has only been observed[7] by low temperature UV-vis and ^{1}H NMR spectroscopies. It was prepared *in situ* (Scheme 12.2) by photolysis of the bis(Dewar benzene) cyclophane derivative **12.3**, which was itself obtained in 15 synthetic steps from diethyl 3,6-dihydroterephthalate. Prolonged irradiation of **12.1a** with light ($\lambda > 335$ nm) leads to an unusual transannular photocycloaddition between the two strained benzene rings of the [1.1]-paracyclophane framework, affording the constitutional isomer **12.4**. More recent research has shown[6] that bulky substituents attached to the phenylene groups of **12.1a** stabilize the [1.1]paracyclophane framework, allowing it to be characterized by single crystal X-ray diffraction.

The molecular square **12.1b** – *i.e.*, [1.1.1.1]paracyclophane – has been synthesized[8] using an adventurous multi-step procedure, which involves flash vacuum pyrolysis at 550 °C and a Diels–Alder cycloaddition to form the

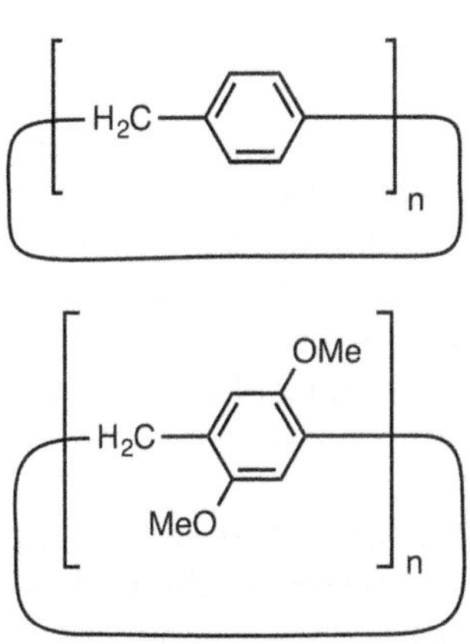

[1n]Paracyclophanes

	n
12.1a	2
12.1b	4
12.1c	5
12.1d	6

Pillar[n]arenes

	n
12.2a	5
12.2b	6
12.2c	7
12.2d	8

Scheme 12.1 Structural formulae illustrating the relationship between the [1*n*]paracyclophanes **12.1a–d** and the pillar[*n*]arene macrocycles **12.2a–d**.

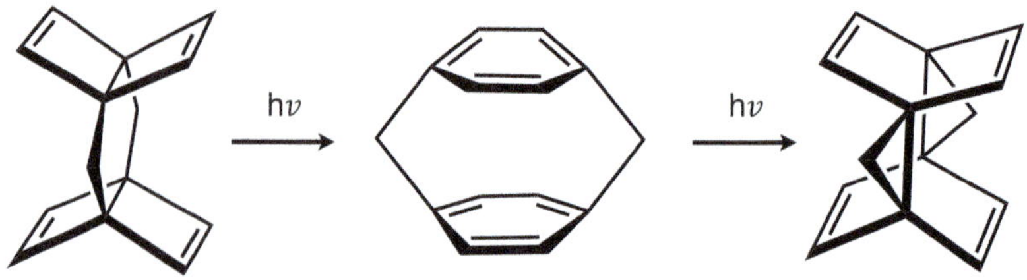

12.3 12.1a 12.4

Scheme 12.2 Synthesis of the highly strained [1.1]paracyclophane **12.1a** from the bis(Dewar benzene) derivative **12.3**.

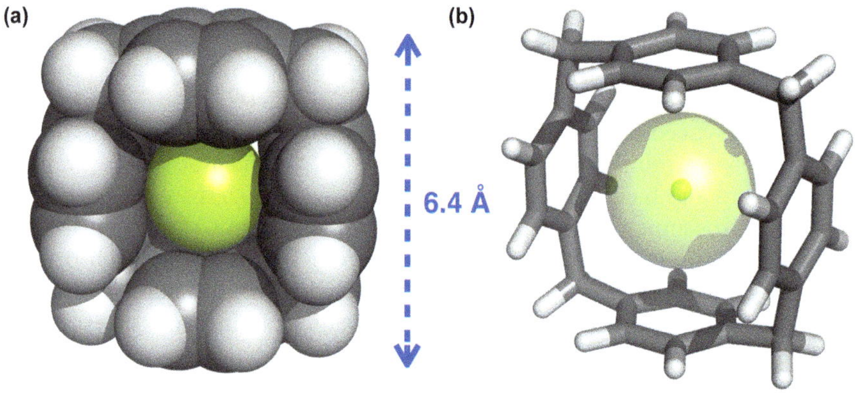

Figure 12.1 Graphical representations – (a) space filling and (b) tubular – of a Cs^+ cation (lemon) binding inside the cavity of [1.1.1.1]paracyclophane **12.1b**. Carbon atoms are illustrated in gray and hydrogens in white.

last phenylene ring assembled in the synthesis. [1.1.1.1]Paracyclophane (**12.1b**) represents a rare example of a hydrocarbon which forms[10] selective adducts with large metal ions through cation–π interactions.[3,11,12] Mass spectrometric data show, for example, that Ag^+, which is known[10] to have a rather strong affinity for carbon=carbon double bonds, binds much more weakly to the molecular square **12.1b** than does Cs^+. This result, which is best explained by the exceptionally good geometrical fit of the $Cs^+ \subset$**12.1b** complex, is surprising because Cs^+ generally shows no special affinity for binding to carbon=carbon double bonds. With the Cs^+ bound inside the square cavity of **12.1b**, the Cs^+ cation is able (Figure 12.1) to exert weakly attractive cation–π interactions with all four walls of the molecular square at the same time, a situation which is not possible with Ag^+ on account of its significantly smaller ionic radius.

The larger [1*n*]paracyclophanes **12.1c** and **12.1d**, with five and six phenylene units per macrocycle, have also been accessed synthetically, for the first time, by Gribble and Nutaitis.[9] The syntheses of **12.1c** and **12.1d** were completed with a Friedel–Crafts alkylation macrocyclization step (Scheme 12.3), catalyzed by trifluoroacetic acid (TFA). While the Friedel–Crafts macrocyclication

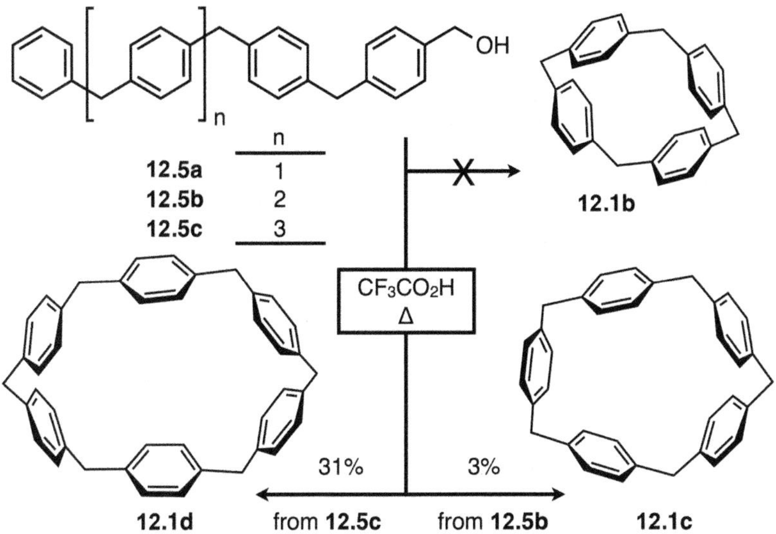

Scheme 12.3 Syntheses of [1.1.1.1.1]paracyclophane **12.1c** and [1.1.1.1.1.1]paracyclophane **12.1d**, using Friedel–Crafts alkylation reactions.

afforded [1.1.1.1.1.1]paracyclophane (**12.1d**) with *six* phenylene units in 31% yield, the yield for the ring-closure step dropped to 3% in the case of [1.1.1.1.1]paracyclophane (**12.1c**) with only *five* phenylene units, presumably because of the increased ring strain associated with the Friedel–Crafts alkylation transition state leading to its formation. Consistent with this hypothesis, analogous Fridel–Crafts macrocyclization conditions did not even afford[9] a trace of [1.1.1.1]paracyclophane (**12.1b**) with *four* phenylene units, presumably as a result of even greater ring strain than that present in **12.1c**.

12.2.2 [1*n*]Cyclophanes with Heteroatom Bridges

Replacing the methylene bridges of [1*n*]paracyclophanes with heteroatoms opens the door to a variety of cylophane derivatives (Scheme 12.4) with tunable structural and electronic properties. Given that a variety of efficient coupling reactions[13,14] between heteroatoms and halogen-substituted phenylene groups have been discovered, heteroatom-substituted paracylophane macrocycles are often more readily accessible, synthetically, than [1*n*]paracyclophanes with methylene bridges. *Vide supra* for some of the heroic and often low-yielding syntheses[4,8,9] used to access all carbon-hydrogen based [1*n*]paracyclophanes.

In an early effort to introduce heteroatoms into [1*n*]paracyclophane frameworks, the close relative **12.6** of the molecular square **12.1b** with all methylene bridges replaced by dimethylsilyl functional groups was synthesized by König *et al.*,[15] using stepwise regioselective ortholithiation reactions followed by coupling with $SiMe_2Cl_2$.

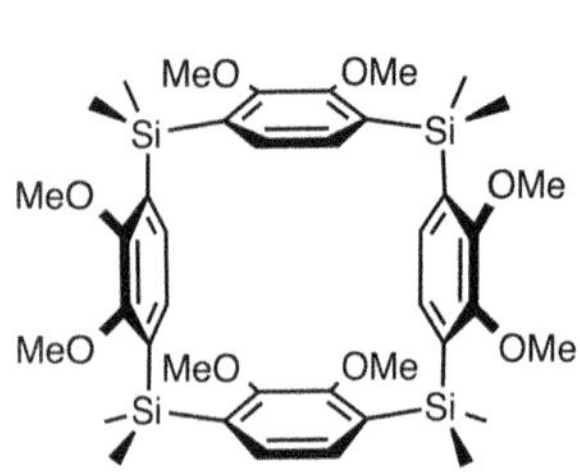

12.6

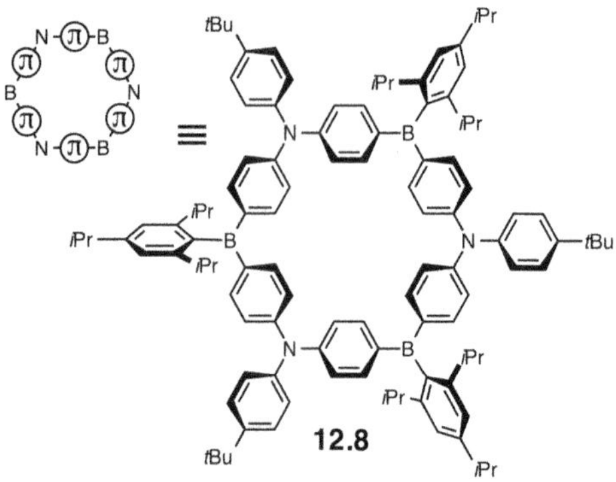

12.8

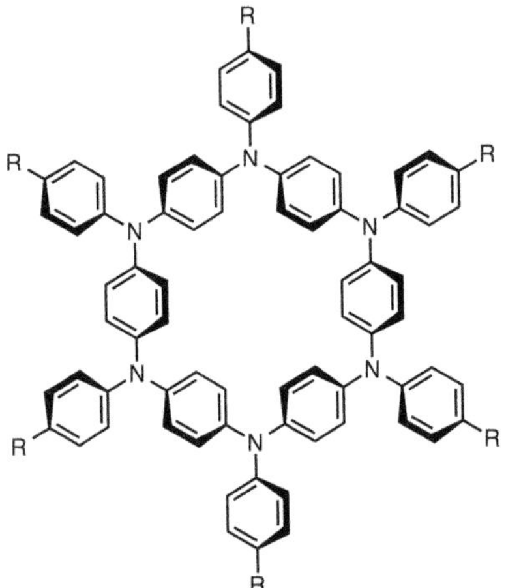

	R
12.7a	OMe
12.7b	H

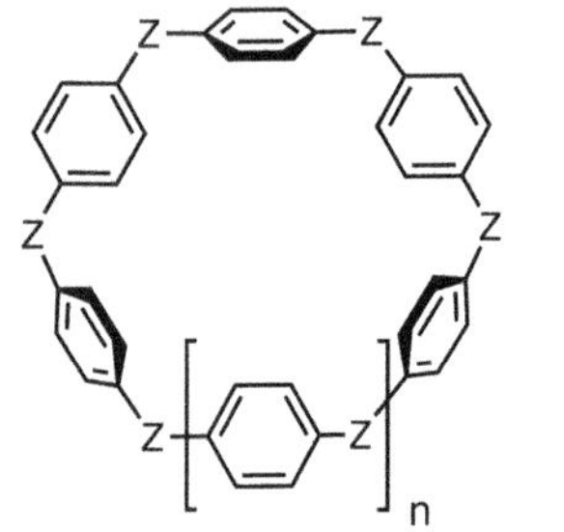

	Z	n
12.9a	O	1
12.9b	O	2
12.9c	O	3
12.9d	O	4
12.9e	O	5
12.10	S	1

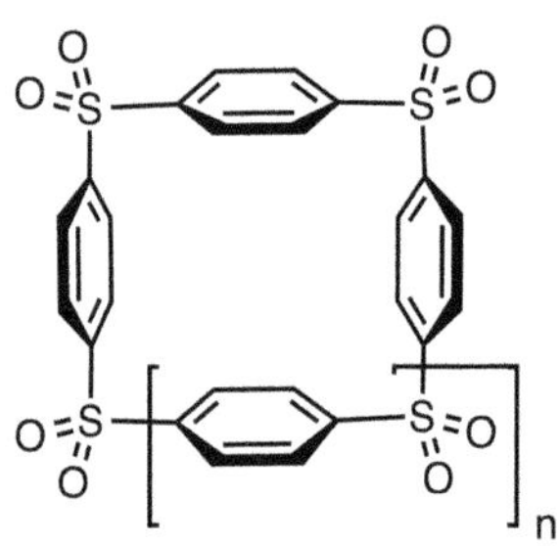

	n
12.11a	1
12.11b	3
12.11c	5
12.11d	7
12.11e	9

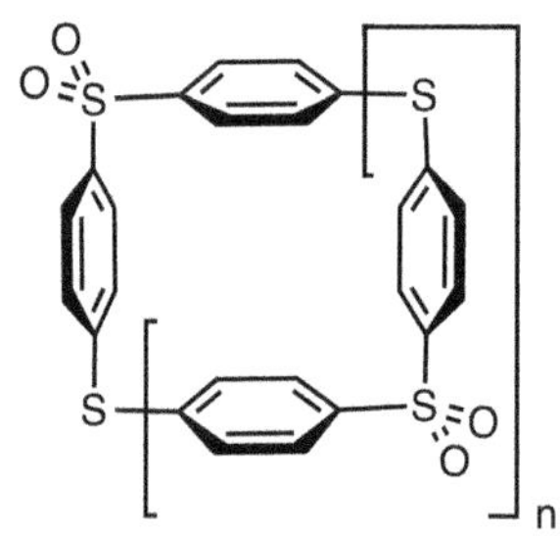

	n
12.12a	1
12.12b	2
12.12c	3
12.12d	4
12.12e	5

Scheme 12.4 Heteroatom-substituted [1*n*]paracyclophane frameworks.

Moving further towards the right in the Periodic Table, the bridging methylene units of [1*n*]paracyclophanes have also been substituted with nitrogens. Tanaka *et al.*[16] have prepared the first example of such a macrocycle, the *N*-anisyl-substituted hexaaza[1.1.1.1.1.1]paracyclophane **12.7a**, employing stepwise Buchwald–Hartwig aryl amination reactions[14] of *tert*-butoxycarbonyl (BOC)-protected aniline intermediates. In a more recent study, Yang *et al.*[17] discovered that, in the synthesis of the phenyl-substituted hexaaza[1.1.1.1.1.1]paracyclophane **12.7b**, no amine protecting groups are required, providing access to **12.7b** in only two synthetic steps. Additionally, Tanaka and co-workers[16] have shown that **12.7a** can be oxidized up to six times to form a hexacationic species, in accord with the six redox-active triarylamine functional groups present in the macrocycle. The potential for the first oxidation of **12.7a** to its radical cationic form, **12.7a**$^{\bullet+}$, is lowered significantly, compared with the first oxidation potential of a corresponding *linear* counterpart to **12.7a**. This finding hints at electron sharing between the six phenylene units of **12.7a**$^{\bullet+}$, a hypothesis which was confirmed by the application of density functional theory (DFT) calculations and electron paramagnetic resonance (EPR) spectroscopy. Interestingly, the spin density distributions (Figure 12.2) of [**12.7a**]$^{\bullet+}$ and [**12.7b**]$^{\bullet+}$ are

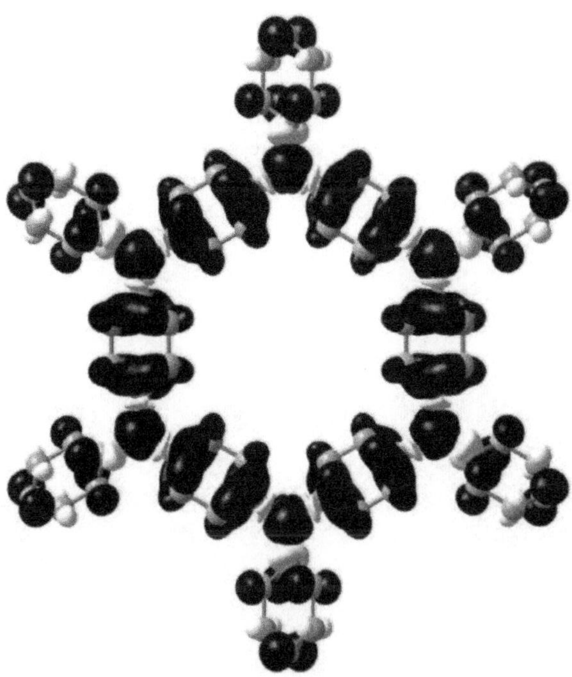

Figure 12.2 Graphical illustration of the delocalized spin density distribution (black: positive spin; white: negative spin; spin isosurface value = 0.0003 electron au^{-3}) of the radical cationic state of the nitrogen-bridged [1*n*]paracyclophane derivative **12.7b** (Scheme 13.4).
Figure reproduced with permission from ref. 16.

predicted to be delocalized completely around the macrocycles at the DFT level of theory, an observation which could be of interest in the study of aromaticity and molecular magnetism in conjugated, toroidal spin systems.[16]

Inspired by the isosterism between CC and BN groupings, Jäkle and co-workers[18] have replaced three of the nitrogen bridges in the hexaaza[1.1.1.1.1.1]paracylophane framework with boron atoms in an alternating manner to create the ambipolar conjugated B–π–N macrocycle **12.8**. Through the *para*-connectivity of the π-conjugated phenylene rings, neighboring nitrogen and boron atoms in **12.8** are coupled with each other electronically. The triaza–tribora macrocycle **12.8**, which shows strong luminescence in solution, can therefore be regarded as an expanded version of borazine. The coordination of anions (*e.g.*, cyanide) to its Lewis acidic boron centers alters the luminescence of **12.8**, rendering this compound's derivatives viable candidates for anion sensing applications. Furthermore, because both electron donating nitrogens and electron withdrawing boron atoms are present within **12.8**, its conjugated cyclic framework might also find future applications in the realm of creating ambipolar semiconducting materials. An X-ray structure (Figure 12.3), obtained from the blue-emissive hexagonally shaped crystals of **12.8**, illustrates that the individual macrocycles are packed loosely on top of each other to form channels which become evident when viewing the superstructure along the *c*-crystallographic axis. The B–N–B–N–B–N– backbone of the macrocycles adopts an almost coplanar conformation, which is only slightly distorted towards a chair-like conformation in the solid state. As a consequence of the large isopropyl substituents in the *ortho* positions, the exocyclic phenylene substituents

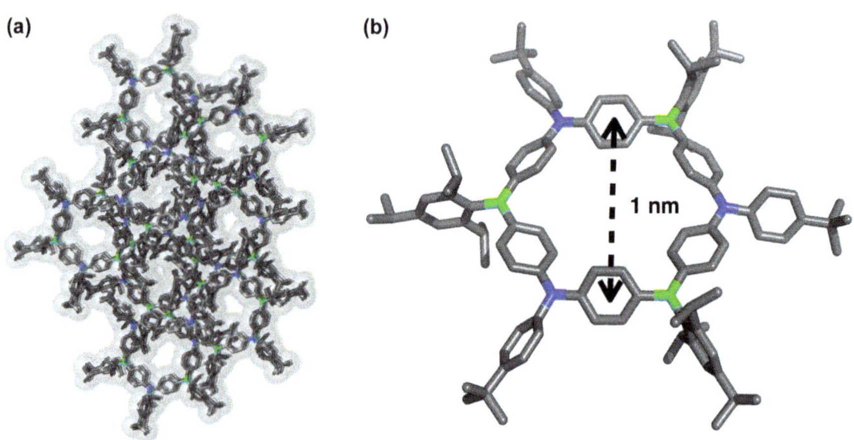

Figure 12.3 A single crystal X-ray structure of the π-expanded borazine macrocycle **12.8** (Scheme 13.4). (a) Sketch of the super-structure, viewed along the *c*-crystallographic axis and (b) graphical illustration of the molecular conformation in the solid state. Carbon atoms are shown in gray, nitrogens in blue, and borons in green. Hydrogen atoms and solvent molecules have been omitted for the sake of clarity.

attached to the boron atoms adopt a near-perpendicular conformation with respect to the boron's vacant *p*-orbitals, hence minimizing the electronic impact of those phenylene groups by π-conjugation.

Besides silicon, nitrogen, and boron, the chalcogen elements oxygen and sulfur have also been incorporated into [1*n*]paracyclophane frameworks. In fact, an early report by Franke and Vögtle[19] describes the syntheses of [1*n*]paracyclophanes with oxygen and sulfur bridges. The incorporation of oxygen into such structures has also been achieved by Osakada *et al.*,[20] who employed Ullman coupling reactions with a CuI/*N*,*N*-dimethylglycine catalyst to access a variety of cylic *p*-phenylene oxide oligomers. Controlled stepwise synthesis afforded *p*-phenylene oxide macrocycles with six to ten phenylene units as pure crystalline materials. Solid-state structures (Figure 12.4) of the *p*-phenylene oxide macrocycles **12.9a–d**, obtained from the analysis of single crystal X-ray diffraction data, illustrate that the neighboring phenylene rings are oriented almost perpendicularly with

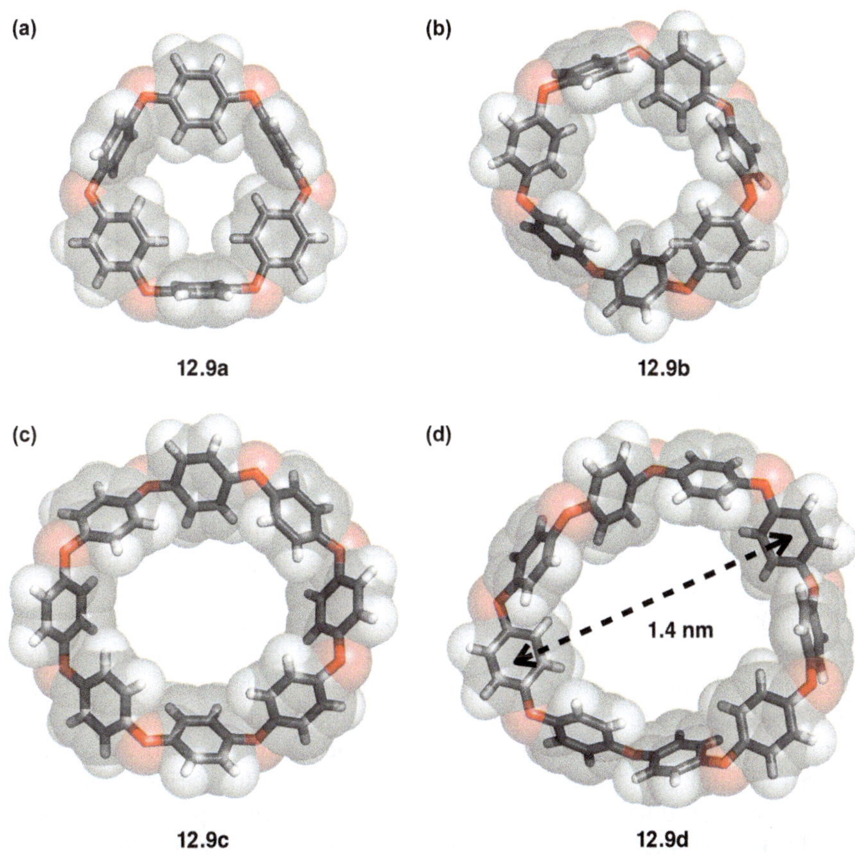

Figure 12.4 Single crystal X-ray structures of the oxygen-bridged [1*n*]paracyclophanes **12.9a–d**. Carbon atoms are illustrated in gray, oxygens in red, and hydrogens in white.

respect to each other. While the *p*-phenylene oxide macrocycles with odd numbers of phenylene units adopt bent conformations in the solid state, the even-numbered ones form almost coplanar structures. Relatively large cavities are formed by the *p*-phenylene oxide macrocycles **12.9a–d**, with diameters on the order of 1.0 to 1.5 nm.

Amongst the most versatile [1*n*]paracyclophane structures are frameworks in which the bridging atoms are sulfurs. Given that sulfur bridges can exist in three different stable oxidation states – *i.e.*, as thioethers, sulfoxides, or sulfones – a great variety of [1*n*]paracyclophanes with tunable redox and guest-binding properties can be envisioned. While cyclic *para*(phenylene sulfides) were first discovered[21] as impurities in samples of commercial poly(paraphenylene sulfides), Franke and Vögtle[19] were able to prepare cyclic *para*(phenylenesulfides) deliberately using Ullman coupling conditions under high dilution. Tsuchida and co-workers[22] later discovered a high-yielding synthesis of cyclic hexakis(thio-1,4-phenylene) (**12.10**), based on the oxidative polymerization of aryl disulfides. Cyclic *para*(phenylenesulfides) are of considerable interest[23] as starting materials for ring-opening polymerizations to access linear poly(*p*-phenylene sulfide) (PPS), a commercial polymer, which has found a multitude of industrial applications[24] as a high performance thermoplastic. Potential benefits of using cyclic oligomers to access PPS include[25] the ability to (1) carry out *in situ* polymerizations to access PPS-based composites with superior adhesion properties, as well as (2) suppress halogenated impurities which are commonly formed during the production of PPS using Ullman coupling and can degrade[23] the electronic and mechanical properties of PPS-based materials.

Oxidation of [1*n*]paracyclophanes with thioether linkages to sulfones has been investigated extensively by Colquhoun *et al.*[26] Reacting 4,4′-sulfonylbis-(benzenethiol) (**13.13**) with 4,4′-dichlorodiphenylsulfone (**12.14**) under pseudo high-dilution conditions – *i.e.*, slow addition of both starting materials to the reaction mixture – leads (Scheme 12.5) to the formation of macrocycles **12.12a**, **12.12b**, and **12.12c** with 4, 6, and 8 phenylene units, respectively. Given that, with a direct nucleophilic aromatic substitution mechanism, only cyclic products with $4n$ ($n = 1, 2, 3, \ldots$) phenylene units should be able to form directly in this reaction, it is surprising that the [1*n*]paracyclophane **12.12b** with six phenylene units is produced as the major macrocycle.

Colquhoun *et al.*[26] explained this unexpected result by proposing a "back-biting" mechanism (Scheme 12.6), wherein a thiophenoxide unit with a *para*-sulfonyl substituent in the reaction intermediate **12.16** can act as a leaving group, leading to scrambling of the thioether units. Oxidation with hydrogen peroxide (H_2O_2) in trifluoroacetic acid (TFA) of the cyclic [1*n*]paracyclophane macrocycles **12.12a–c**, with alternating phenylene sulfide/sulfone units, affords the highly symmetrical polysulfone rings **12.11a**, **12.11b**, and **12.11c**. Interestingly, if the oxidation reaction is carried out in acetic acid (AcOH) rather than TFA as the solvent, then the last thioether unit in **12.12c** is arrested at the sulfoxide stage to afford (***RS***)-**12.15**. This incomplete oxidation is presumably caused by the very low solubility of the hepta-sulfone-mono-sulfoxide (***RS***)-**12.15** in acetic acid, a physical

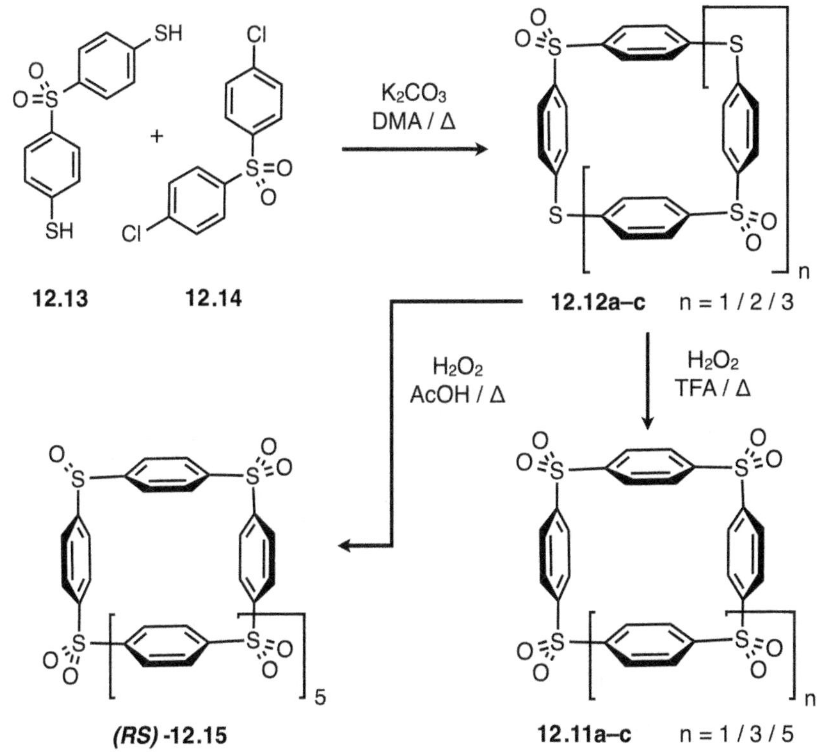

Scheme 12.5 Synthesis of sulfur-bridged [1*n*]paracyclophanes.

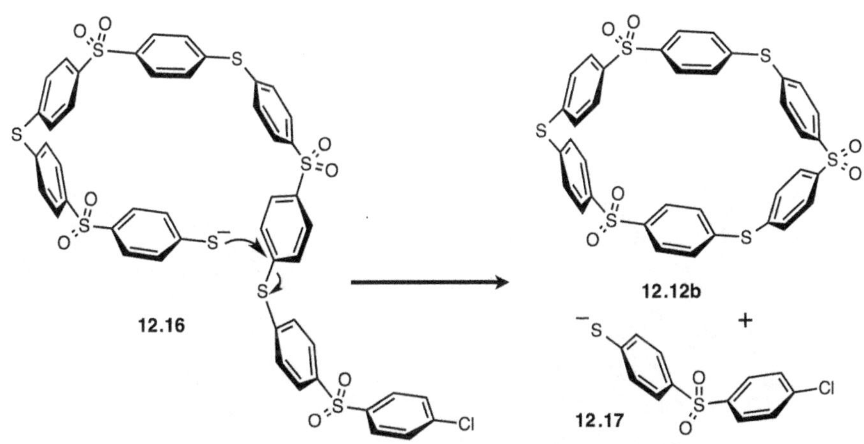

Scheme 12.6 The "back-biting" reaction mechanism, which is likely responsible for the formation of the sulfur-bridged [1*n*]paracyclophane **12.12b** with *six* phenylene units.

situation which leads to fast crystallization of (***RS***)-**12.15**, thereby inhibiting further oxidation of the last sulfoxide unit to a sulfonyl group kinetically in the solid state. Colquhoun *et al.*[26] were able to grow single crystals of compound

(***RS***)-**12.15**, suitable for single crystal X-ray diffraction measurements. Remarkably, analysis of the X-ray diffraction data showed (Figure 12.5) that the heptasulfonyl-mono-sulfoxyl derivative (***RS***)-**12.15** adopts a solid-state structure very similar to that of the octasulfone **12.11c**. The sulfoxide unit of (***RS***)-**12.15** is disordered over all eight possible positions of the macrocycle, with an occupancy value for each sulfoxide oxygen of 0.96. This occupancy value agrees well with the theoretically expected value of 0.94 (15/16) for this situation. The paracyclophane (***RS***)-**12.15**, which adopts a folded "figure-of-eight" conformation in the solid state, is stabilized by weak [C–H···O] hydrogen bonds formed between the inwards-directed sulfone oxygen atoms and aromatic hydrogen substituents.

A rather dramatic change of the cyclic backbone conformation was observed in the solid state for compound **12.12c** which has eight phenylene units joined together by alternating thioether/sulfoxide bridges. It crystallized in two different forms (Figure 12.5b and d), neither of which resembles the "figure-of-eight" conformation of the heptasulfone–monosulfoxide macrocycle (***RS***)-**12.15**. The backbone of **12.12c** adopts a chair-like

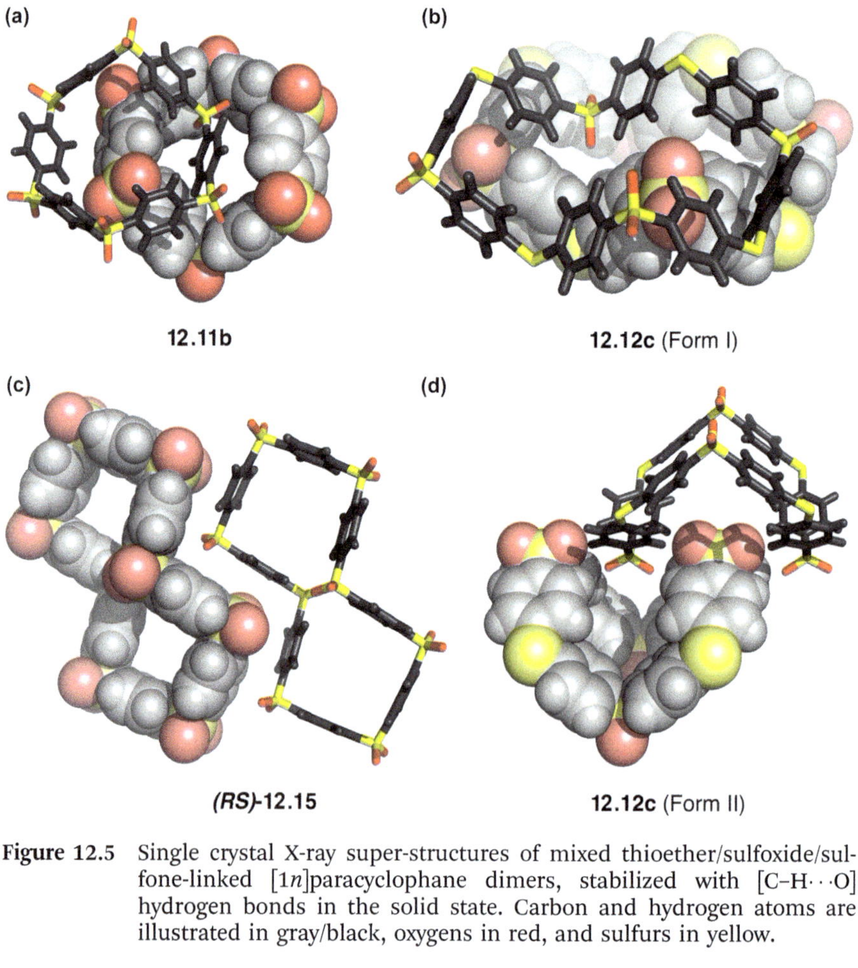

Figure 12.5 Single crystal X-ray super-structures of mixed thioether/sulfoxide/sulfone-linked [1*n*]paracyclophane dimers, stabilized with [C–H···O] hydrogen bonds in the solid state. Carbon and hydrogen atoms are illustrated in gray/black, oxygens in red, and sulfurs in yellow.

conformation in form I (crystallized from benzene), while in form II (crystallized from dichloromethane/chloroform) the macrocycle backbones resemble the pattern of a tennis ball. Intermolecular [C–H···O] interactions also lead to the formation of dimers (Figure 12.5) in the solid-state superstructures of the compounds **12.11b**, **12.12c**, and (***RS***)-**12.15**.

Those examples highlight how replacing the methylene bridges in pillararene-related [1*n*]paracyclophanes can be employed to fine-tune the electronic and supramolecular properties of such macrocycles. The molecular structures of heteroatom-bridged cyclophanes are affected in complex ways by the fine balance between (1) conjugation of heteroatom orbitals with the phenylene groups and (2) noncovalent bonding interactions between neighboring/transannular phenylene groups. Replacing methylene groups with heteroatoms therefore represents a promising approach to engineering novel macrocycles with enhanced properties for applications in chemistry and materials science.

12.3 Rigidified Relatives of Pillararenes

12.3.1 Collarene Molecular Belts

By adding (Scheme 12.7) an additional methylene bridge connection between all neighboring phenylene units of the [1*n*]cylophane backbone of

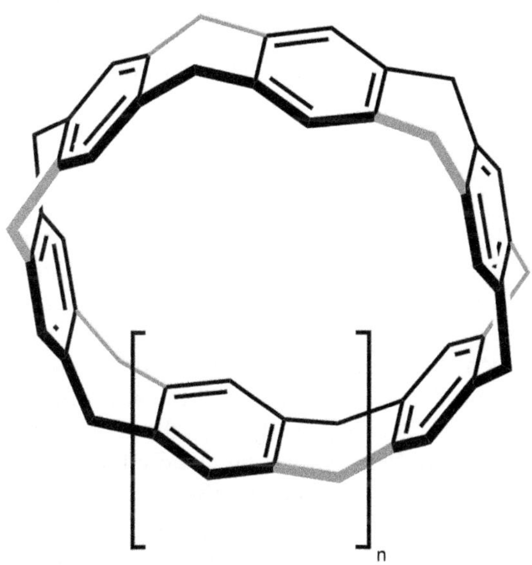

	n
12.18a	0
12.18b	1

Scheme 12.7 Structural formulae of collarene macrocycles. The [1*n*]paracyclophane backbone of pillararenes – contained within the collarene framework – is illustrated in black, while the additional bridging methylene units are illustrated in gray.

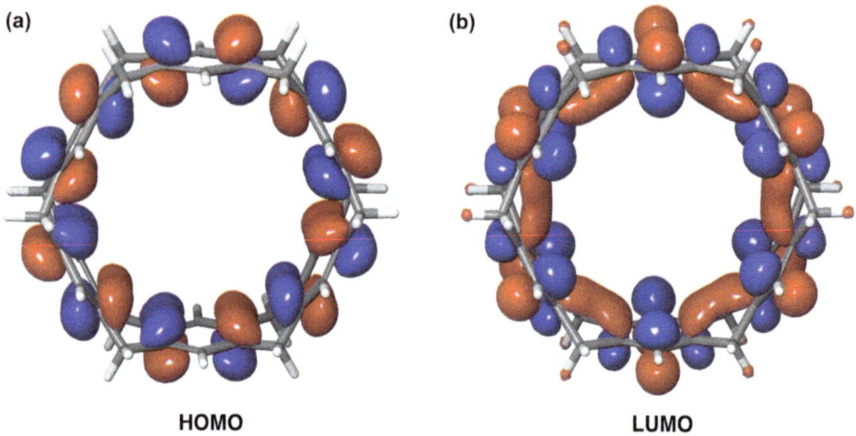

Figure 12.6 Graphical illustration of the DFT (B3LYP/6-31G**) frontier molecular orbitals of the [12]collarene molecular belt **12.18b**.

pillar[*n*]arenes, a new type of rigid, belt-like molecular structure (**12.18a–b**), which we have dubbed[27] [*n*]collarene, is obtained. Belt-like molecular structures have attracted a considerable amount of interest during recent years[28] on account of their special opto-electronic and guest-binding properties. Many of the special properties associated with these structures are caused[29] by the enhanced orbital overlap between neighboring aromatic units, which are arranged in a toroidal fashion. In the case of [12]collarene (**12.18b**), for example, the frontier molecular orbitals (Figure 12.6) are completely delocalized around the cavity of the cyclic molecular belt.

12.3.2 Synthesis of Collarenes

The synthesis (Scheme 12.8) of the first [12]collarene molecular belt (**12.18b**) was achieved[27] in our laboratory through the repeated application of Diels–Alder (DA) cycloadditions between the dienophile **12.20** and the diene **12.21**, which proceeds[30] with extraordinarily high diastereoselectivity. While the DA reaction between **12.20** and **12.21** could, potentially, lead to *four* different diastereoisomeric products, only a *single* diastereoisomer was actually observed in this reaction. This especially high diastereoselectivity is dictated primarily by electronic effects, which kinetically favor attack on the diene **12.21** from the *endo* face, while attack on the dieneophile **12.20** is preferred from the *exo* side. Density functional theory (DFT) calculations (Figure 12.7) reveal that the observed diastereoselectivity has its cause in the selective stabilization of the *endo–exo* transition state (**12.19a**).

As a consequence of the very high stereoelectronic control, the diastereoselectivity of the DA reaction between **12.20** and **12.21** was unaffected when carried out numerous times, making it possible to isolate (Scheme 12.8) the belt-like [12]cyclacene derivative Kohnkene (**12.24**) selectively in only three synthetic steps starting from **12.20** and **12.21**. Partial

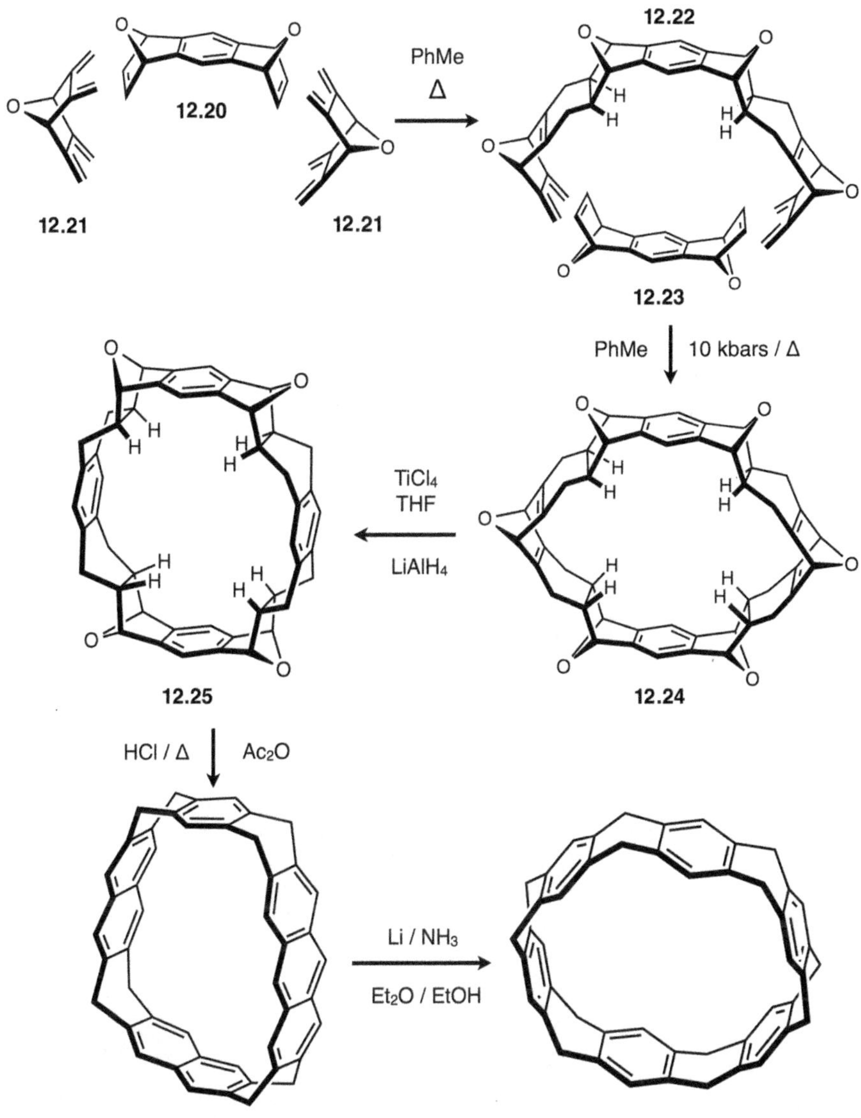

Scheme 12.8 Synthesis of the [12]collarene molecular belt **12.18b**.

deoxygenation of Kohnkene (**12.24**), using a low-valent titanium reagent, led subsequently to the formation of the [12]cyclacene derivative **12.25**, which adopts a structure resembling a "Celtic cross". Most remarkably, the hydrophobic cavity of **12.25** is filled with a disordered water molecule in the solid state, as illustrated by single crystal X-ray diffraction data. Furthermore, in the solid-state superstructure (Figure 12.8), two macrocycles are stacked side-on with the hydrophobic parts of their outer faces pointing

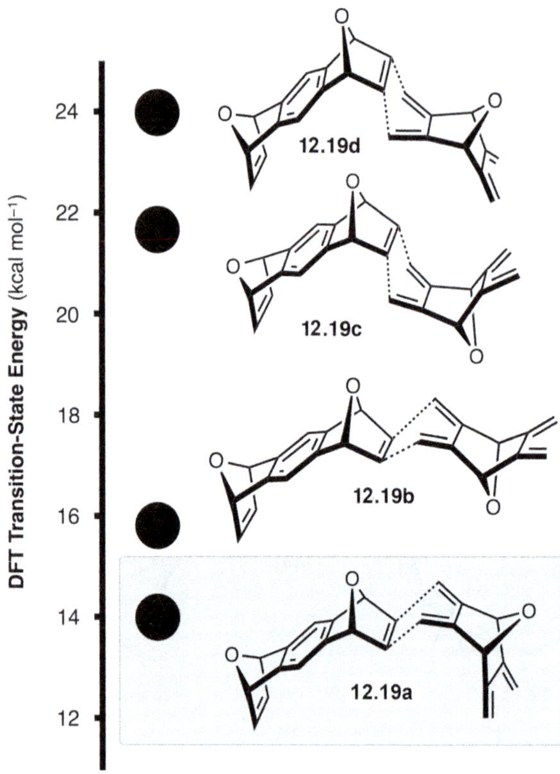

Figure 12.7 All four possible transition states of the highly diastereoselective Diels–Alder (DA) cycloaddition reaction employed for the synthesis of the [12]collarene **12.18b**. The superb diastereoselectivity of the reaction is explained by the DFT (B3LYP/6-31G**) transition-state energies, which have been plotted relative to the energy of the starting materials.

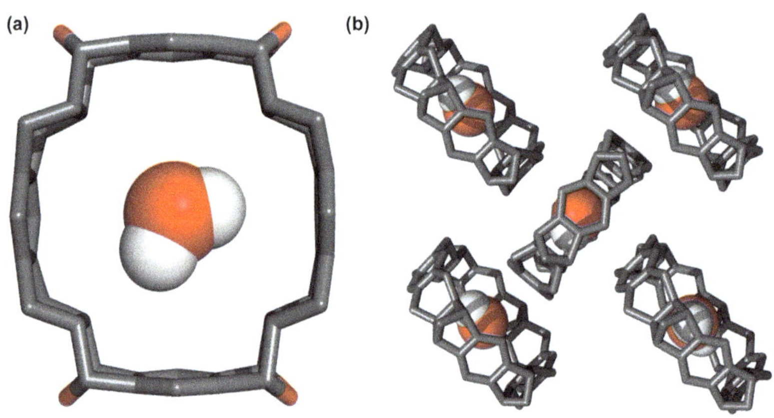

Figure 12.8 Single crystal X-ray structure (a) and superstructure (b) of Kohnkene (**12.25**), showing a water molecule bound inside the macrocycle's hydrophobic cavity.

towards the water molecule encapsulated inside the cavity of the central [12]cyclacene. The encapsulated water molecule is therefore not forming any strong hydrogen bonds with other water molecules in the solid-state superstructure, an unusual result which highlights the special supramolecular properties of cyclacene molecular belts. Compound **12.25** was then dehydrated under acid catalysis, leading to the isolation of the partially aromatized macrocycle **12.26**. Finally, Birch reduction of **12.26** afforded the highly symmetrical molecular belt **12.8b**, which we named [12]collarene.

12.3.3 Collarenes for Selective Cation Binding and Membrane Transport

The highly symmetrical arrangement of phenylene units in collarenes – with the faces of the aromatic groups pointing inward – maximizes[12] cation–π interactions when cations are included[31] inside the cavity of a collarene. Kim and co-workers[32] have predicted, therefore, that collarenes of different sizes will act as selective hosts for cations. Using *ab initio* calculations, as well as statistical perturbation theory in conjunction with classical Monte Carlo and molecular dynamics simulations in aqueous solutions, these authors predict that (1) [6]collarene will act as a selective host for Li^+ and Mg^{2+}, (2) [8]collarene will complex K^+ and Sr^{2+} and (3) [10]collarene (**12.8a**) will bind Cs^+ and Ba^{2+} selectively. As pointed out, this excellent size selectivity for cation complexation could be of interest in cleaning up radioactive waste, where in particular the long-lived, radioactive isotopes ^{137}Cs and ^{90}Sr pose significant environmental problems. Furthermore, [12]collarene (**12.8b**), which has been prepared[27] (*vide supra*) in our laboratory, was predicted[32] to form complexes with organic cations (*e.g.*, acetylcholine) of biological relevance.

12.3.4 Bridge-substituted Collarene Derivatives

Radially substituted derivatives of collarenes, wherein all neighboring methylene bridges are linked together with *ortho*-phenylene substituents, are part of the family of cyclic iptycenes. Yet, to the present day, only two members of this class of compounds, the cyclodeciptycene quinones **12.27a** and **12.27b**, have been isolated,[33] following multi-step synthetic procedures which involved Diels–Alder cycloadditions (Scheme 12.9) between the bis(quinone) **12.28** and the bis(anthracene) derivatives **12.29a** and **12.29b**. Hua and co-workers[33] were able to obtain a single crystal X-ray structure of the cyclodeciptycene quinone **12.27a**, which reveals that it has an inner ring diameter of ~0.9 nm. The four quinone and two methoxyl groups of compound **12.27a** are rotationally disordered within the unit cell of the solid-state superstructure with the averaged structure displaying occupancies of 33.3% (2 : 6) and 66.7% (4 : 6) for the methoxyl and quinone moieties, respectively. These occupancies are in good agreement with the compound's chemical formula.

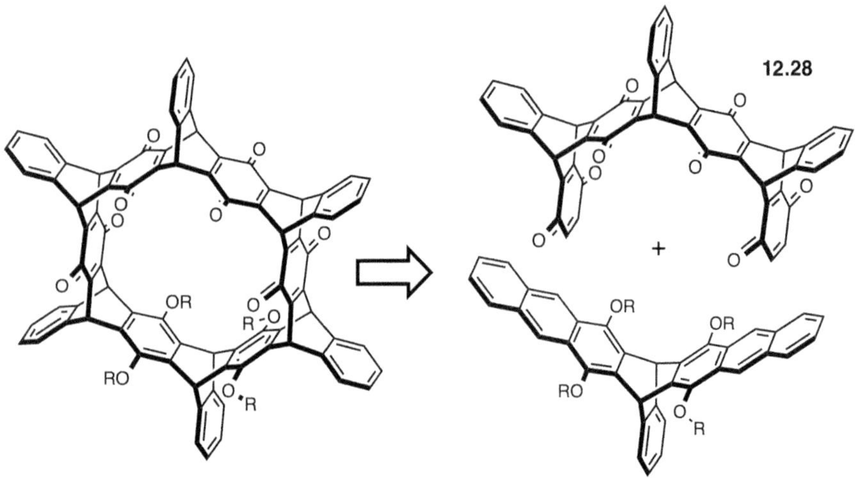

	R
12.27a	Me
12.27b	COMe

	R
12.29a	Me
12.29b	COMe

Scheme 12.9 Retrosynthetic strategy employed in the syntheses of cyclodeciptycene derivatives **12.27a** and **12.27b**.
Reproduced with permission from ref. 33. Copyright 2010 American Chemical Society.

By analogy with the [1*n*]paracyclophanes (*vide supra*), it should also be possible to substitute the methylene bridges of collarenes with heteroatoms. While – to the best of our knowledge – such structures have not been accessed synthetically to date, DFT calculations performed by Gleiter *et al.*[34] predict highly symmetrical and nearly strain-free structures for such heteroatom-substituted [10]collarene derivatives with methylene bridges substituted by sulfur and nitrogen, respectively.

12.3.5 Freezing the Conformation of Pillar[6]arene with Silyl Bridges

Rigid, doubly bridged pillar[*n*]arene-based frameworks, which are structurally related to collarenes (*vide supra*), have been isolated recently by Ogoshi and co-workers.[35] By simply reacting perhydroxylated pillar[6]arene with dialkyldichlorosilanes, they were able to link up all neighboring hydroxyl groups of the pillar[6]arene in a highly diastereoselective manner to form the rigid, belt-like derivatives **12.30a** and **12.30b** in good yield. The silyl bridges of **12.30a** and **12.30b** are oriented (Scheme 12.10) in an alternating "up–down" manner, which adds an element of inversion symmetry to the conformations of **12.30a** and **12.30b**. With an inversion center, **12.30a** and **12.30b** are achiral and therefore represent isolated *meso* forms of pillar[*n*]arene derivatives.

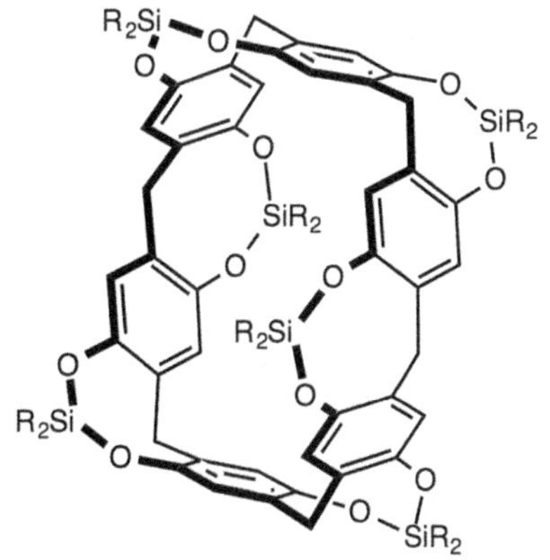

	R
12.30a	Me
12.30b	Et

Scheme 12.10 Structural formulae of the silyl-bridged, rigidified pillar[6]arene derivatives **12.30a** and **12.30b**.

12.4 Asararenes

12.4.1 Macrocycles with Fully Substituted Phenylene Units

The substituents attached to molecular frameworks can play the role of "adjustable knobs", which can be used to fine-tune the geometrical and supramolecular characteristics of macrocyclic compounds in order to achieve desired optoelectronic properties and self-assembly behavior. Inspired by the previous work (*vide supra*) involving pillar[*n*]arene related structures, we envisioned that attaching additional heteroatom substituents to the phenylene units of the pillar[*n*]arene [1*n*]paracyclophane backbone would provide access to a new family of macrocyclic structures with the potential to display enhanced supramolecular recognition and assembly properties. Our first attempt to move forward in this research direction involved the design of pillar[*n*]arene-related macrocycles, which contain *four* – rather than *two* as in the case of pillar[*n*]arenes – oxygen-based substituents attached to each phenylene ring. Consequently, the basic monomeric building unit for this new type of macrocycles is tetramethoxybenzene (**12.31**) and not *para*-dimethoxybenzene as in the case of the pillar[*n*]arenes. Given that tetramethoxybenzene (**12.31**) is also called "asarol methyl ether" colloquially, based on the fact that tetramethoxybenzene derivatives are commonly found[36] in the oils of wild ginger (*asarum*), we have chosen[37] the name asar[*n*]arenes for this new class of asarol methyl ether-derived macrocycle. Analogous to its use in pillar[*n*]arenes, the parameter *n* simply represents the number of phenylene repeating units present in the macrocycle.

12.4.2 Asararene Synthesis

The starting material for assembly of asar[*n*]arene macrocycles is tetramethoxybenzene (**12.31**), which can be prepared[37] from commercial dihydroxybenzoquinone (**12.33**) in abundant quantities. We subjected (Scheme 12.11) a mixture of tetramethoxybenzene and paraformaldehyde to Friedel–Crafts alkylation conditions at 80 °C in chlorinated solvents, with $BF_3 \cdot OEt_2$ as the Lewis acid. Influenced by the prior work on the synthesis of pillar[*n*]arenes – where pillar[5]arene is formed[1] primarily under similar reaction conditions – we were expecting this reaction mixture to form asar[5]arene as the major product. To our surprise, we found, however, that the reaction mixture produced only asar[6]arene and not even a trace of asar[5]arene. It is most likely that the increased steric demand imposed on the macrocyclic framework by the two additional methoxyl groups not present in pillar[*n*]arenes is responsible for this striking difference in reactivity between the asar[*n*]arene and pillar[*n*]arene families of macrocycles. Soxhlet extraction of the crude reaction mixture with acetonitrile as the solvent was then used as a scalable method of purification to access pure asar[6]arene (**12.32a**) in bulk quantities.

Asar[*n*]arene rings with more than six phenylene units ($n = 6$–15) can be accessed when conducting the synthesis of asar[*n*]arenes at lower temperatures (~36 °C). Over time, this reaction mixture starts slowly to become enriched in asar[6]arene (**12.32a**). This finding suggests that, as in the synthesis of pillar[*n*]arenes,[38] formation of the asar[*n*]arene macrocycles

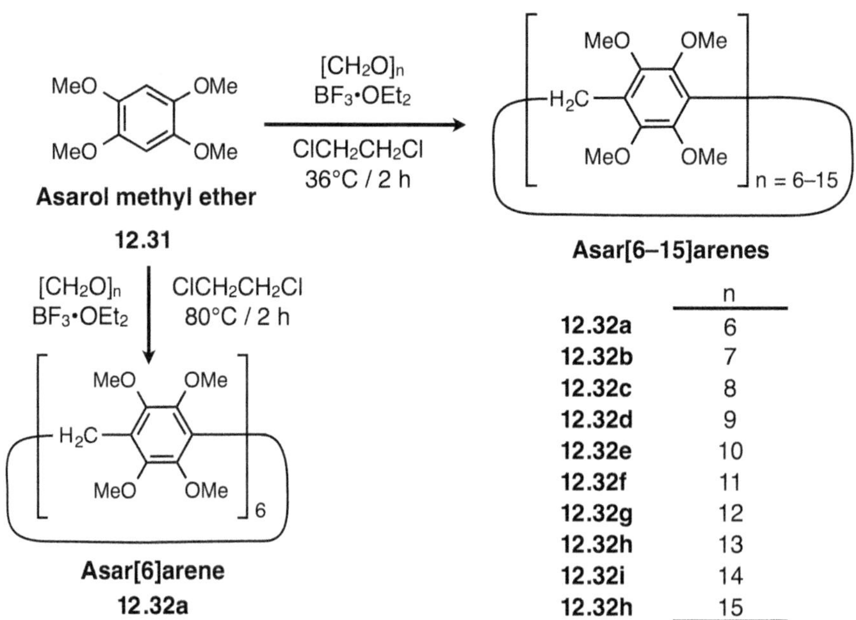

Scheme 12.11 Synthesis of differently sized asar[*n*]arene macrocycles. Note that asar[6]arene (**12.32a**) is the thermodynamically favored product. Reproduced with permission from ref. 37.

12.32a–h is a reversible process in the presence of the BF_3 Lewis acid. Furthermore, asar[6]arene (**12.32a**) seems to be the thermodynamically preferred product of this reversible reaction, with a lower free energy of formation than any of the larger asar[*n*]arenes. The differently sized asar[*n*]arenes were found to have very distinct solubilities in acetonitrile, a property which allowed us to isolate several of the large asar[*n*]arenes selectively through the application of fractional Soxhlet extraction followed by crystallization. Interestingly, we found that the asar[*n*]arenes **12.32b, 12.32d,** and **12.32f** containing an odd number of phenylene units generally display higher solubilities in organic solvents than the even-numbered asararenes **12.32a, 12.32c, 12.32e,** and **12.32g**. This distribution of solubilities is most likely caused by the more symmetrical structures (*vide infra*) adopted by the even-numbered asar-[*n*]arene macrocycles in the solid-state.

12.4.3 The Structures of Asararenes

We were able to solve the single crystal X-ray structures of asar[6]- (**12.32a**), asar[7]- (**12.32b**), asar[8]- (**12.32c**), asar[9]- (**12.32d**), asar[10]- (**12.32e**), and asar[11]arene (**12.32f**). The asar[*n*]arenes adopt highly diverse shapes (Figure 12.9) in the solid state, with the structures of the even-numbered homologs being more symmetrical than those of the odd-numbered ones. The solid-state structures of asar[6]- and asar[8]arene, for example, both contain an *n*-fold improper axis of rotational symmetry (S_n) with $n=6$ and 4, respectively, while the symmetry elements of asar[10]arene include an inversion center. All the odd-numbered asar[*n*]arenes, on the other hand, do not contain any symmetry elements other than a trivial C_1 axis. The steric requirements of the four methoxyl groups attached to each phenylene unit force neighboring phenylenes in the macrocycles to adopt twisted conformations with respect to each other. The twisted arrangement of phenylene units renders the backbone conformations of the asar[*n*]arenes – constructed (Figure 12.10) by linking all topologically connected methylene units together directly – reminiscent of conformations adopted by cyclic hydrocarbons. Along that line of thinking, the backbone conformation of asar[6]arene, for example, resembles the chair conformation of an inflated cyclohexane, while the spine geometries of asar[8]- and asar[10]arene resemble giant cyclooctatetrane or cyclodexane rings, respectively.

Asar[10]arene (**12.32e**) contains the largest cavity of all asararenes obtained in the crystalline form to date, with a diameter of ~1 nm. In the solid-state superstructure (Figure 12.11), the asar[10]arene rings are stacked on top of one another in an eclipsed fashion, leading to continuous channels extending infinitely through the crystals along the *c*-crystallographic axis. The individual stacks formed by asar[10]arene are packed together in a regular checkerboard-like fashion.

While the structures of the methoxyl-protected asar[*n*]arenes are dominated by the steric demands of the methoxyl substituents, DFT calculations predict very different structural features for the de-protected,

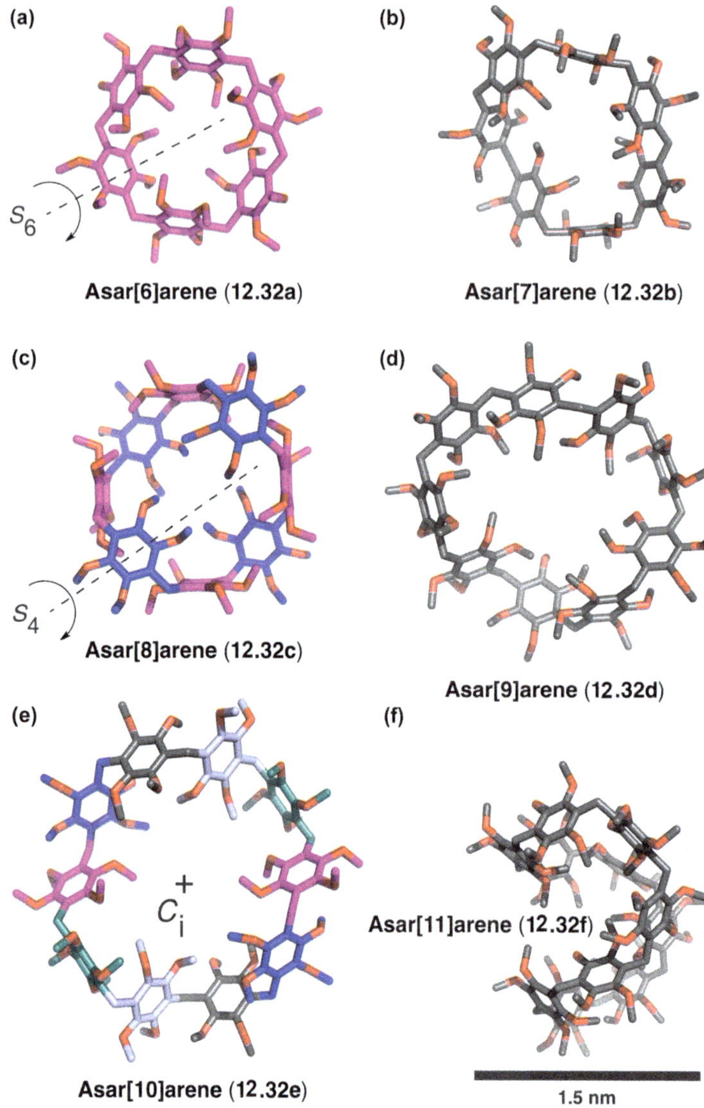

Figure 12.9 Tubular representations of the solid-state conformations of the asar[*n*]arene macrocycles **12.32a–f**, obtained from single crystal X-ray diffraction data. Oxygen atoms are illustrated in red, while carbon atoms are illustrated in all other colors – with symmetry-equivalent phenylene units being colored identically.
Reproduced with permission from ref. 37.

perhydroxylated asararenes. The structure (Figure 12.12) of perhydroxy-asar[6]arene, for example, is anticipated to adopt an "open-book" conformation, which is stabilized by a continuous network of hydrogen bonds extending along both rims of the perhydroxy-asar[6]arene.

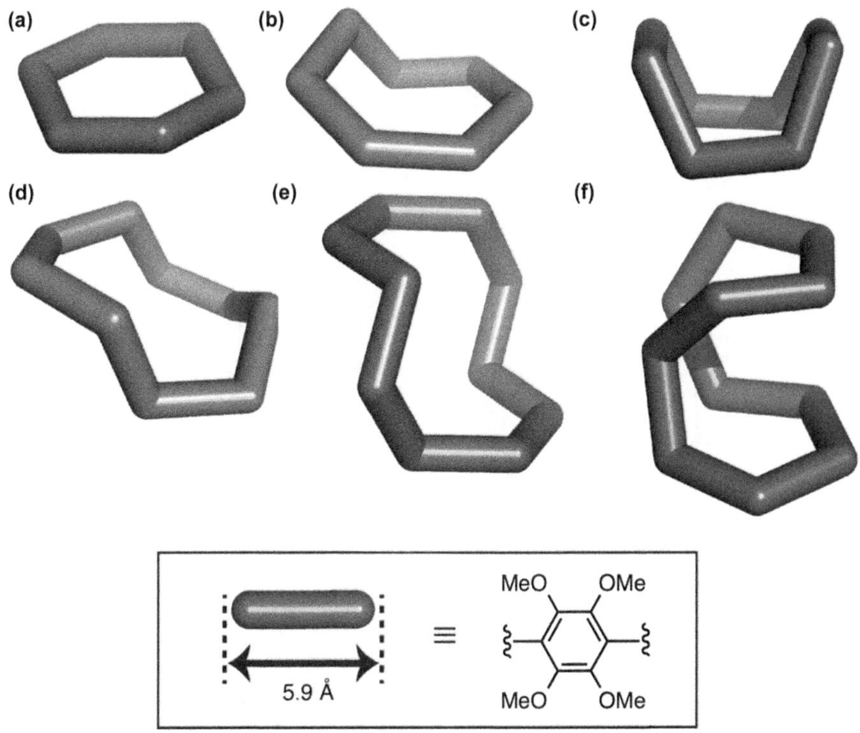

Figure 12.10 The solid-state backbone conformations – obtained from the single crystal X-ray structures and constructed by directly linking topologically connected methylene units – of (a) asar[6]arene (**12.32a**), (b) asar[7]arene (**12.32b**), (c) asar[8]arene (**12.32c**), (d) asar[9]arene (**12.32d**), (e) asar[10]arene (**12.32e**), and (f) asar[11]arene (**12.32f**). Reproduced with permission from ref. 37.

12.4.4 Applications of Asararene Macrocycles

With their diverse structures, the asar[*n*]arenes might find applications as selective supramolecular hosts[39] and components of molecular machines[40] in the fullness of time. On the other hand, the tubular channels formed by asar[10]arene molecules in the solid state could potentially absorb gases in an efficient manner, a process which might be assisted by the many methoxyl groups covering the walls of these channels. Furthermore, we envision that after de-protection of the asararenes' methoxyl groups, oxidation to hydroxyquinone moieties will occur spontaneously in air. Based on the well-known[41] coordination chemistry of hydroxybenzoquinone derivatives, which form infinite linear sheets with a variety of metal ions, the hydroxyquinone forms of the perhydroxylated asararenes could then serve as preprogrammed molecular building blocks to create tubular, metal-directed assemblies with tunable shapes and sizes.

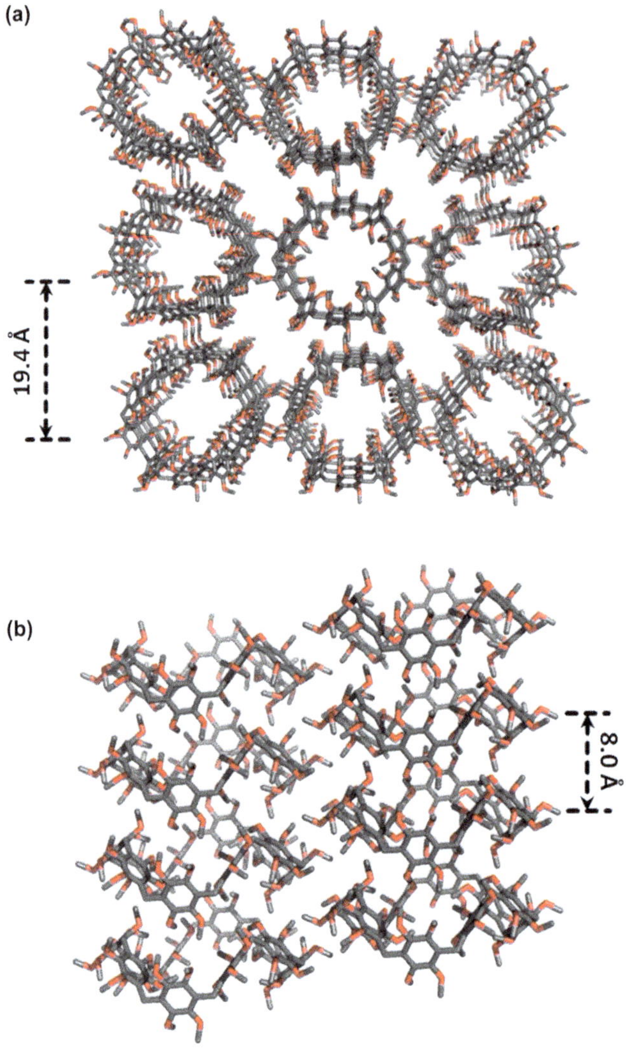

Figure 12.11 The solid-state superstructure of asar[10]arene (**12.32e**), viewed along (a) the *c*- and (b) the *a*-crystallographic axis. Oxygen atoms are illustrated in red and carbons in gray. Hydrogen atoms have been omitted for the sake of clarity.
Reproduced with permission from ref. 37.

12.4.5 Amino-functionalized Pillararenes

In going from pillar[*n*]arenes to asar[*n*]arenes,[37] two hydroxyl groups are substituted onto each of the phenylene units in the host. Analogous pillar[*n*]arene derivatives can be prepared using different heteroatoms substituted at these positions to produce hosts with unique properties. Amino groups directly substituted on the pillar[*n*]arenes have the potential

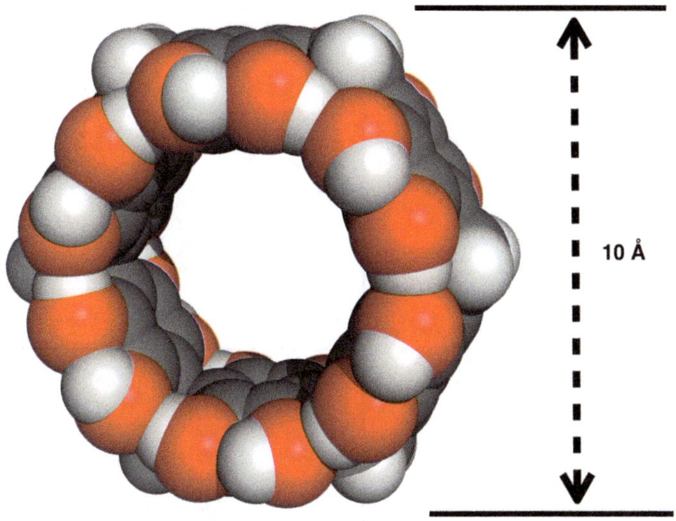

Figure 12.12 The structure (predicted) of perhydroxy-asar[6]arene (**12.33a**), stabilized by hydrogen-bonded rims.

to influence the electron density in the host's cavity, allowing for stronger associations with electron-poor guests. This added functionality can also allow novel chemical modifications of the hosts, enabling chemists to prepare structures unobtainable with previously synthesized pillar[*n*]arene derivatives.

The pillar[5]arene **12.34**, composed of one diamine-benzoquinone unit and four 1,4-dimethoxybenzene units, was synthesized (Scheme 12.12) by us[42] in order to investigate the potential applications of an amino-functionalized host. The synthetic pathway to this amino-functionalized pillar[5]-arene involved the conjugate addition of 2 equiv. of allylamine onto a benzoquinone unit within the host, followed by de-protection. The X-ray crystal structure of **12.34** (Scheme 12.12) was obtained by the slow evaporation of CH_2Cl_2 and indicates that intramolecular hydrogen bonds are formed between the protons of the amino group and the oxygen atoms of the neighboring 1,4-dimethoxybenzene units. Dynamic ^1H NMR spectroscopy was employed to investigate the energy barrier to stereochemical inversion in **12.34**, which was calculated to be 16.6 kcal mol^{-1} using the Eyring equation.

The major motivation for making pillar[*n*]arene derivatives containing diamino-benzoquinones is that these units can act as the starting material in the synthesis of tubular structures utilizing a growth-from-template approach. Diamino-benzoquinones react[43] with two equivalents of an aldehyde to afford a benzdioxazole, allowing for substitution at positions which are almost exactly orthogonal to the plane encompassing the five methylene carbons of the pillar[5]arene backbone. This unique substitution geometry allows for the growth of a one-dimensional nanostructure in a direction

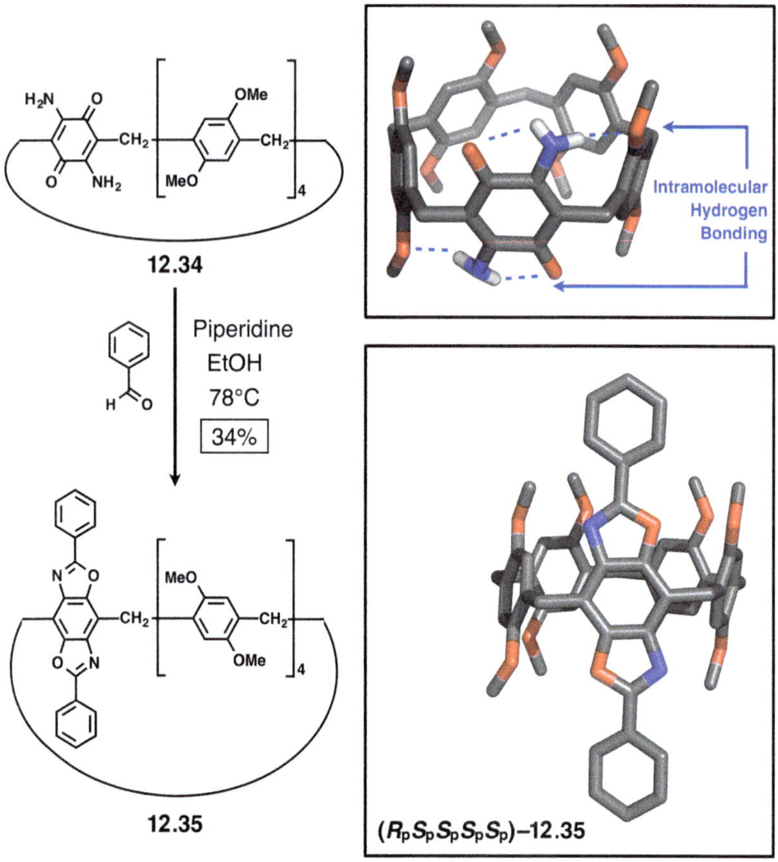

Scheme 12.12 Synthesis and single crystal X-ray structures of the amino-substituted pillar[5]arene derivatives **12.34** and **12.35**. Carbon atoms are illustrated in gray, oxygens in red, and nitrogens in blue.

which places all of the pillar[5]arene rings "in register". In order to investigate the reactivity of diamine-benzoquinones within pillar[5]arene towards aldehydes, **12.34** was reacted with benzaldehyde in the presence of a catalytic amount of base to afford the pillar[5]arene derivative **12.35** containing one benzdioxazole unit. The solid-state structure of **12.35** (Scheme 12.12) illustrates the desired orthogonal relationship between the plane of the methylene carbon atoms and the phenylene-substituents at the 2- and 6-positions of the benzdioxazole.

Rigid oligomers composed of pillar[5]arene repeating units connected through oxazole and phenylene spacers have been prepared (Scheme 12.13) through the polymerization of the A–B type monomer **12.36**, which is derived from **12.34**. These rigid arrays consist of nanostructures from the **2-Mer** oligomer – two repeating units – up to the **9-*Mer***. Modeling of the **9-*Mer*** indicates that the length of this nanostructure is approximately 11 nm. Further investigations are underway currently to produce the analogous

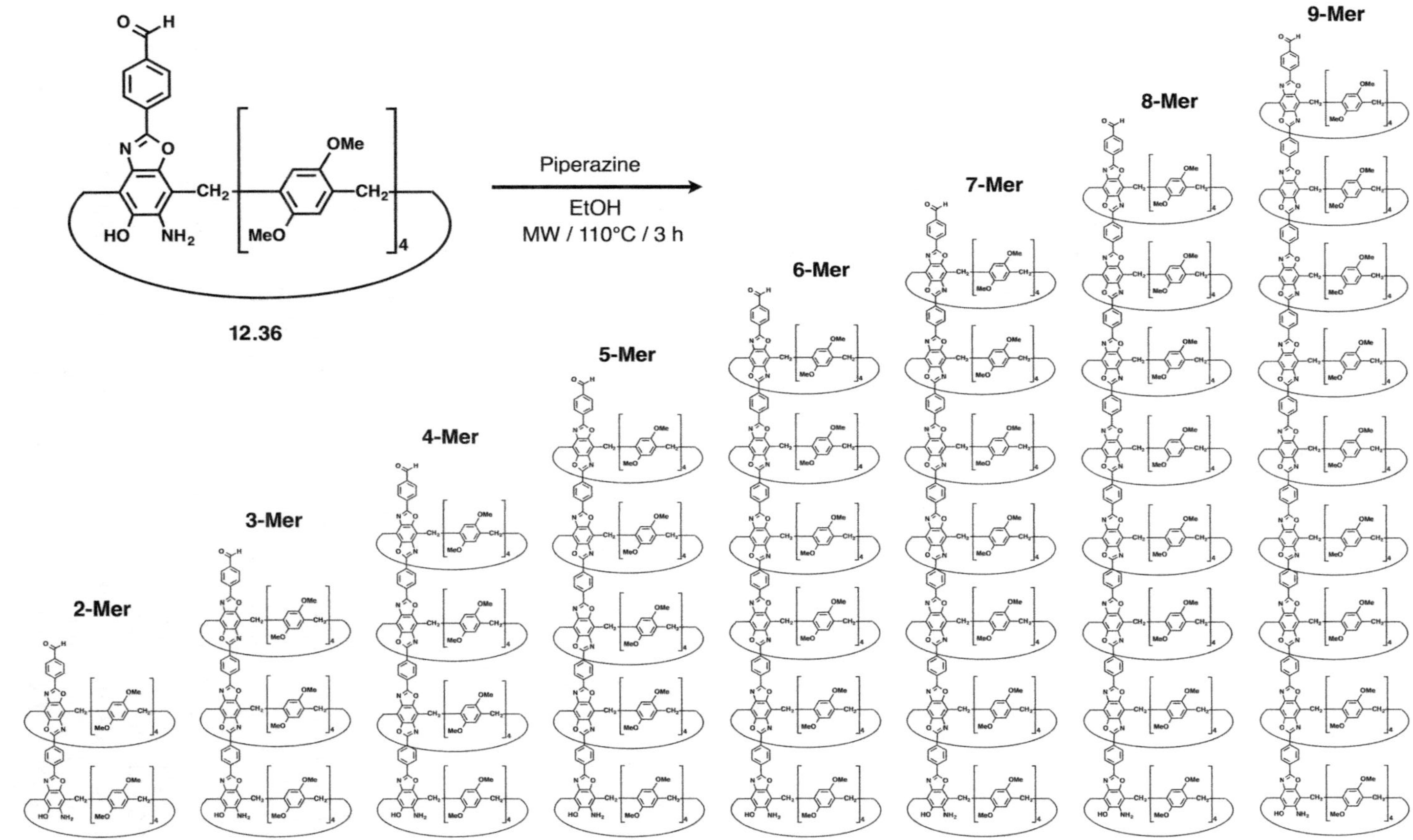

Scheme 12.13 Synthesis of tube-shaped oligomers of the amino-substituted pillar[5]arene derivative **12.36**.

covalently linked tubular arrays from pillar[5]arene precursors which are connected at all five phenylene units.

12.5 Conclusions and Outlook

In this chapter, we have shown how pillar[*n*]arene macrocycles can serve as an inspiration to create more general structural frameworks with enhanced substitution of the phenylene rings. These additional phenylene substituents provide access to pillararene-related structures, which include heteroatom-bridged [1*n*]cyclophanes, collarene-like molecular belts, cyclic iptycenes, and giant asar[*n*]arene rings, as well as amino-substituted pillararenes. Thanks to the discovery of Diels–Alder cycloadditions with superb diastereoselectivities and reversible Fridel–Crafts alkylations, efficient methods of synthesis of collarenes and asararenes have been found, which opens the door for the application of such structurally complex, programmable macrocycles in the exploration of selective host–guest binding, directed supramolecular/metal-directed/covalent assembly, and catalysis.

References

1. T. Ogoshi, S. Kanai, S. Fujinami, T. A. Yamagishi and Y. Nakamoto, *J. Am. Chem. Soc.*, 2008, **130**, 5022–5023.
2. D. Cao, Y. Kou, J. Liang, Z. Chen, L. Wang and H. Meier, *Angew. Chem., Int. Ed.*, 2009, **48**, 9721–9723; C. Han, F. Ma, Z. Zhang, B. Xia, Y. Yu and F. Huang, *Org. Lett.*, 2010, **12**, 4360–4363; T. Ogoshi, K. Kitajima, T. Aoki, S. Fujinami, T. A. Yamagishi and Y. Nakamoto, *J. Org. Chem.*, 2010, **75**, 3268–3273; Z. Zhang, B. Xia, C. Han, Y. Yu and F. Huang, *Org. Lett.*, 2010, **12**, 3285–3287; T. Ogoshi, *J. Inclusion Phenom. Macrocyclic Chem.*, 2011, **72**, 247–262; N. L. Strutt, R. S. Forgan, J. M. Spruell, Y. Y. Botros and J. F. Stoddart, *J. Am. Chem. Soc.*, 2011, **133**, 5668–5671; P. J. Cragg and K. Sharma, *Chem. Soc. Rev.*, 2012, **41**, 597–607; Y. Ma, X. Chi, X. Yan, J. Liu, Y. Yao, W. Chen, F. Huang and J. L. Hou, *Org. Lett.*, 2012, **14**, 1532–1535; T. Ogoshi, D. Yamafuji, D. Kotera, T. Aoki, S. Fujinami and T. A. Yamagishi, *J. Org. Chem.*, 2012, **77**, 11146–11152; N. L. Strutt, D. Fairen-Jimenez, J. Iehl, M. B. Lalonde, R. Q. Snurr, O. K. Farha, J. T. Hupp and J. F. Stoddart, *J. Am. Chem. Soc.*, 2012, **134**, 17436–17439; M. Xue, Y. Yang, X. Chi, Z. Zhang and F. Huang, *Acc. Chem. Res.*, 2012, **45**, 1294–1308; G. Yu, M. Xue, Z. Zhang, J. Li, C. Han and F. Huang, *J. Am. Chem. Soc.*, 2012, **134**, 13248–13251; Y. Jin, C. Yu, R. J. Denman and W. Zhang, *Chem. Soc. Rev.*, 2013, **42**, 6634–6654; H. Zhang and Y. Zhao, *Chem. – Eur. J.*, 2013, **19**, 16862–16879; D. Cao and H. Meier, *Asian J. Org. Chem.*, 2014, **3**, 244–262.
3. J. C. Ma and D. A. Dougherty, *Chem. Rev.*, 1997, **97**, 1303–1324.
4. H. Takemura, *Curr. Org. Chem.*, 2009, **13**, 1633–1653.
5. T. Tsuji, *Highly Strained Cyclophanes*, Wiley-VCH Verlag GmbH & Co, 2005.

6. H. Kawai, T. Suzuki, M. Ohkita and T. Tsuji, *Angew. Chem., Int. Ed.*, 1998, **37**, 817–819; H. Kawai, T. Suzuki, M. Ohkita and T. Tsuji, *Chem. - Eur. J.*, 2000, **6**, 4177–4187.
7. T. Tsuji, M. Ohkita, T. Konno and S. Nishida, *J. Am. Chem. Soc.*, 1997, **119**, 8425–8431.
8. Y. Miyahara, T. Inazu and T. Yoshino, *Tetrahedron Lett.*, 1983, **24**, 5277–5280.
9. G. W. Gribble and C. F. Nutaitis, *Tetrahedron Lett.*, 1985, **26**, 6023–6026.
10. F. Inokuchi, Y. Miyahara, T. Inazu and S. Shinkai, *Angew. Chem., Int. Ed. Engl.*, 1995, **34**, 1364–1366.
11. E. A. Meyer, R. K. Castellano and F. Diederich, *Angew. Chem., Int. Ed.*, 2003, **42**, 1210–1250; A. Frontera, D. Quinonero and P. M. Deya, *Wiley Interdiscip. Rev.: Comput. Mol. Sci.*, 2011, **1**, 440–459; K. D. Daze and F. Hof, *Acc. Chem. Res.*, 2013, **46**, 937–945; A. S. Mahadevi and G. N. Sastry, *Chem. Rev.*, 2013, **113**, 2100–2138; M. M. Watt, M. S. Collins and D. W. Johnson, *Acc. Chem. Res.*, 2013, **46**, 955–966.
12. D. A. Dougherty, *Acc. Chem. Res.*, 2013, **46**, 885–893.
13. J. F. Hartwig, *Acc. Chem. Res.*, 1998, **31**, 852–860; A. R. Muci and S. L. Buchwald, *Top. Curr. Chem.*, 2002, **219**, 131–209.
14. D. S. Surry and S. L. Buchwald, *Chem. Sci.*, 2011, **2**, 27–50.
15. B. König, M. Rödel, P. Bubenitschek, P. G. Jones and I. Thondorf, *J. Org. Chem.*, 1995, **60**, 7406–7410; B. König, M. Rödel, P. Bubenitschek and P. G. Jones, *Angew. Chem., Int. Ed. Engl.*, 1995, **34**, 661–662.
16. A. Ito, Y. Yokoyama, R. Aihara, K. Fukui, S. Eguchi, K. Shizu, T. Sato and K. Tanaka, *Angew. Chem., Int. Ed.*, 2010, **49**, 8205–8208.
17. T. F. Yang, K. Y. Chiu, H. C. Cheng, Y. W. Lee, M. Y. Kuo and Y. O. Su, *J. Org. Chem.*, 2012, **77**, 8627–8633.
18. P. Chen, R. A. Lalancette and F. Jäkle, *Angew. Chem., Int. Ed.*, 2012, **51**, 7994–7998.
19. J. Franke and F. Vögtle, *Tetrahedron Lett.*, 1984, **25**, 3445–3448.
20. D. Takeuchi, I. Asano and K. Osakada, *J. Org. Chem.*, 2006, **71**, 8614–8617.
21. M. L. Kaplan and W. D. Reents, *Tetrahedron Lett.*, 1982, **23**, 373–374.
22. E. Tsuchida, K. Miyatake, K. Yamamoto and A. S. Hay, *Macromolecules*, 1998, **31**, 6469–6475.
23. D. A. Zimmerman, J. L. König and H. Ishida, *Polymer*, 1996, **37**, 3111–3116.
24. D. Parker, J. Bussink, H. T. van de Grampel, G. W. Wheatley, E.-U. Dorf, E. Ostlinning and K. Reinking, *Polymers, High-Temperature*, 2000.
25. K. Miyatake, Y. Yokoi, K. Yamamoto, E. Tsuchida and A. S. Hay, *Macromolecules*, 1997, **30**, 4502–4503.
26. I. Baxter, D. J. Williams, H. M. Colquhoun, P. Hodge and F. H. Kohnke, *Chem. Commun.*, 1998, 1991–1992; I. Baxter, A. Ben-Haida, H. M. Colquhoun, P. Hodge, F. H. Kohnke and D. J. Williams, *Chem. - Eur. J.*, 2000, **6**, 4285–4296; A. Ben-Haida, H. M. Colquhoun, P. Hodge,

J. Raftery, A. J. White and D. J. Williams, *Org. Biomol. Chem.*, 2009, **7**, 5229–5235.
27. P. R. Ashton, G. R. Brown, N. S. Isaacs, D. Giuffrida, F. H. Kohnke, J. P. Mathias, A. M. Z. Slawin, D. R. Smith, J. F. Stoddart and D. J. Williams, *J. Am. Chem. Soc.*, 1992, **114**, 6330–6353.
28. W. D. Neudorff, D. Lentz, M. Anibarro and A. D. Schlüter, *Chem. – Eur. J.*, 2003, **9**, 2745–2757; R. Herges, *Chem. Rev.*, 2006, **106**, 4820–4842; T. Kawase and H. Kurata, *Chem. Rev.*, 2006, **106**, 5250–5273; K. Tahara and Y. Tobe, *Chem. Rev.*, 2006, **106**, 5274–5290; P. Evans and R. Jasti, *Molecular Belts*, Springer Berlin Heidelberg, 2013.
29. M. Stuparu, D. Lentz, H. Rüegger and A. D. Schlüter, *Eur. J. Org. Chem.*, 2007, **2007**, 88–100; T. Yao, H. Yu, R. J. Vermeij and G. J. Bodwell, *Pure Appl. Chem.*, 2008, **80**, 533–546; R. Gleiter, B. Esser and S. C. Kornmayer, *Acc. Chem. Res.*, 2009, **42**, 1108–1116; D. Eisenberg, R. Shenhar and M. Rabinovitz, *Chem. Soc. Rev.*, 2010, **39**, 2879–2890.
30. F. H. Kohnke, A. M. Z. Slawin, J. F. Stoddart and D. J. Williams, *Angew. Chem., Int. Ed. Engl.*, 1987, **26**, 892–894; P. R. Ashton, N. S. Isaacs, F. H. Kohnke, A. M. Z. Slawin, C. M. Spencer, J. F. Stoddart and D. J. Williams, *Angew. Chem., Int. Ed. Engl.*, 1988, **27**, 966–969; P. Ellwood, J. P. Mathias, J. F. Stoddart and F. H. Kohnke, *Bull. Soc. Chim. Belg.*, 1988, **97**, 669–678; P. R. Ashton, N. S. Isaacs, F. H. Kohnke, J. P. Mathias and J. F. Stoddart, *Angew. Chem., Int. Ed. Engl.*, 1989, **28**, 1258–1261; F. H. Kohnke, J. P. Mathias and J. F. Stoddart, *Angew. Chem., Int. Ed. Engl.*, 1989, **28**, 1103–1110; F. H. Kohnke and J. F. Stoddart, *Pure Appl. Chem.*, 1989, **61**, 1581–1586.
31. J. Benkhoff, R. Boese, F.-G. Klärner and A. E. Wigger, *Tetrahedron Lett.*, 1994, **35**, 73–76; J. Benkhoff, R. Boese and F.-G. Klärner, *Liebigs Ann.*, 1997, **1997**, 501–516.
32. H. S. Choi, S. B. Suh, S. J. Cho and K. S. Kim, *Proc. Natl. Acad. Sci. U. S. A.*, 1998, **95**, 12094–12099.
33. K. Lou, A. M. Prior, B. Wiredu, J. Desper and D. H. Hua, *J. Am. Chem. Soc.*, 2010, **132**, 17635–17641.
34. S. C. Kornmayer, B. Esser and R. Gleiter, *Org. Lett.*, 2009, **11**, 725–728.
35. T. Ogoshi, K. Demachi, K. Masaki and T. A. Yamagishi, *Chem. Commun.*, 2013, **49**, 3952–3954.
36. E. J. Parry, *The Chemistry of Essential Oils and Artificial Perfumes*, Scott, Greenwood & Son, London, UK, 1908.
37. S. T. Schneebeli, C. Cheng, K. J. Hartlieb, N. L. Strutt, A. A. Sarjeant, C. L. Stern and J. F. Stoddart, *Chem. – Eur. J.*, 2013, **19**, 3860–3868.
38. M. Holler, N. Allenbach, J. Sonet and J. F. Nierengarten, *Chem. Commun.*, 2012, **48**, 2576–2578.
39. H. J. Schneider and A. K. Yatsimirsky, *Chem. Soc. Rev.*, 2008, **37**, 263–277.

40. E. R. Kay, D. A. Leigh and F. Zerbetto, *Angew. Chem., Int. Ed.*, 2007, **46**, 72–191; A. Coskun, M. Banaszak, R. D. Astumian, J. F. Stoddart and B. A. Grzybowski, *Chem. Soc. Rev.*, 2012, **41**, 19–30.
41. T. Yamada, S. Morikawa and H. Kitagawa, *Bull. Chem. Soc. Jpn.*, 2010, **83**, 42–48.
42. N. L. Strutt, H. Zhang, S. T. Schneebeli and J. F. Stoddart, *Chem. - Eur. J.*, 2014, **20**, 10996–11004.
43. A.-M. Osman, *J. Am. Chem. Soc.*, 1957, **79**, 966–968.

Subject Index